Oxford Graduate Texts in Mathematics

Series Editors

R. Cohen S. K. Donaldson S. Hildebrandt
T. J. Lyons M. J. Taylor

OXFORD GRADUATE TEXTS IN MATHEMATICS

Books in the series

1. Keith Hannabuss: *An introduction to quantum theory*
2. Reinhold Meise and Dietmar Vogt: *Introduction to functional analysis*
3. James G. Oxley: *Matroid theory*
4. N. J. Hitchin, G. B. Segal, and R. S. Ward: *Integrable systems: twistors, loop groups, and Riemann surfaces*
5. Wulf Rossmann: *Lie groups: An introduction through linear groups*
6. Qing Liu: *Algebraic geometry and arithmetic curves*
7. Martin R. Bridson and Simon M. Salamon (eds): *Invitations to geometry and topology*
8. Shmuel Kantorovitz: *Introduction to modern analysis*
9. Terry Lawson: *Topology: A geometric approach*
10. Meinolf Geck: *An introduction to algebraic geometry and algebraic groups*
11. Alastair Fletcher and Vladimir Markovic: *Quasiconformal maps and Teichmüller theory*
12. Dominic Joyce: *Riemannian holonomy groups and calibrated geometry*
13. Fernando Villegas: *Experimental Number Theory*
14. Péter Medvegyev: *Stochastic Integration Theory*
15. Martin A. Guest: *From Quantum Cohomology to Integrable Systems*
16. Alan D. Rendall: *Partial Differential Equations in General Relativity*
17. Yves Félix, John Oprea and Daniel Tanré: *Algebraic Models in Geometry*
18. Jie Xiong: *Introduction to Stochastic Filtering Theory*
19. Maciej Dunajski: *Solitons, Instantons, and Twistors*

Solitons, Instantons, and Twistors

Maciej Dunajski
Department of Applied Mathematics and Theoretical Physics, University of Cambridge

OXFORD
UNIVERSITY PRESS

OXFORD
UNIVERSITY PRESS

Great Clarendon Street, Oxford OX2 6DP

Oxford University Press is a department of the University of Oxford.
It furthers the University's objective of excellence in research, scholarship,
and education by publishing worldwide in

Oxford New York

Auckland Cape Town Dar es Salaam Hong Kong Karachi
Kuala Lumpur Madrid Melbourne Mexico City Nairobi
New Delhi Shanghai Taipei Toronto

With offices in

Argentina Austria Brazil Chile Czech Republic France Greece
Guatemala Hungary Italy Japan South Korea Poland Portugal
Singapore Switzerland Thailand Turkey Ukraine Vietnam

Oxford is a registered trade mark of Oxford University Press
in the UK and in certain other countries

Published in the United States
by Oxford University Press Inc., New York

© Maciej Dunajski, 2010

The moral rights of the author have been asserted

Database right Oxford University Press (maker)

Reprinted 2010

All rights reserved. No part of this publication may be reproduced,
stored in a retrieval system, or transmitted, in any form or by any means,
without the prior permission in writing of Oxford University Press,
or as expressly permitted by law, or under terms agreed with the appropriate
reprographics rights organization. Enquiries concerning reproduction
outside the scope of the above should be sent to the Rights Department,
Oxford University Press, at the address above

You must not circulate this book in any other binding or cover
And you must impose this same condition on any acquirer

ISBN 978-0-19-857063-9

Printed and bound in Great Britain by
CPI Antony Rowe, Chippenham and Eastbourne

Preface

This book grew out of lecture courses I have given to mathematics Part II, Part III, and graduate students in Cambridge between the years 2003 and 2008.

The first four chapters could form a basis of a one-term lecture course on integrable systems covering the Arnold–Liouville theorem, inverse scattering transform, Hamiltonian methods in soliton theory, and Lie point symmetries. The additional, more advanced topics are covered in Chapters 7 and 8. They include the anti-self-dual Yang–Mills equations, their symmetry reductions, and twistor methods. Chapters 5, 6, and 9 provide material for an advanced course on field theory where particular emphasis is paid to non-perturbative solutions to classical field equations. We shall discuss scalar kinks, sigma-model lumps, non-abelian magnetic monopoles in $\mathbb{R}^3$, instanton solutions to pure Yang–Mills equations in $\mathbb{R}^4$, and finally gravitational instantons. Although the material is entirely 'classical', the motivation comes from quantum field theories, including gauge theories, where it is necessary to consider solutions of the non-linear field equations that are topologically distinct from the vacuum. Chapter 10 contains a discussion of the anti-self-dual conformal structures, some of which have not been presented in literature before. This chapter links with the rest of the book as the anti-self-dual conformal structures provide a unifying framework for studying the dispersionless integrable systems. There are three appendices. The first two provide a mathematical background in topology of manifolds and Lie groups (Appendix A), and complex analysis required by twistor theory (Appendix B). Appendix C is self-contained and can form a basis of a one-term lecture course on overdetermined partial differential equations. This appendix gives an elementary introduction to the subject known as the exterior differential system.

Although the term *soliton* – a localized non-singular lump of energy – plays a central role in the whole book, its precise meaning changes as the reader progresses through the chapters. In Chapters 1, 2, 3, and 8 solitons arise as solutions to completely integrable equations. They are time-dependent localized waves which scatter without emitting radiation, and owe their stability to the existence of infinitely many dynamically conserved currents. In Chapters 5 and 6 the solitons are topological – they are characterized by discrete homotopy invariants (usually in the form of Chern numbers) which are conserved as the continuous field configurations have finite energy. Finally in Chapter 9 the gravitational solitons arise as lifts of Riemannian gravitational instantons to higher dimensions.

The book should be of use to advanced undergraduate and research students, as well as experts in soliton theory who want to broaden their techniques. It is aimed at both mathematicians and those physicists who are willing to go beyond perturbation theory. The revived interest in twistor theory in recent years can be largely attributed to Witten's twistor-string theory [185]. It is hoped that those researchers who come to twistor strings with the string theory or quantum field theory (QFT) background will find this book an accessible introduction to twistor theory.

There are some excellent text books which treat the material presented here in great depth. Readers should consult [122] for inverse scattering transform, [124] for the symmetry methods, and [114] for topological solitons. The twistor approach to integrability is a subject of the monograph [118], while [83, 132, 175] are books on twistors which concentrate on aspects of the theory other than integrability. The full treatment of exterior differential systems can be found in [23].

The twistor approach to integrability used in the second half of the book has been developed over the last thirty years by the Oxford school of Sir Roger Penrose with a particular input from Richard Ward. I am grateful to Sir Roger for sharing his inspirational ideas with the rest of us. His original motivation was to unify general relativity and quantum mechanics in a non-local theory based on complex numbers. The application of twistor theory to integrability has been an unexpected spin-off from the twistor programme.

While preparing the manuscript I have benefited from many valuable discussions with my colleagues, collaborators, and research students. In particular I would like to thank Robert Bryant, David Calderbank, Mike Eastwood, Jenya Ferapontov, Gary Gibbons, Sean Hartnoll, Nigel Hitchin, Marcin Kaźmierczak, Piotr Kosinski, Nick Manton, Lionel Mason, Vladimir Matveev, Paweł Nurowski, Roger Penrose, Prim Plansangkate, Maciej Przanowski, George Sparling, David Stuart, Paul Tod, Simon West, and Nick Woodhouse. I am especially grateful to Paul Tod for carefully reading the manuscript.

Finally, I thank my wife Asia and my sons Adam and Nico not least for making me miss several submission deadlines. I dedicate this book to the three of them with gratitude.

Cambridge
January 2009

Maciej Dunajski

Contents

List of Figures

List of Abbreviations

ASDYM	Anti-self-dual Yang-Mills
AHS	Atiyah–Hitchin–Singer
ALE	Asymptotically locally Euclidean
ALF	Asymptotically locally flat
ASD	Anti-self-dual
BGPP	Belinskii–Gibbons–Page–Pope
BPS	Bogomolny–Prasad–Sommerfeld
CR	Cauchy–Riemann
DE	Differential equation
dKP	Dispersionless Kadomtsev–Petviashvili
EDS	Exterior differential system
EW	Einstein–Weyl
GLM	Gelfand–Levitan–Marchenko
IST	Inverse scattering transform
KdV	Korteweg–de Vries
LHC	Large hadron collider
LHS	Left-hand side
ODE	Ordinary differential equation
PDE	Partial differential equation
PP	Painlevé property
QM	Quantum mechanics
RHS	Right-hand side
SD	Self-dual
WZW	Wess–Zumino–Witten
YM	Yang–Mills

1 Integrability in classical mechanics

Integrable systems are non-linear differential equations (DEs) which 'in principle' can be solved analytically. This means that the solution can be reduced to a finite number of algebraic operations and integrations. Such systems are very rare – most non-linear DEs admit chaotic behaviour and no explicit solutions can be written down. Integrable systems nevertheless lead to very interesting mathematics ranging from differential geometry and complex analysis to quantum field theory and fluid dynamics. In this chapter we shall introduce the integrability of ordinary differential equations (ODEs). This is a fairly clear concept based on existence of sufficiently many well-behaved first integrals, or, as a physicist would put it, constants of the motion.

1.1 Hamiltonian formalism

Motion of a system with n degrees of freedom is described by a trajectory in a $2n$ dimensional phase space M (locally think of an open set in $\mathbb{R}^{2n}$ but globally it can be a topologically non-trivial manifold – for example, a sphere or a torus. See Appendix A) with local coordinates

$$(p_j, q_j), \quad j = 1, 2, \ldots, n.$$

The dynamical variables are functions $f : M \times \mathbb{R} \longrightarrow \mathbb{R}$, so that $f = f(p, q, t)$ where t is called 'time'. Let $f, g : M \times \mathbb{R} \longrightarrow \mathbb{R}$. Define the Poisson bracket of f, g to be the function

$$\{f, g\} := \sum_{k=1}^{n} \frac{\partial f}{\partial q_k}\frac{\partial g}{\partial p_k} - \frac{\partial f}{\partial p_k}\frac{\partial g}{\partial q_k}. \tag{1.1.1}$$

It satisfies

$$\{f, g\} = -\{g, f\}, \qquad \{f, \{g, h\}\} + \{g, \{h, f\}\} + \{h, \{f, g\}\} = 0.$$

These two properties are called the *skew-symmetry* and the *Jacobi identity*, respectively. One says that two functions f, g are *in involution* if $\{f, g\} = 0$.

The coordinate functions (p_j, q_j) satisfy the canonical commutation relations

$$\{p_j, p_k\} = 0, \quad \{q_j, q_k\} = 0, \text{ and } \{q_j, p_k\} = \delta_{jk}.$$

Given a Hamiltonian $H = H(p, q, t)$ (usually $H(p, q)$) the dynamics is determined by

$$\frac{df}{dt} = \frac{\partial f}{\partial t} + \{f, H\}, \qquad \text{for any} \qquad f = f(p, q, t).$$

Setting $f = p_j$ or $f = q_j$ yields Hamilton's equations of motion

$$\dot{p}_j = -\frac{\partial H}{\partial q_j} \quad \text{and} \quad \dot{q}_j = \frac{\partial H}{\partial p_j}. \tag{1.1.2}$$

The system (1.1.2) of $2n$ ODEs is *deterministic* in the sense that $(p_j(t), q_j(t))$ are uniquely determined by $2n$ initial conditions $(p_j(0), q_j(0))$. Equations (1.1.2) also imply that volume elements in phase space are conserved. This system is essentially equivalent to Newton's equations of motion. The Hamiltonian formulation allows a more geometrical insight into classical mechanics. It is also the starting point to quantization.

Definition 1.1.1 *A function $f = f(p_j, q_j, t)$ which satisfies $\dot{f} = 0$ when equations* (1.1.2) *hold is called a first integral or a constant of motion. Equivalently,*

$$f(p(t), q(t), t) = const$$

if $p(t), q(t)$ are solutions of (1.1.2).

In general the system (1.1.2) will be solvable if it admits 'sufficiently many' first integrals and the reduction of order can be applied. This is because any first integral eliminates one equation.

- **Example.** Consider a system with one degree of freedom with $M = \mathbb{R}^2$ and the Hamiltonian

  $$H(p, q) = \frac{1}{2}p^2 + V(q).$$

  Hamilton's equations (1.1.2) give

  $$\dot{q} = p \text{ and } \dot{p} = -\frac{dV}{dq}.$$

  The Hamiltonian itself is a first integral as $\{H, H\} = 0$. Thus

  $$\frac{1}{2}p^2 + V(q) = E$$

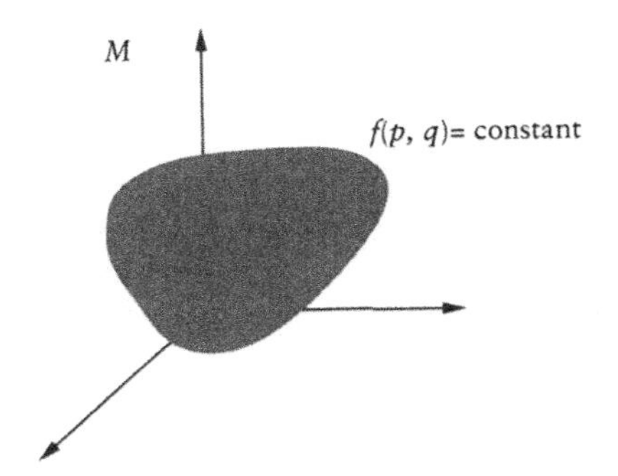

Figure 1.1 *Level surface*

where E is a constant called energy. Now

$$\dot{q} = p, \quad p = \pm\sqrt{2[E - V(q)]}$$

and one integration gives a solution in the implicit form

$$t = \pm \int \frac{dq}{\sqrt{2[E - V(q)]}}.$$

The explicit solution could be found if we can perform the integral on the right-hand side (RHS) and invert the relation $t = t(q)$ to find $q(t)$. These two steps are not always possible but nevertheless we would certainly regard this system as integrable.

It is useful to adopt a more geometrical approach. Assume that a first integral f does not explicitly depend on time, and that it defines a hypersurface $f(p, q) = \text{const}$ in M (Figure.1.1). Two hypersurfaces corresponding to two independent first integrals generically intersect in a surface of co-dimension 2 in M. In general the trajectory lies on a surface of dimension $2n - L$ where L is the number of independent first integrals. If $L = 2n - 1$ this surface is a curve – a solution to (1.1.2).

How may we find first integrals? Given two first integrals which do not explicitly depend on time their Poisson bracket will also be a first integral if it is not zero. This follows from the Jacobi identity and the fact all first integrals Poisson commute with the Hamiltonian. More generally, the Noether theorem gives some first integrals (this theorem relates the symmetries that Hamilton's equation (1.1.2) may possess, for example, time translation, rotations, etc., to first integrals) but not enough. The difficulty with finding the first integrals has deep significance. For assume we use some existence theorem for ODEs and apply it to (1.1.2). Now solve the algebraic equations

$$q_k = q_k(p^0, q^0, t) \text{ and } p_k = p_k(p^0, q^0, t),$$

for the initial conditions (p^0, q^0) thus giving

$$q^0{}_k = q^0{}_k(p, q, t) \text{ and } p^0{}_k = p^0{}_k(p, q, t).$$

This gives $2n$ first integrals as obviously (p^0, q^0) are constants which we can freely specify. One of these integrals determines the time parametrizations and others could perhaps be used to construct the trajectory in the phase space. However for some of the integrals the equation

$$f(p, q) = \text{const}$$

may not define a 'nice' surface in the phase space. Instead it defines a pathological (at least from the applied mathematics point of view) set which densely covers the phase space. Such integrals do not separate points in M.

One first integral – energy – always exist for Hamiltonian systems giving the energy surface $H(p, q) = E$, but often it is the only first integral. Sufficiently complicated, deterministic systems may behave according to the laws of thermodynamics: the probability that the system is contained in some element of the energy surface is proportional to the normalized volume of this element. This means that the time evolution covers uniformly the entire region of the constant energy surface in the phase space. It is not known whether this ergodic postulate can be derived from Hamilton's equations.

Early computer simulations in the 1960s revealed that some non-linear systems (with infinitely many degrees of freedom!) are not ergodic. Soliton equations

$$u_t = 6uu_x - u_{xxx}, \quad u = u(x, t), \qquad \text{KdV}$$

or

$$\phi_{xx} - \phi_{tt} = \sin\phi, \quad \phi = \phi(x, t), \qquad \text{Sine-Gordon}$$

are examples of such systems. Both possess infinitely many first integrals. We shall study them in Chapter 2.

1.2 Integrability and action–angle variables

Given a system of Hamilton's equations (1.1.2) it is often sufficient to know n (rather than $2n - 1$) first integrals as each of them reduces the order of the system by two. This underlies the following definition of an *integrable system.*

Definition 1.2.1 *An integrable system consists of a $2n$-dimensional phase-space M together with n globally defined independent functions (in the sense that the gradients ∇f_j are linearly independent vectors on the tangent space at* *any point in M)* $f_1, \ldots, f_n : M \to \mathbb{R}$ *such that*

$$\{f_j, f_k\} = 0, \qquad j, k = 1, \ldots, n. \tag{1.2.3}$$

The vanishing of Poisson brackets (1.2.3) means that the first integrals are in involution. We shall show that integrable systems lead to completely solvable Hamilton's equations of motion. Let us first explore the freedom in (1.1.2) given by a coordinate transformation of phase space

$$Q_k = Q_k(p, q) \text{ and } P_k = P_k(p, q).$$

This transformation is called *canonical* if it preserves the Poisson bracket

$$\sum_{k=1}^{n} \frac{\partial f}{\partial q_k}\frac{\partial g}{\partial p_k} - \frac{\partial f}{\partial p_k}\frac{\partial g}{\partial q_k} = \sum_{k=1}^{n} \frac{\partial f}{\partial Q_k}\frac{\partial g}{\partial P_k} - \frac{\partial f}{\partial P_k}\frac{\partial g}{\partial Q_k}$$

for all $f, g : M \longrightarrow \mathbb{R}$. Canonical transformations preserve Hamilton's equations (1.1.2).

Given a function $S(q, P, t)$ such that

$$\det\left(\frac{\partial^2 S}{\partial q_j \partial P_k}\right) \neq 0,$$

we can construct a canonical transformation by setting

$$p_k = \frac{\partial S}{\partial q_k}, \qquad Q_k = \frac{\partial S}{\partial P_k}, \text{ and } \widetilde{H} = H + \frac{\partial S}{\partial t}.$$

The function S is an example of a generating function [5, 102, 187]. The idea behind the following theorem is to seek a canonical transformation such that in the new variables $H = H(P_1, \ldots, P_n)$ so that

$$P_k(t) = P_k(0) = \text{const} \quad \text{and} \quad Q_k(t) = Q_k(0) + t\frac{\partial H}{\partial P_k}.$$

Finding the generating function for such canonical transformation is in practice very difficult, and deciding whether a given Hamiltonian system is integrable (without a priori knowledge of n Poisson commuting integrals) is still an open problem.

Theorem 1.2.2 (Arnold–Liouville theorem [5]) *Let*

$$(M, f_1, \ldots, f_n)$$

be an integrable system with Hamiltonian $H = f_1$, *and let*

$$M_f := \{(p, q) \in M; f_k(p, q) = c_k\}, \qquad c_k = const, \qquad k = 1, \ldots, n$$

be an n-dimensional level surface of first integrals f_k. *Then*

- *If M_f is compact and connected then it is diffeomorphic to a torus*

$$T^n := S^1 \times S^1 \times \cdots \times S^1,$$

and (in a neighbourhood of this torus in M) one can introduce 'action–angle' coordinates

$$I_1, \ldots, I_n, \phi_1, \ldots, \phi_n, \qquad 0 \le \phi_k \le 2\pi,$$

such that the angles ϕ_k *are coordinates on* M_f *and the* actions $I_k = I_k(f_1, \ldots, f_n)$ *are first integrals.*
- *The canonical equations of motion* (1.1.2) *become*

$$\dot{I}_k = 0 \text{ and } \dot{\phi}_k = \omega_k(I_1, \ldots, I_n), \qquad k = 1, \ldots, n \tag{1.2.4}$$

and so the integrable systems are solvable by quadratures (a finite number of algebraic operations and integrations of known functions).

Proof We shall follow the proof given in [5], but try to make it more accessible by avoiding the language of differential forms.

- The motion takes place on the surface

$$f_1(p, q) = c_1, \; f_2(p, q) = c_2, \ldots, \; f_n(p, q) = c_n$$

of dimension $2n - n = n$. The first part of the theorem says that this surface is a torus.[1] For each point in M there exists precisely one torus T^n passing through that point. This means that M admits a foliation by n-dimensional leaves. Each leaf is a torus and different tori correspond to different choices of the constants $c_1, \ldots, c_n$.

Assume

$$\det\left(\frac{\partial f_j}{\partial p_k}\right) \neq 0$$

so that the system $f_k(p, q) = c_k$ can be solved for the momenta p_i

$$p_i = p_i(q, c)$$

and the relations $f_i(q, p(q, c)) = c_i$ hold identically. Differentiate these identities with respect to q_j

$$\frac{\partial f_i}{\partial q_j} + \sum_k \frac{\partial f_i}{\partial p_k}\frac{\partial p_k}{\partial q_j} = 0$$

[1] This part of the proof requires some knowledge of Lie groups and Lie algebras. It is given in Appendix A.

and multiply the resulting equations by $\partial f_m/\partial p_j$

$$\sum_j \frac{\partial f_m}{\partial p_j}\frac{\partial f_i}{\partial q_j} + \sum_{j,k} \frac{\partial f_m}{\partial p_j}\frac{\partial f_i}{\partial p_k}\frac{\partial p_k}{\partial q_j} = 0.$$

Now swap the indices and subtract $(mi) - (im)$. This yields

$$\{f_i, f_m\} + \sum_{j,k}\left(\frac{\partial f_m}{\partial p_j}\frac{\partial f_i}{\partial p_k}\frac{\partial p_k}{\partial q_j} - \frac{\partial f_i}{\partial p_j}\frac{\partial f_m}{\partial p_k}\frac{\partial p_k}{\partial q_j}\right) = 0.$$

The first term vanishes as the first integrals are in involution. Rearranging the indices in the second term gives

$$\sum_{j,k}\frac{\partial f_i}{\partial p_k}\frac{\partial f_m}{\partial p_j}\left(\frac{\partial p_k}{\partial q_j} - \frac{\partial p_j}{\partial q_k}\right) = 0$$

and, as the matrices $\partial f_i/\partial p_k$ are invertible,

$$\frac{\partial p_k}{\partial q_j} - \frac{\partial p_j}{\partial q_k} = 0. \qquad (1.2.5)$$

This condition implies that

$$\oint \sum_j p_j dq_j = 0$$

for any closed contractible curve on the torus T^n. This is a consequence of Stokes' theorem. To see it recall that in $n = 3$

$$\oint_{\delta D} \mathbf{p}\cdot d\mathbf{q} = \int_D (\nabla \times \mathbf{p}) \cdot d\mathbf{q}$$

where δD is the boundary of the surface D and

$$(\nabla \times \mathbf{p})_m = \frac{1}{2}\epsilon_{jkm}\left(\frac{\partial p_k}{\partial q_j} - \frac{\partial p_j}{\partial q_k}\right).$$

- There are n closed curves which cannot be contracted down to a point, so that the corresponding integrals do not automatically vanish.

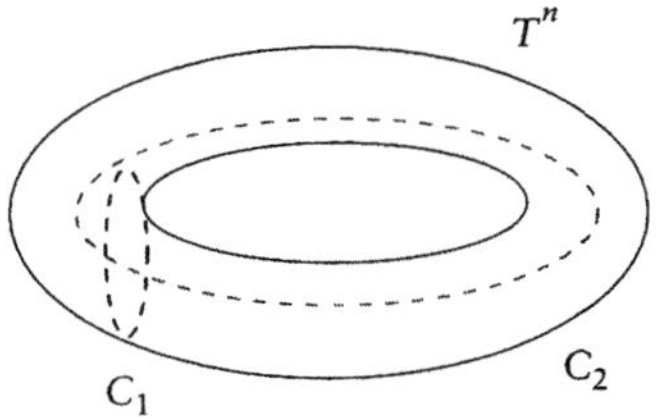

Cycles on a torus

Therefore we can define the action coordinates

$$I_k := \frac{1}{2\pi}\oint_{\Gamma_k}\sum_j p_j dq_j, \tag{1.2.6}$$

where the closed curve Γ_k is the kth basic cycle (the term 'cycle' in general means 'submanifold without boundary') of the torus T^n

$$\Gamma_k = \{(\tilde{\phi}_1, \ldots, \tilde{\phi}_n) \in T^n; 0 \le \tilde{\phi}_k \le 2\pi, \tilde{\phi}_j = \text{const} \ \text{ for } j \neq k\},$$

where $\tilde{\phi}$ are some coordinates[2] on T^n.
Stokes' theorem implies that the actions (1.2.6) are independent of the choice of Γ_k.

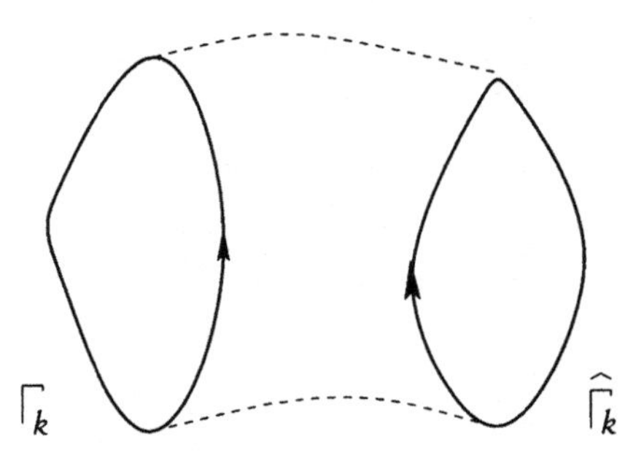

Stokes' theorem

This is because

$$\oint_{\Gamma_k}\sum_j p_j dq_j + \oint_{\hat{\Gamma}_k}\sum_j p_j dq_j = \int\left(\frac{\partial p_i}{\partial q_j} - \frac{\partial p_j}{\partial q_i}\right) dq_j \wedge dq_i = 0$$

where we have chosen Γ and $\hat{\Gamma}$ to have opposite orientations.

- The actions (1.2.6) are also first integrals as $\oint p(q, c)dq$ only depends on $c_k = f_k$ and the f_k's are first integrals. The actions are Poisson commuting

$$\{I_i, I_j\} = \sum_{r,s,k}\frac{\partial I_i}{\partial f_r}\frac{\partial f_r}{\partial q_k}\frac{\partial I_j}{\partial f_s}\frac{\partial f_s}{\partial p_k} - \frac{\partial I_i}{\partial f_r}\frac{\partial f_r}{\partial p_k}\frac{\partial I_j}{\partial f_s}\frac{\partial f_s}{\partial q_k} = \sum_{r,s}\frac{\partial I_i}{\partial f_r}\frac{\partial I_j}{\partial f_s}\{f_r, f_s\} = 0$$

and in particular $\{I_k, H\} = 0$.

The torus M_f can be equivalently represented by

$$I_1 = \tilde{c}_1, \ldots, \ I_1 = \tilde{c}_n,$$

[2] This is a non-trivial step. In practice it is unclear how to explicitly describe the n-dimensional torus and the curves Γ_k in $2n$-dimensional phase space. Thus, to some extend the Arnold–Liouville theorem has the character of an existence theorem.

for some constants $\tilde{c}_1, \ldots, \tilde{c}_n$. (We might have been tempted just to define $I_k = f_k$, but then the transformation $(p, q) \to (I, \phi)$ would not be canonical in general.)

- We shall construct the angle coordinates ϕ_k canonically to conjugate to the actions using a generating function

$$S(q, I) = \int_{q_0}^{q} \sum_j p_j dq_j,$$

where q_0 is some chosen point on the torus. This definition does not depend on the path joining q_0 and q as a consequence of (1.2.5) and Stokes' theorem. Choosing a different q_0 just adds a constant to S thus leaving the *angles*

$$\phi_i = \frac{\partial S}{\partial I_i}$$

invariant.

- The angles are periodic coordinates with a period 2π. To see this consider two paths C and $C \cup C_k$ (where C_k represents the kth cycle) between q_0 and q and calculate

$$S(q, I) = \int_{C \cup C_k} \sum_j p_j dq_j = \int_C \sum_j p_j dq_j + \int_{C_k} \sum_j p_j dq_j = S(q, I) + 2\pi I_k$$

so

$$\phi_k = \frac{\partial S}{\partial I_k} = \phi_k + 2\pi.$$

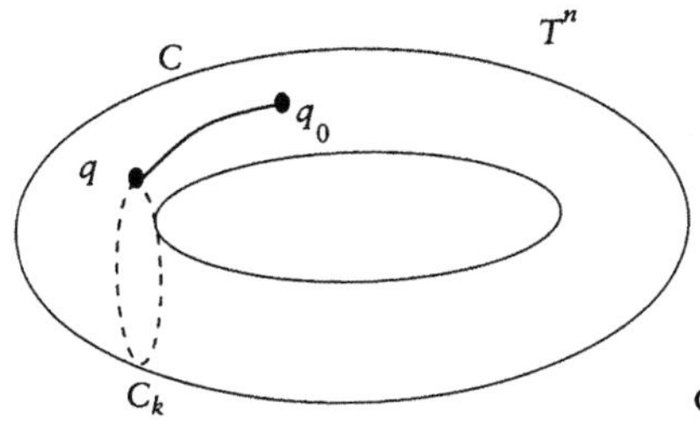

Generating function

- The transformations

$$q = q(\phi, I), \quad p = p(\phi, I) \qquad \text{and} \qquad \phi = \phi(q, p), \quad I = I(q, p)$$

are canonical (as they are defined by a generating function) and invertible. Thus,

$$\{I_j, I_k\} = 0, \quad \{\phi_j, \phi_k\} = 0, \ \text{and} \ \{\phi_j, I_k\} = \delta_{jk}$$

and the dynamics is given by

$$\dot{\phi}_k = \{\phi_k, \widetilde{H}\} \text{ and } \dot{I}_k = \{I_k, \widetilde{H}\},$$

where

$$\widetilde{H}(\phi, I) = H(q(\phi, I), p(\phi, I)).$$

The I_k's are first integrals, therefore

$$\dot{I}_k = -\frac{\partial \widetilde{H}}{\partial \phi_k} = 0$$

so $\widetilde{H} = \widetilde{H}(I)$ and

$$\dot{\phi}_k = \frac{\partial \widetilde{H}}{\partial I_k} = \omega_k(I)$$

where the ω_k's are also first integrals. This proves (1.2.4). Integrating these canonical equations of motion yields

$$\phi_k(t) = \omega_k(I)t + \phi_k(0) \quad \text{and} \quad I_k(t) = I_k(0). \tag{1.2.7}$$

These are n circular motions with constant angular velocities. □

The trajectory (1.2.7) may be closed on the torus or it may cover it densely. This depends on the values of the angular velocities. If $n = 2$ the trajectory will be closed if ω_1/ω_2 is rational and dense otherwise.

Interesting things happen to the tori under a small perturbation of the integrable Hamiltonian

$$H(I) \longrightarrow H(I) + \epsilon K(I, \phi).$$

In some circumstances the motion is still periodic and most tori do not vanish but become deformed. This is governed by the Kolmogorov–Arnold–Moser theorem – not covered in this book. Consult the popular book by Schuster [145], or read the complete account given by Arnold [5].

- **Example.** All time-independent Hamiltonian systems with two-dimensional phase spaces are integrable. Consider the harmonic oscillator with the Hamiltonian

$$H(p, q) = \frac{1}{2}(p^2 + \omega^2 q^2).$$

Different choices of the energy E give a foliation of M by ellipses

$$\frac{1}{2}(p^2 + \omega^2 q^2) = E.$$

For a fixed value of E we can take $\Gamma = M_f$. Therefore

$$I = \frac{1}{2\pi}\oint_{M_f} pdq = \frac{1}{2\pi}\int\int_S dpdq = \frac{E}{\omega}$$

where we used the Stokes' theorem to express the line integral in terms of the area enclosed by M_f.

The Hamiltonian expressed in the new variables is $\widetilde{H} = \omega I$ and

$$\dot{\phi} = \frac{\partial \widetilde{H}}{\partial I} = \omega \quad \text{and} \quad \phi = \omega t + \phi_0.$$

To complete the picture we need to express (I, ϕ) in terms of (p, q). We already know

$$I = \frac{1}{2}\left(\frac{1}{\omega}p^2 + \omega q^2\right).$$

Thus the generating function is

$$S(q, I) = \int pdq = \pm\int\sqrt{2I\omega - \omega^2 q^2}dq$$

and (choosing a sign)

$$\phi = \frac{\partial S}{\partial I} = \int\frac{\omega dq}{\sqrt{2I\omega - \omega^2 q^2}} = \arcsin\left(q\sqrt{\frac{\omega}{2I}}\right) - \phi_0.$$

This gives

$$q = \sqrt{\frac{2I}{\omega}}\sin(\phi + \phi_0)$$

and finally we recover the familiar solution

$$p = \sqrt{2E}\cos(\omega t + \phi_0) \quad \text{and} \quad q = \sqrt{2E/\omega^2}\sin(\omega t + \phi_0).$$

- **Example.** The Kepler problem is another tractable example. Here the four-dimensional phase space is coordinatized by $(q_1 = \phi, q_2 = r, p_1 = p_\phi, p_2 = p_r)$ and the Hamiltonian is

$$H = \frac{p_\phi{}^2}{2r^2} + \frac{p_r{}^2}{2} - \frac{\alpha}{r}$$

where $\alpha > 0$ is a constant. One readily verifies that

$$\{H, p_\phi\} = 0$$

so the system is integrable in the sense of Definition 1.2.1. The level set M_f of first integrals is given by

$$H = E \quad \text{and} \quad p_\phi = \mu$$

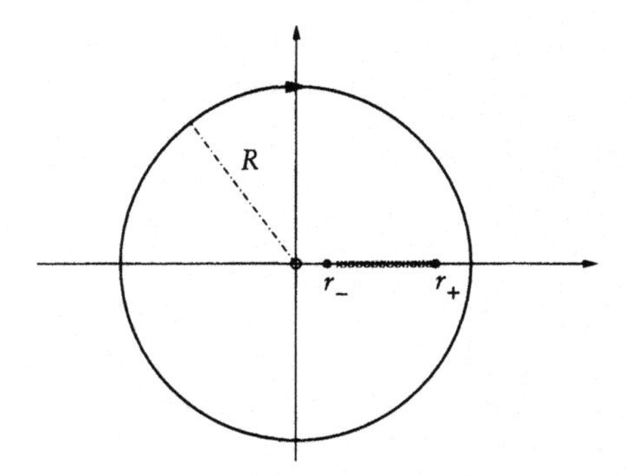

Figure 1.2 *Branch cut for the Kepler integral*

which gives

$$p_\phi = \mu \quad \text{and} \quad p_r = \pm\sqrt{2E - \frac{\mu^2}{r^2} + \frac{2\alpha}{r}}.$$

This leaves ϕ arbitrary and gives one constraint on (r, p_r). Thus, ϕ and one function of (r, p_r) parameterize M_f. Varying ϕ and fixing the other coordinate gives one cycle $\Gamma_\phi \subset M_f$ and

$$I_\phi = \frac{1}{2\pi}\oint_{\Gamma_\phi} p_\phi d\phi + p_r dr = \frac{1}{2\pi}\int_0^{2\pi} p_\phi d\phi = p_\phi.$$

To find the second action coordinate fix ϕ (as well as H and p_ϕ). This gives another cycle Γ_r and

$$\begin{aligned} I_r &= \frac{1}{2\pi}\oint_{\Gamma_r} p_r dr \\ &= 2\frac{1}{2\pi}\int_{r_-}^{r_+}\sqrt{2E - \frac{\mu^2}{r^2} + \frac{2\alpha}{r}}dr \\ &= \frac{\sqrt{-2E}}{\pi}\int_{r_-}^{r_+}\frac{\sqrt{(r-r_-)(r_+-r)}}{r}dr \end{aligned}$$

where the periodic orbits have $r_- \le r \le r_+$ and

$$r_\pm = \frac{\alpha \pm \sqrt{\alpha^2 + 2\mu^2 E}}{2E}.$$

The integral can be performed using the residue calculus and choosing a contour with a branch cut (Figure 1.2) from r_- to r_+ on the real axis.[3] Consider

[3] The following method is taken from Max Born's *The Atom* published in 1927. I thank Gary Gibbons for pointing out this reference to me.

a branch of

$$f(z) = \sqrt{(z - r_-)(r_+ - z)}$$

defined by a branch cut from r_- to r_+ with $f(0) = i\sqrt{(r_+ r_-)}$ on the top side of the cut. We evaluate the integral over a large circular contour $|z| = R$ integrating the Laurent expansion

$$\int_{|z|=R} z^{-1} f(z)dz = \int_0^{2\pi} \sqrt{-1}\left(1 - \frac{r_-}{R}e^{-i\theta}\right)^{1/2}\left(1 - \frac{r_+}{R}e^{-i\theta}\right)^{1/2} i Re^{i\theta} d\theta$$
$$= \pi(r_+ + r_-) \qquad \text{when} \quad R \to \infty,$$

since all terms containing powers of $\exp(i\theta)$ are periodic and do not contribute to the integral. The same value must arise from the residue at 0 and collapsing the contour onto the branch cut (when calculating the residue remember that $z = 0$ is on the left-hand side (LHS) of the cut and thus $\sqrt{-1} = -i$. Integration along the big circle is equivalent to taking a residue at ∞ which is on the right side of the cut where $\sqrt{-1} = i$). Thus

$$\pi(r_+ + r_-) = 2\pi\sqrt{r_+ r_-} + \int_{r_-}^{r_+} \frac{\sqrt{(r - r_-)(r_+ - r)}}{r} dr$$
$$- \int_{r_+}^{r_-} \frac{\sqrt{(r - r_-)(r_+ - r)}}{r} dr$$

and

$$I_r = \frac{\sqrt{-2E}}{\pi}\frac{\pi}{2}(r_+ + r_- - 2\sqrt{r_+ r_-})$$
$$= \alpha\sqrt{\frac{1}{2|E|}} - \mu.$$

The Hamiltonian becomes

$$\widetilde{H} = -\frac{\alpha^2}{2(I_r + I_\phi)^2}$$

and we conclude that the absolute values of frequencies are equal and given by

$$\frac{\partial \widetilde{H}}{\partial I_r} = \frac{\partial \widetilde{H}}{\partial I_\phi} = \frac{\alpha^2}{(I_r + I_\phi)^3} = \left(\frac{r_+ + r_-}{2}\right)^{-3/2}\sqrt{\alpha}.$$

This is a particular case when the ratio of two frequencies is a rational number (here it is equal to 1). The orbits are therefore closed – a remarkable result known to Kepler.

1.3 Poisson structures

There is a natural way to extend the Hamiltonian formalism by generalizing the notion of Poisson bracket (1.1.1). A geometric approach is given by symplectic geometry [5]. We shall take a lower level (but a slightly more general) point of view and introduce Poisson structures. The phase space M is m dimensional with local coordinates $(\xi^1, \ldots, \xi^m)$. In particular we do not distinguish between positions and momenta.

Definition 1.3.1 *A skew-symmetric matrix $\omega^{ab} = \omega^{ab}(\xi)$ is called a Poisson structure if the Poisson bracket defined by*

$$\{f, g\} = \sum_{a,b=1}^{m} \omega^{ab}(\xi) \frac{\partial f}{\partial \xi^a} \frac{\partial g}{\partial \xi^b} \tag{1.3.8}$$

satisfies

$$\{f, g\} = -\{g, f\},$$

$$\{f, \{g, h\}\} + \{h, \{f, g\}\} + \{g, \{h, f\}\} = 0.$$

The second property is called the Jacobi identity. It puts restrictions on $\omega^{ab}(\xi)$ which can be seen noting that

$$\omega^{ab}(\xi) = \{\xi^a, \xi^b\}$$

and evaluating the Jacobi identity on coordinate functions which yields

$$\sum_{d=1}^{m} \omega^{dc} \frac{\omega^{ab}}{\partial \xi^d} + \omega^{db} \frac{\omega^{ca}}{\partial \xi^d} + \omega^{da} \frac{\omega^{bc}}{\partial \xi^d} = 0.$$

Given a Hamiltonian $H : M \times \mathbb{R} \longrightarrow \mathbb{R}$ the dynamics is governed by

$$\frac{df}{dt} = \frac{\partial f}{\partial t} + \{f, H\}$$

and Hamilton's equations generalizing (1.1.2) become

$$\dot{\xi}^a = \sum_{b=1}^{m} \omega^{ab}(\xi) \frac{\partial H}{\partial \xi^b}. \tag{1.3.9}$$

- **Example.** Let $M = \mathbb{R}^3$ and $\omega^{ab} = \sum_{c=1}^{3} \varepsilon^{abc} \xi^c$, where ε^{abc} is the standard totally antisymmetric tensor. Thus

$$\{\xi^1, \xi^2\} = \xi^3, \quad \{\xi^3, \xi^1\} = \xi^2, \text{ and } \{\xi^2, \xi^3\} = \xi^1.$$

This Poisson structure admits a *Casimir* – any function $f(r)$ where

$$r = \sqrt{(\xi^1)^2 + (\xi^2)^2 + (\xi^3)^2}.$$

Poisson commutes with the coordinate functions

$$\{f(r), \xi^a\} = 0.$$

This is independent of the choice of Hamiltonian. With the choice

$$H = \frac{1}{2}\left[\frac{(\xi^1)^2}{a_1} + \frac{(\xi^2)^2}{a_2} + \frac{(\xi^3)^2}{a_3}\right]$$

where a_1, a_2, and a_3 are constants, and Hamilton's equations (1.3.9) become the equations of motion of a rigid body fixed at its centre of gravity

$$\dot{\xi}^1 = \frac{a_3 - a_2}{a_2 a_3}\xi^2\xi^3, \qquad \dot{\xi}^2 = \frac{a_1 - a_3}{a_1 a_3}\xi^1\xi^3, \quad \text{and} \quad \dot{\xi}^3 = \frac{a_2 - a_1}{a_1 a_2}\xi^1\xi^2.$$

Assume that $m = 2n$ is even and the matrix ω is invertible with $W_{ab} := (\omega^{-1})_{ab}$. The Jacobi identity implies that the antisymmetric matrix $W_{ab}(\xi)$ is closed, that is,

$$\partial_a W_{bc} + \partial_c W_{ab} + \partial_b W_{ca} = 0, \quad \forall a, b, c = 1, \ldots, m.$$

In this case W_{ab} is called a *symplectic structure*. The Darboux theorem [5] states that in this case there locally exists a coordinate system

$$\xi^1 = q_1, \ldots, \xi^n = q_n, \xi^{n+1} = p_1, \ldots, \xi^{2n} = p_n$$

such that

$$\omega = \begin{pmatrix} 0 & 1_n \\ -1_n & 0 \end{pmatrix}$$

and the Poisson bracket reduces to the standard form (1.1.1). A simple proof can be found in [5]. One constructs a local coordinate system (p, q) by induction with respect to half of the dimension of M. Choose a function p_1, and find q_1 by solving the equation $\{q_1, p_1\} = 1$. Then consider a level set of (p_1, q_1) in M which is locally a symplectic manifold. Now look for (p_2, q_2), etc.

- **Example.** The Poisson structure in the last example is degenerate as the matrix ω^{ab} is not invertible. This degeneracy always occurs if the phase space is odd dimensional or/and there exists a non-trivial Casimir. Consider the restriction of $\omega^{ab} = \sum_{c=1}^{3} \varepsilon^{abc}\xi^c$ to a two-dimensional sphere $r = C$. This gives a symplectic structure on the sphere given by

$$\{\xi^1, \xi^2\} = \sqrt{C^2 - (\xi^1)^2 - (\xi^2)^2}$$

or

$$W = \frac{1}{\sqrt{C^2 - (\xi^1)^2 - (\xi^2)^2}} \begin{pmatrix} 0 & 1 \\ -1 & 0 \end{pmatrix}.$$

This of course has no Casimir functions apart from constants. It is convenient to choose a different parameterization of the sphere: if

$$\xi^1 = C \sin\theta \cos\phi, \quad \xi^2 = C \sin\theta \sin\phi, \text{ and } \xi^3 = C \cos\theta$$

then in the local coordinates (θ, ϕ) the symplectic structure is given by $\{\theta, \phi\} = \sin^{-1}\theta$ or

$$W = \sin\theta \begin{pmatrix} 0 & 1 \\ -1 & 0 \end{pmatrix}$$

which is equal to the volume form on the two-sphere. The radius C is arbitrary. Therefore the Poisson phase space $\mathbb{R}^3$ is foliated by symplectic phase spaces S^2 as there is exactly one sphere centred at the origin through any point of $\mathbb{R}^3$. This is a general phenomenon: fixing the values of the Casimir functions on Poisson spaces gives the foliations by symplectic spaces. The local Darboux coordinates on S^2 are given by $q = -\cos\theta$, $p = \phi$ as then

$$\{q, p\} = 1.$$

The Poisson generalization is useful to set up the Hamiltonian formalism in the infinite-dimensional case. Formally one can think of replacing the coordinates on the trajectory $\xi^a(t)$ by a dynamical variable $u(x, t)$. Thus the discrete index a becomes the continuous independent variable x (think of m points on a string versus the whole string). The phase space $M = \mathbb{R}^m$ is replaced by a space of smooth functions on a line with appropriate boundary conditions (decay or periodic). The whole formalism may be set up making the following replacements

$$\text{ODEs} \longrightarrow \text{PDEs}$$

$$\xi^a(t), a = 1, \dots, m \longrightarrow u(x, t), x \in \mathbb{R}$$

$$\sum_a \longrightarrow \int_{\mathbb{R}} dx$$

$$\text{function } f(\xi) \longrightarrow \text{functional } F[u]$$

$$\frac{\partial}{\partial \xi^a} \longrightarrow \frac{\delta}{\delta u}.$$

The functionals are given by integrals

$$F[u] = \int_{\mathbb{R}} f(u, u_x, u_{xx}, \dots) dx$$

(we could in principle allow t derivatives, but we will not for reasons to become clear shortly). Recall that the functional derivative is

$$\frac{\delta F}{\delta u(x)} = \frac{\partial f}{\partial u} - \frac{\partial}{\partial x}\frac{\partial f}{\partial (u_x)} + \left(\frac{\partial}{\partial x}\right)^2 \frac{\partial f}{\partial (u_{xx})} + \cdots$$

and

$$\frac{\delta u(y)}{\delta u(x)} = \delta(y-x)$$

where the δ on the RHS is the Dirac delta which satisfies

$$\int_{\mathbb{R}} \delta(x)dx = 1, \qquad \delta(x) = 0 \text{ for } x \neq 0.$$

The presence of the Dirac delta will constantly remind us that we have entered a territory which is rather slippery from a pure mathematics perspective. We should be reassured that the formal replacements made above can nevertheless be given a solid functional-analytic foundation. This will not be done in this book.

The analogy with finite-dimensional situation (1.3.8) suggests the following definition of a Poisson bracket:

$$\{F, G\} = \int_{\mathbb{R}^2} \omega(x, y, u)\frac{\delta F}{\delta u(x)}\frac{\delta G}{\delta u(y)}dxdy$$

where the Poisson structure $\omega(x, y, u)$ should be such that the bracket is antisymmetric and the Jacobi identity holds. A canonical (but not the only) choice is

$$\omega(x, y, u) = \frac{1}{2}\frac{\partial}{\partial x}\delta(x-y) - \frac{1}{2}\frac{\partial}{\partial y}\delta(x-y).$$

This is analogous to the Darboux form in which ω^{ab} is a constant and antisymmetric matrix and the Poisson bracket reduces to (1.1.1). This is because the differentiation operator $\partial/\partial x$ is anti-self-adjoint with respect to the inner product

$$< u, v >= \int_{\mathbb{R}} u(x)v(x)dx$$

which is analogous to a matrix being antisymmetric. With this choice

$$\{F, G\} = \int_{\mathbb{R}} \frac{\delta F}{\delta u(x)}\frac{\partial}{\partial x}\frac{\delta G}{\delta u(x)}dx \tag{1.3.10}$$

and Hamilton's equations become

$$\frac{\partial u}{\partial t} = \{u, H[u]\} = \int_{\mathbb{R}} \frac{\delta u(x)}{\delta u(y)}\frac{\partial}{\partial y}\frac{\delta H}{\delta u(y)}dy$$

$$= \frac{\partial}{\partial x} \frac{\delta H[u]}{\delta u(x)}. \tag{1.3.11}$$

- **Example.** The KdV (Korteweg–de Vries) equation mentioned earlier is a Hamiltonian system with the Hamiltonian given by the functional

$$H[u] = \int_{\mathbb{R}} \left(\frac{1}{2} u_x^2 + u^3 \right) dx.$$

It is assumed that u belongs to the space of functions decaying sufficiently fast when $x \to \pm\infty$.

Exercises

1. Assume that (p_j, q_j) satisfy Hamilton's equations and show that any function $f = f(p, q, t)$ satisfies

$$\frac{df}{dt} = \frac{\partial f}{\partial t} + \{f, H\},$$

where H is the Hamiltonian.

Show that the Jacobi identity

$$\{f_1, \{f_2, f_3\}\} + \{f_3, \{f_1, f_2\}\} + \{f_2, \{f_3, f_1\}\} = 0 \tag{1.3.12}$$

holds for Poisson brackets.

Deduce that if functions f_1 and f_2 which do not explicitly depend on time are first integrals of a Hamiltonian system then so is $f_3 = \{f_1, f_2\}$.

2. • Find the canonical transformation generated by

$$S = \sum_{k=1}^{n} q_k P_k.$$

- Show that the canonical transformations preserve volume in the two-dimensional phase space, that is,

$$\frac{\partial(P, Q)}{\partial(p, q)} = 1.$$

[This result also holds in phase spaces of arbitrary dimension.]

- Show that the transformations

$$Q = \cos(\beta) q - \sin(\beta) p \quad \text{and} \quad P = \sin(\beta) q + \cos(\beta) p$$

are canonical for any constant $\beta \in \mathbb{R}$. Find the corresponding generating functions. Are they defined for all β?

3. Demonstrate that the system of n coupled harmonic oscillators with the Hamiltonian

$$H = \frac{1}{2} \sum_{k=1}^{n} (p_k^2 + \omega_k^2 q_k^2),$$

where $\omega_1, \ldots, \omega_n$ are constants, is completely integrable. Find the action variables for this system.

[Hint: Consider a function $F_1 = (p_1^2 + \omega_1^2 q_1^2)/2$.]

4. Consider the Poisson structure ω^{ab} on $\mathbb{R}^{2n}$ defined by (1.3.8). Show that

$$\{fg, h\} = f\{g, h\} + \{f, h\}g.$$

Assume that the matrix ω is invertible with $W := (\omega^{-1})$ and show that the antisymmetric matrix $W_{ab}(\xi)$ satisfies

$$\partial_a W_{bc} + \partial_c W_{ab} + \partial_b W_{ca} = 0.$$

Deduce that if $n = 1$ then any antisymmetric invertible matrix $\omega(\xi^1, \xi^2)$ gives rise to a Poisson structure (i.e. show that the Jacobi identity holds automatically in this case).

2 Soliton equations and the inverse scattering transform

A universally accepted definition of integrability does not exist for partial differential equations (PDEs). The phase space is infinite dimensional but having 'infinitely many' first integrals may not be enough – we could have missed every second one. One instead focuses on properties of solutions and solution-generation techniques. We shall study solitons – solitary non-linear waves which preserve their shape (and other characteristics) in the evolution. These soliton solutions will be constructed by means of an inverse problem: recovering a potential from the scattering data.

2.1 The history of two examples

Soliton equations originate in the nineteenth century. Some of them appeared in the study of non-linear wave phenomena and others arose in the differential geometry of surfaces in $\mathbb{R}^3$:

- The KdV equation

$$u_t - 6uu_x + u_{xxx} = 0, \quad \text{where} \quad u = u(x, t) \tag{2.1.1}$$

has been written down, and solved in the simplest case, by Korteweg and de Vries in 1895 to explain the following account of J. Scott Russell. Russell observed a soliton while riding on horseback beside a narrow barge channel. The following passage has been taken from J. Scott Russell. Report on waves, 14th meeting of the British Association for the Advancement of Science, 1844.

I was observing the motion of a boat which was rapidly drawn along a narrow channel by a pair of horses, when the boat suddenly stopped – not so the mass of water in the channel which it had put in motion; it accumulated round the prow of the vessel in a state of violent agitation, then suddenly leaving it behind, rolled forward with great velocity, assuming the form of a large solitary elevation, a rounded, smooth and well-defined heap of water, which continued its course along the channel apparently without change of form or diminution of speed. I followed it on horseback, and overtook it still rolling on at a rate of some eight or nine miles an hour, preserving its original figure some thirty feet long and a foot to a foot and a half in height. Its height gradually diminished, and after a chase of one or two miles I lost it in the windings of the channel. Such, in the month of August 1834, was my first chance interview with that singular and beautiful phenomenon which I have called the Wave of Translation.

- The Sine-Gordon equation

$$\phi_{xx} - \phi_{tt} = \sin\phi \quad \text{where} \quad \phi = \phi(x, t) \tag{2.1.2}$$

locally describes the isometric embeddings of surfaces with constant negative Gaussian curvature in the Euclidean space $\mathbb{R}^3$. The function $\phi = \phi(x, t)$ is the angle between two asymptotic directions $\tau = (x + t)/2$ and $\rho = (x - t)/2$ on the surface along which the second fundamental form is zero. If the first fundamental form of a surface parameterized by (ρ, τ) is

$$ds^2 = d\tau^2 + 2\cos\phi \, d\rho d\tau + d\rho^2, \quad \text{where} \quad \phi = \phi(\tau, \rho)$$

then the Gaussian curvature is constant and equal to -1 provided that

$$\phi_{\tau\rho} = \sin\phi$$

which is equivalent to (2.1.2).

The integrability of the Sine-Gordon equation has been used by Bianchi, Bäcklund, Lie, and other classical differential geometers to construct new embeddings.

2.1.1 A physical derivation of KdV

Consider the linear wave equation

$$\Psi_{xx} - \frac{1}{v^2}\Psi_{tt} = 0$$

where $\Psi_{xx} = \partial_x^2\Psi$, etc. which describes a propagation of waves travelling with a constant velocity v. Its derivation is based on three simplifying assumptions:

- There is no dissipation, that is, the equation is invariant with respect to time inversion $t \to -t$.
- The amplitude of oscillation is small and so the non-linear terms (like Ψ^2) can be omitted.
- There is no dispersion, that is, the group velocity is constant.

In the derivation of the KdV we follow [122] and relax these assumptions.

The general solution of the wave equation is a superposition of two waves travelling in opposite directions

$$\Psi = f(x - vt) + g(x + vt)$$

where f and g are arbitrary functions of one variable. Each of these two waves is characterized by a linear first order PDE, for example,

$$\Psi_x + \frac{1}{v}\Psi_t = 0 \longrightarrow \Psi = f(x - vt).$$

To introduce the dispersion consider a complex wave

$$\Psi = e^{i[kx - \omega(k)t]}$$

where $\omega(k) = vk$ and so the group velocity $d\omega/dk$ equals to the phase velocity v. We change this relation by introducing the dispersion

$$\omega(k) = v(k - \beta k^3 + \cdots)$$

where the absence of even terms in this expansion guarantees real dispersion relations. Let us assume that the dispersion is small and truncate this series keeping only the first two terms. The equation satisfied by

$$\Psi = e^{i[kx - v(kt - \beta k^3 t)]}$$

is readily found to be

$$\Psi_x + \beta\Psi_{xxx} + \frac{1}{v}\Psi_t = 0.$$

This can be rewritten in a form of a conservation law

$$\rho_t + j_x = 0,$$

where the density ρ and the current j are given by

$$\rho = \frac{1}{v}\Psi \quad \text{and} \quad j = \Psi + \beta\Psi_{xx}.$$

To introduce non-linearity modify the current

$$j = \Psi + \beta\Psi_{xx} + \frac{\alpha}{2}\Psi^2.$$

The resulting equation is

$$\frac{1}{v}\Psi_t + \Psi_x + \beta\Psi_{xxx} + \alpha\Psi\Psi_x = 0.$$

The non-zero constants (v, β, α) can be eliminated by a simple change of variables $x \to x - vt$ and rescaling Ψ. This leads to the standard form of the KdV equation

$$u_t - 6uu_x + u_{xxx} = 0.$$

The simplest one-soliton solution found by Korteweg and de-Vires is

$$u(x,t) = -\frac{2\chi^2}{\cosh^2 \chi(x - 4\chi^2 t - \phi_0)}. \tag{2.1.3}$$

The KdV is not a linear equation therefore multiplying this solution by a constant will not give another solution. The constant ϕ_0 determines the location of the extremum at $t = 0$. We should therefore think of a one-parameter family of solutions labelled by $\chi \in \mathbb{R}$.

The one-soliton (2.1.3) was the only regular solution of KdV such that $u, u_x \to 0$ as $|x| \to \infty$ known until 1965 when Gardner, Green, Kruskal, and Miura analysed KdV numerically. They took two waves with different amplitudes as their initial profile. The computer simulations revealed that the initially separated waves approached each other distorting their shapes, but eventually the larger wave overtook the smaller wave and both waves re-emerged with their sizes and shapes intact. The relative phase shift was the only result of the non-linear interaction. This behaviour resembles what we usually associate with particles and not waves. Thus Zabruski and Kruskal named these waves 'solitons' (like electrons, protons, baryons, and other particles ending with 'ons'). In this chapter we shall construct more general N-soliton solutions describing the interactions of one-solitons.

To this end we note that the existence of a stable solitary wave is a consequence of cancellations of effects caused by non-linearity and dispersion.

- If the dispersive terms were not present, the equation would be

$$u_t - 6uu_x = 0$$

and the resulting solution would exhibit a discontinuity of first derivatives at some $t_0 > 0$ (a shock or 'breaking wave'). This solution can be easily found using the method of characteristics (see formula (C33)).

Shock

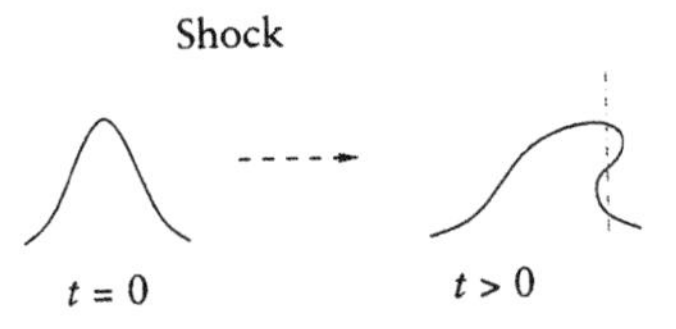

- If the non-linear terms were not present the initial wave profile would disperse in the evolution $u_t + u_{xxx} = 0$.

Dispersion

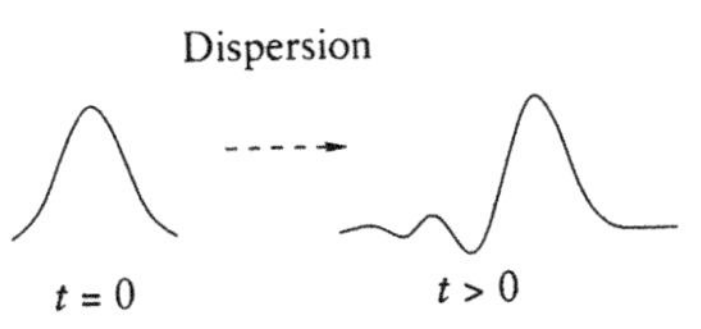

- The presence of both terms allows smooth localized soliton solutions

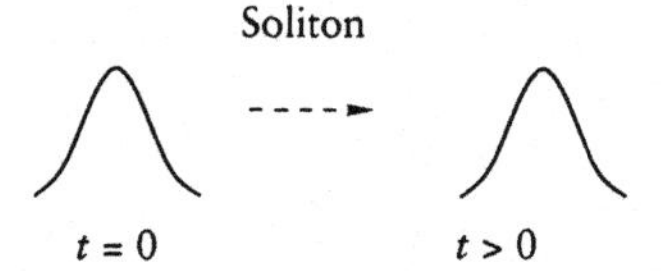

of which (2.1.3) is an example (the plot gives $-u(x,t)$).

2.1.2 Bäcklund transformations for the Sine-Gordon equation

Let us consider the Sine-Gordon equation – the other soliton equation mentioned in the introduction to this chapter. The simplest solution-generating technique is the Bäcklund transformation. Set $\tau = (x+t)/2$ and $\rho = (x-t)/2$ so that the equation (2.1.2) becomes

$$\phi_{\tau\rho} = \sin\phi.$$

Now define the Bäcklund relations

$$\partial_\rho(\phi_1 - \phi_0) = 2b\sin\left(\frac{\phi_1+\phi_0}{2}\right) \quad \text{and} \quad \partial_\tau(\phi_1+\phi_0) = 2b^{-1}\sin\left(\frac{\phi_1-\phi_0}{2}\right),$$

where $b = \text{const}$.

Differentiating the first equation with respect to τ, and using the second equation yields

$$\begin{aligned}\partial_\tau\partial_\rho(\phi_1-\phi_0) &= 2b\,\partial_\tau\sin\left(\frac{\phi_1+\phi_0}{2}\right) = 2\sin\left(\frac{\phi_1-\phi_0}{2}\right)\cos\left(\frac{\phi_1+\phi_0}{2}\right)\\ &= \sin\phi_1 - \sin\phi_0.\end{aligned}$$

Therefore ϕ_1 is a solution to the Sine-Gordon equation if ϕ_0 is. Given ϕ_0 we can solve the first order Bäcklund relations for ϕ_1 and generate new solutions from the ones we know. The trivial solution $\phi_0 = 0$ yields the one-soliton solution of Sine-Gordon

$$\phi_1(x,t) = 4\arctan\left[\exp\left(\frac{x-vt}{\sqrt{1-v^2}} - x_0\right)\right]$$

where v is a constant with $|v| < 1$. This solution is called a kink (Figure 2.1). A static kink corresponds to a special case $v = 0$.

One can associate a topological charge

$$N = \frac{1}{2\pi}\int_{\mathbb{R}} d\phi = \frac{1}{2\pi}\Big[\phi(x=\infty,t) - \phi(x=-\infty,t)\Big]$$

with any solution of the Sine-Gordon equation. It is an integral of a total derivative which depends only on boundary conditions. It is conserved if one

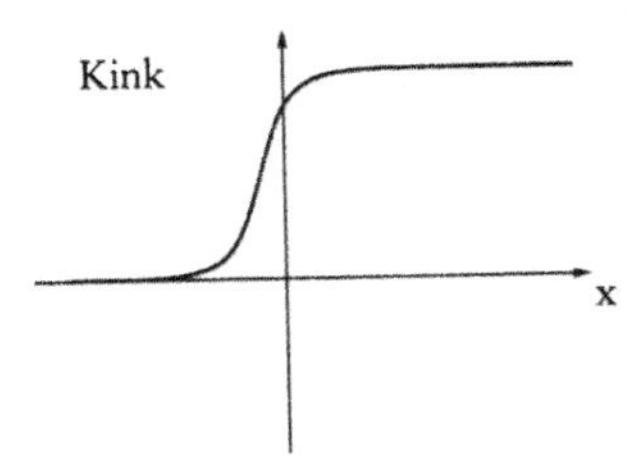

Figure 2.1 *Sine-Gordon kink*

insists on finiteness of the energy

$$E = \int_{\mathbb{R}} \left\{ \frac{1}{2}\left(\phi_t^2 + \phi_x^2\right) + [1 - \cos(\phi)] \right\} dx.$$

Note that the Sine-Gordon equation did not enter the discussion at this stage. Topological charges, like N, are in this sense different from first integrals like E which satisfy $\dot{E} = 0$ as a consequence of (2.1.2). For the given kink solution $N(\phi) = 1$ and the kink is stable[1] as it would take infinite energy to change this solution into a constant solution $\phi = 0$ with $E = 0$.

There exist interesting solutions with $N = 0$: a soliton–antisoliton pair has $N = 0$ but is non-trivial:

$$\phi(x, t) = 4 \arctan \left(\frac{v \cosh \frac{x}{\sqrt{1-v^2}}}{\sinh \frac{vt}{\sqrt{1-v^2}}} \right).$$

At $t \to -\infty$, this solution represents widely separated pair of kink and anti-kink approaching each other with velocity v. A non-linear interaction takes place at $t = 0$ and as $t \to \infty$ kink and anti-kink re-emerge unchanged.

2.2 Inverse scattering transform for KdV

One of the most spectacular methods of solving soliton equations comes from quantum mechanics (QM). It is quite remarkable, as the soliton equations we have discussed so far have little to do with the quantum world.

Recall that the mathematical arena of QM is the infinite-dimensional complex vector space $\mathcal{H}$ of functions [144]. Elements Ψ of this space are referred to as wave functions, or state vectors. In the case of one-dimensional QM we have $\Psi : \mathbb{R} \to \mathbb{C}$, $\Psi = \Psi(x) \in \mathbb{C}$. The space $\mathcal{H}$ is equipped with a unitary

[1] The physical interpretation of kinks within the framework of field theory is discussed in Section 5.3.

inner product

$$(\Psi, \Phi) = \int_{\mathbb{R}} \overline{\Psi(x)}\Phi(x)dx. \tag{2.2.4}$$

Functions which are square integrable, that is, $(\Psi, \Psi) < \infty$ like $\Psi = e^{-x^2}$, are called bound states. Other functions, like e^{-ix}, are called the scattering states.

Given a real-valued function $u = u(x)$ called the potential, the time-independent Schrödinger equation

$$-\frac{\hbar^2}{2m}\frac{d^2\Psi}{dx^2} + u\Psi = E\Psi$$

determines the x dependence of the wave function. Here $\hbar$ and m are constants which we shall not worry about and E is the energy of the quantum system. The energy levels can be discrete for bound states or continuous for scattering states. This depends on the potential $u(x)$. We shall regard the Schrödinger equation as an eigen-value problem and refer to Ψ and E as eigenvector and eigenvalue, respectively.

According to the Copenhagen interpretation of QM the probability density for the position of a quantum particle is given by $|\Psi|^2$, where Ψ is a solution to the Schrödinger equation. The time evolution of the wave function is governed by a time-dependent Schrödinger equation

$$i\hbar\frac{\partial\Psi}{\partial t} = -\frac{\hbar^2}{2m}\frac{\partial^2\Psi}{\partial x^2} + u\Psi.$$

This equation implies that for bound states the QM probability is conserved in the sense that

$$\frac{d}{dt}\int_{\mathbb{R}} |\Psi|^2 dx = 0.$$

The way physicists discover new elementary particles is by scattering experiments. Huge accelerators collide particles through targets and, by analysing the changes to momenta of scattered particles, a picture of a target is built.[2] Given a potential $u(x)$ one can use the Schrödinger equation to find Ψ, the associated energy levels, and the scattering data in the form of so-called reflection and transmission coefficients. Experimental needs are however different: the scattering data is measured in the accelerator but the potential (which gives the internal structure of the target) needs to be recovered. This comes down to the following mathematical problem

- Recover the potential from the scattering data.

[2] These kind of experiments will take place in the Large Hadron Collider (LHC) opened in September 2008 at CERN. The LHC is located in a 27-km long tunnel under the Swiss/French border outside Geneva. It is hoped that the elusive Higgs particle and a whole bunch of other exotic forms of matter will be discovered.

This problem was solved in the 1950s by Gelfand, Levitan, and Marchenko [70, 115] who gave a linear algorithm for reconstructing $u(x)$. Gardner, Green, Kruskal, and Miura [67] used this algorithm to solve the Cauchy problem for the KdV equation. Their remarkable idea was to regard the initial data in the solution of KdV as a potential in the Schrödinger equation.

Set $\hbar^2/(2m) = 1$ and write the one-dimensional Schrödinger equation as an eigenvalue problem

$$\left[-\frac{d^2}{dx^2} + u(x)\right]\Psi = E\Psi.$$

We allow u to depend on x as well as t which at this stage should be regarded as a parameter.

In the scattering theory one considers the beam of free particles incident from $+\infty$. Some of the particles will be reflected by the potential (which is assumed to decay sufficiently fast as $|x| \to \infty$) and some will be transmitted. There may also be a number of bound states with discrete energy levels. The Gelfand–Levitan–Marchenko (GLM) theory shows that given

- the energy levels, E,
- the transmission probability, T, and
- the reflection probability, R,

one can find the potential u. Given $u_0(x)$ one finds the scattering data at $t = 0$. If $u(x, t)$ is a solution to the KdV equation (2.1.1) with $u(x, 0) = u_0(x)$ then the scattering data $(E(t), T(t), R(t))$ satisfies simple linear ODEs determining their time evolution. In particular E does not depend on t. Once this has been determined, $u(x, t)$ is recovered by solving a linear integral equation. The Gardner, Green, Kruskal, and Miura scheme for solving KdV is summarized in the following table:

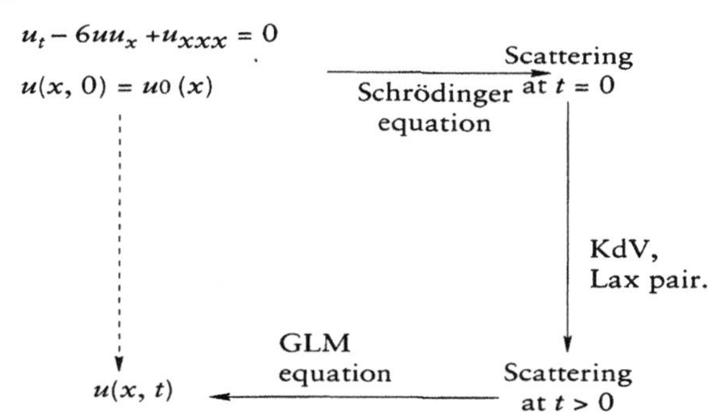

We should stress that in this method the time evolution of the scattering data is governed by the KdV and not by the time-dependent Schrödinger equation. In fact the time-dependent Schrödinger equation will not play any role in the following discussion.

2.2.1 Direct scattering

The following discussion summarizes the basic one-dimensional QM of a particle scattering on a potential [122, 144].

- Set $E = k^2$ and rewrite the Schrödinger equation as

$$Lf := \left[-\frac{d^2}{dx^2} + u(x)\right] f = k^2 f \tag{2.2.5}$$

where L is called the Schrödinger operator. Consider the class of potentials $u(x)$ such that $|u(x)| \rightarrow 0$ as $x \rightarrow \pm\infty$ and

$$\int_{\mathbb{R}} (1 + |x|)|u(x)|dx < \infty.$$

This integral condition guarantees that there exist only a finite number of discrete energy levels (thus it rules out both the harmonic oscillator and the hydrogen atom).

- At $x \rightarrow \pm\infty$ the problem (2.2.5) reduces to a 'free particle'

$$f_{xx} + k^2 f = 0$$

with the general solution

$$f = C_1 e^{ikx} + C_2 e^{-ikx}.$$

The pair of constants (C_1, C_2) is in general different at $+\infty$ and $-\infty$.

- For each $k \neq 0$ the set of solutions to (2.2.5) forms a two-dimensional complex vector space G_k. The reality of $u(x)$ implies that if f satisfies (2.2.5) then so does $\bar{f}$.

Consider two bases $(\psi, \overline{\psi})$ and $(\phi, \overline{\phi})$ of G_k determined by the asymptotic

$$\psi(x, k) \cong e^{-ikx}, \quad \overline{\psi}(x, k) \cong e^{ikx} \quad \text{as} \quad x \longrightarrow \infty$$

and

$$\phi(x, k) \cong e^{-ikx}, \quad \overline{\phi}(x, k) \cong e^{ikx} \quad \text{as} \quad x \longrightarrow -\infty.$$

Any solution can be expanded in the first basis, so in particular

$$\phi(x, k) = a(k)\psi(x, k) + b(k)\overline{\psi}(x, k).$$

Therefore, if $a \neq 0$, we can write

$$\frac{\phi(x, k)}{a(k)} = \begin{cases} \dfrac{e^{-ikx}}{a(k)}, & \text{for } x \rightarrow -\infty \\ e^{-ikx} + \dfrac{b(k)}{a(k)} e^{ikx}, & \text{for } x \rightarrow \infty. \end{cases} \tag{2.2.6}$$

- Consider a particle incident from ∞ with the wave function e^{-ikx} (Figure 2.2). The transmission coefficient $t(k)$ and the reflection coefficient

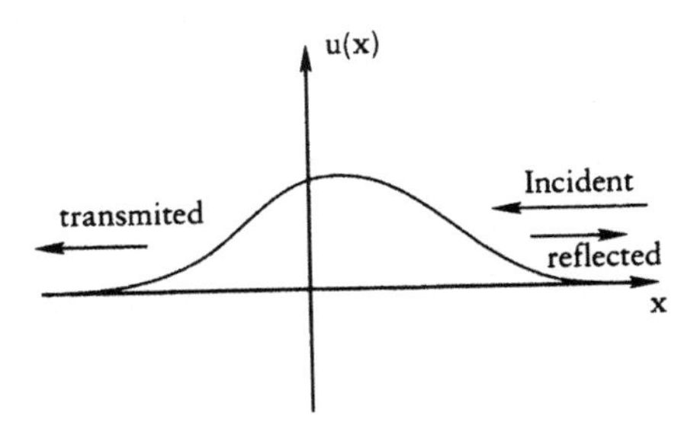

Figure 2.2 *Reflection and transmission*

$r(k)$ are defined by

$$t(k) = \frac{1}{a(k)} \text{ and } r(k) = \frac{b(k)}{a(k)}.$$

They satisfy

$$|t(k)|^2 + |r(k)|^2 = 1 \tag{2.2.7}$$

which is intuitively clear as the particle is 'either reflected or transmitted'. To prove it recall that given the Wronskian

$$W(f, g) = fg_x - gf_x$$

of any two functions we have

$$W_x = fg_{xx} - gf_{xx} = 0$$

if f, g both satisfy the Schrödinger equation (2.2.5). Thus $W(\phi, \overline{\phi})$ is a constant which can be calculated for $x \to -\infty$

$$W(\phi, \overline{\phi}) = e^{-ikx}(e^{ikx})_x - e^{ikx}(e^{-ikx})_x = 2ik.$$

Analogous calculation at $x \to \infty$ gives $W(\psi, \overline{\psi}) = 2ik$. On the other hand

$$\begin{aligned} W(\phi, \overline{\phi}) &= W(a\psi + b\overline{\psi}, \overline{a}\overline{\psi} + \overline{b}\psi) \\ &= |a|^2 W(\psi, \overline{\psi}) + a\overline{b}W(\psi, \psi) + b\overline{a}W(\overline{\psi}, \overline{\psi}) - |b|^2 W(\psi, \overline{\psi}) \\ &= 2ik(|a|^2 - |b|^2). \end{aligned}$$

Thus $|a(k)|^2 - |b(k)|^2 = 1$ or equivalently (2.2.7) holds.

2.2.2 Properties of the scattering data

Assume that $k \in \mathbb{C}$. In scattering theory (see, e.g., [122]) one proves the following:

- $a(k)$ is holomorphic in the upper half-plane $\mathrm{Im}(k) > 0$.
- $\{\mathrm{Im}(k) \geq 0, |k| \to \infty\} \quad \longrightarrow \quad |a(k)| \to 1.$

- Zeros of $a(k)$ in the upper half-plane lie on the imaginary axis. The number of these zeros is finite if

$$\int_{\mathbb{R}} (1 + |x|)|u(x)| < \infty.$$

Thus $a(i\chi_1) = \cdots = a(i\chi_N) = 0$ where $\chi_n \in \mathbb{R}$ can be ordered as

$$\chi_1 > \chi_2 > \cdots > \chi_N > 0.$$

- Consider the asymptotics of ϕ at these zeros. Formula (2.2.6) gives

$$\phi(x, i\chi_n) = \begin{cases} e^{-i(i\chi_n)x}, & \text{for } x \to -\infty \\ a(i\chi_n)e^{-i(i\chi_n)x} + b(i\chi_n)e^{i(i\chi_n)}, & \text{for } x \to \infty. \end{cases}$$

Thus

$$\phi(x, i\chi_n) = \begin{cases} e^{\chi_n x}, & \text{for } x \to -\infty \\ b(i\chi_n)e^{-\chi_n x}, & \text{for } x \to \infty. \end{cases}$$

Moreover

$$\left[-\frac{d^2}{dx^2} + u(x)\right]\phi(x, i\chi_n) = -\chi_n^2\phi(x, i\chi_n)$$

so ϕ is square integrable with energy $E = -\chi_n^2$.
- Set $b_n = b(i\chi_n)$. Then $b_n \in \mathbb{R}$ and

$$b_n = (-1)^n|b_n|.$$

Also $ia'(i\chi_n)$ has the same sign as b_n.

2.2.3 Inverse scattering

We want to recover the potential $u(x)$ from the scattering data which consists of the reflection coefficients and the energy levels

$$r(k), \{\chi_1, \ldots, \chi_N\}$$

so that $E_n = -\chi_n^2$ and

$$\phi(x, i\chi_n) = \begin{cases} e^{\chi_n x}, & \text{for } x \to -\infty \\ b_n e^{-\chi_n x}, & \text{for } x \to \infty. \end{cases}$$

The inverse scattering transform (IST) of GLM consists of the following steps:

- Set

$$F(x) = \sum_{n=1}^{N} \frac{b_n e^{-\chi_n x}}{ia'(i\chi_n)} + \frac{1}{2\pi}\int_{-\infty}^{\infty} r(k)e^{ikx}dk. \tag{2.2.8}$$

- Consider the GLM integral equation

$$K(x, y) + F(x + y) + \int_{x}^{\infty} K(x, z)F(z + y)dz = 0 \tag{2.2.9}$$

  and solve it for $K(x, y)$.
- Then

$$u(x) = -2\frac{d}{dx}K(x, x) \tag{2.2.10}$$

  is the potential in the corresponding Schrödinger equation.

These formulae are given in the t-independent way, but t can be introduced as a parameter. If the time dependence of the scattering data is known, the solution of the GLM integral equation $K(x, y, t)$ will also depend on t and so will the potential $u(x, t)$.

2.2.4 Lax formulation

If the potential $u(x)$ in the Schrödinger equation depends on a parameter t, its eigenvalues will in general change with t. The IST is an example of an isospectral problem, when this does not happen.

Proposition 2.2.1 *If there exist a differential operator A such that*

$$\dot{L} = [L, A] \tag{2.2.11}$$

where

$$L = -\frac{d^2}{dx^2} + u(x, t),$$

then the spectrum of L does not depend on t.

Proof Consider the eigenvalue problem

$$Lf = Ef.$$

Differentiating gives

$$L_t f + Lf_t = E_t f + Ef_t.$$

Note that $ALf = EAf$ and use the representation (2.2.11) to find

$$(L - E)(f_t + Af) = E_t f. \tag{2.2.12}$$

Take the inner product (2.2.4) of this equation with f and use the fact that L is self-adjoint

$$E_t\|f\|^2 =< f, (L-E)(f_t + Af) >=< (L-E)f, f_t + Af >= 0.$$

Thus $E_t = 0$. This derivation also implies that if $f(x,t)$ is an eigenfunction of L with eigenvalue $E = k^2$ then so is $(f_t + Af)$. □

What makes the method applicable to KdV equation (2.1.1) is that KdV is equivalent to (2.2.11) with

$$L = -\frac{d^2}{dx^2} + u(x,t) \quad \text{and} \quad A = 4\frac{d^3}{dx^3} - 3\left(u\frac{d}{dx} + \frac{d}{dx}u\right). \tag{2.2.13}$$

To prove this statement it is enough to compute both sides of (2.2.11) on a function and verify that $[L, A]$ is multiplication by $6uu_x - u_{xxx}$ (also $\dot{L} = u_t$). This is the *Lax representation* of KdV [103]. Such representations (for various choices of operators L, A) underlie integrability of PDEs and ODEs.

2.2.5 Evolution of the scattering data

We will now use the Lax representation to determine the time evolution of the scattering data. Assume that the potential $u(x,t)$ in the Schrödinger equation satisfies the KdV equation (2.1.1). Let $f(x,t)$ be an eigenfuction of the Schrödinger operator $Lf = k^2 f$ defined by the asymptotic behaviour

$$f = \phi(x,k) \longrightarrow e^{-ikx}, \quad \text{as } x \to -\infty.$$

Equation (2.2.12) implies that if $f(x,t)$ is an eigenfunction of L with eigenvalue k^2 then so is $(f_t + Af)$. Moreover $u(x) \to 0$ as $|x| \to \infty$ therefore

$$\dot{\phi} + A\phi \longrightarrow 4\frac{d^3}{dx^3}e^{-ikx} = 4ik^3e^{-ikx} \quad \text{as } x \to -\infty.$$

Thus $4ik^3\phi(x,k)$ and $\dot{\phi} + A\phi$ are eigenfunctions of the Schrödinger operator with the same asymptotics and we deduce that they must be equal. Their difference is in the kernel of L and so must be a linear combination of ψ and $\overline{\psi}$. But this combination vanishes at ∞ so, using the independence of ψ and $\overline{\psi}$, it must vanish everywhere. Thus the ODE

$$\dot{\phi} + A\phi = 4ik^3\phi$$

holds for all $x \in \mathbb{R}$. We shall use this ODE and the asymptotics at $+\infty$ to find ODEs for $a(k)$ and $b(k)$. Recall that

$$\phi(x,k) = a(k,t)e^{-ikx} + b(k,t)e^{ikx} \quad \text{as } x \to \infty.$$

Substituting this into the ODE gives

$$\dot{a}e^{-ikx} + \dot{b}e^{ikx} = \left(-4\frac{d^3}{dx^3} + 4ik^3\right)(ae^{-ikx} + be^{ikx})$$
$$= 8ik^3be^{ikx}.$$

Equating the exponentials gives

$$\dot{a} = 0, \quad \dot{b} = 8ik^3b$$

and

$$a(k,t) = a(k,0), \quad b(k,t) = b(k,0)e^{8ik^3t}.$$

In the last section we have shown that k does not depend on t and so the zeros $i\chi_n$ of a are constant. The evolution of the scattering data is thus given by the following:

$$\begin{aligned}
a(k,t) &= a(k,0),\\
b(k,t) &= b(k,0)e^{8ik^3t},\\
r(k,t) &= \frac{b(k,t)}{a(k,t)} = r(k,0)e^{8ik^3t},\\
\chi_n(t) &= \chi_n(0),\\
b_n(t) &= b_n(0)e^{8\chi_n^3t},\\
a_n(t) &= 0, \quad \text{and}\\
\beta_n(t) &= \frac{b_n(t)}{ia'(i\chi_n)} = \beta_n(0)e^{8\chi_n^3t}.
\end{aligned} \tag{2.2.14}$$

2.3 Reflectionless potentials and solitons

The formula (2.2.14) implies that if the reflection coefficient is initially zero, it is zero for all t. In this case the inverse scattering procedure can be carried out explicitly. The resulting solutions are called N-solitons, where N is the number of zeros $i\chi_1, \ldots, i\chi_N$ of $a(k)$. These solutions describe collisions of one-solitons (2.1.3) without any non-elastic effects. The one-solitons generated after collisions are 'the same' as those before the collision. This fact was discovered numerically in the 1960s and boosted the interest in the whole subject.

Assume $r(k, 0) = 0$ so that (2.2.14) implies

$$r(k, t) = 0.$$

2.3.1 One-soliton solution

We shall first derive the one-soliton solution. The formula (2.2.8) with $N = 1$ gives

$$F(x, t) = \beta(t)e^{-\chi x}.$$

This depends on x as well as t because $\beta(t) = \beta(0)e^{8\chi^3 t}$ from (2.2.14). We shall suppress this explicit t dependence in the following calculation and regard t as a parameter. The GLM equation (2.2.9) becomes

$$K(x, y) + \beta e^{-\chi(x+y)} + \int_x^\infty K(x, z)\beta e^{-\chi(z+y)}dz = 0.$$

Look for solutions of the form

$$K(x, y) = K(x)e^{-\chi y}.$$

This gives

$$K(x) + \beta e^{-\chi x} + K(x)\beta \int_x^\infty e^{-2\chi z}dz = 0,$$

and after a simple integration

$$K(x) = -\frac{\beta e^{-\chi x}}{1 + \frac{\beta}{2\chi}e^{-2\chi x}}.$$

Thus

$$K(x, y) = -\frac{\beta e^{-\chi(x+y)}}{1 + \frac{\beta}{2\chi}e^{-2\chi x}}.$$

This function also depends on t because β does. Finally the formula (2.2.10) gives

$$\begin{aligned} u(x, t) &= -2\frac{\partial}{\partial x}K(x, x) = -\frac{4\beta\chi e^{-2\chi x}}{(1 + \frac{\beta}{2\chi}e^{-2\chi x})^2} \\ &= -\frac{8\chi^2}{\hat{\beta}^{-1}e^{\chi x} + \hat{\beta}e^{-\chi x}}, \quad \text{where} \quad \hat{\beta} = \sqrt{\beta/(2\chi)} \\ &= -\frac{2\chi^2}{\cosh\left[\chi(x - 4\chi^2 t - \phi_0)\right]^2}, \quad \phi_0 = \frac{1}{2\chi}\log\left(\frac{\beta_0}{2\chi}\right) \end{aligned}$$

which is the one-soliton solution (2.1.3).

The energy of the corresponding solution to the Schrödinger equation determines the amplitude and the velocity of the soliton. The soliton is of the form $u = u(x - 4\chi^2 t)$ so it represents a wave moving to the right with velocity $4\chi^2$ and phase ϕ_0.

2.3.2 *N*-soliton solution

Suppose there are N energy levels which we order $\chi_1 > \chi_2 > \cdots > \chi_N > 0$. The function (2.2.8) is

$$F(x) = \sum_{n=1}^{N} \beta_n e^{-\chi_n x}$$

and the GLM equation (2.2.9) becomes

$$K(x, y) + \sum_{n=1}^{N} \beta_n e^{-\chi_n(x+y)} + \int_x^\infty K(x, z) \sum_{n=1}^{N} \beta_n e^{-\chi_n(z+y)} dz = 0.$$

The kernel of this integral equation is degenerate in the sense that

$$F(z + y) = \sum_{n=1}^{N} k_n(z) h_n(y),$$

so we seek solutions of the form

$$K(x, y) = \sum_{n=1}^{N} K_n(x) e^{-\chi_n y}.$$

After one integration this gives

$$\sum_{n=1}^{N} [K_n(x) + \beta_n e^{-\chi_n x}] e^{-\chi_n y} + \sum_{n=1}^{N} \left[\beta_n \sum_{m=1}^{N} \frac{K_m(x)}{\chi_m + \chi_n} e^{-(\chi_n + \chi_m)x} \right] e^{-\chi_n y} = 0.$$

The functions $e^{-\chi_n y}$ are linearly independent, so

$$K_n(x) + \beta_n e^{-\chi_n x} + \sum_{m=1}^{N} \beta_n K_m(x) \frac{1}{\chi_m + \chi_n} e^{-(\chi_n + \chi_m)x} = 0.$$

Define a matrix

$$A_{nm}(x) = \delta_{nm} + \frac{\beta_n e^{-(\chi_n + \chi_m)x}}{\chi_n + \chi_m}.$$

The linear system becomes

$$\sum_{m=1}^{N} A_{nm}(x) K_m(x) = -\beta_n e^{-\chi_n x},$$

or

$$AK + B = 0,$$

where B is a column vector

$$B = [\beta_1 e^{-\chi_1 x}, \beta_2 e^{-\chi_2 x}, \dots, \beta_n e^{-\chi_n x}]^T.$$

The solution of this system is

$$K = -A^{-1} B.$$

Using the relation

$$\frac{dA_{mn}(x)}{dx} = -B_m e^{-\chi_n x}$$

we can write

$$\begin{aligned}
K(x, x) &= \sum_{m=1}^{N} e^{-\chi_m x} K_m(x) = -\sum_{m,n=1}^{N} e^{-\chi_m x} (A^{-1})_{mn} B_n \\
&= \sum_{m,n=1}^{N} (A^{-1})_{mn} \frac{dA_{nm}(x)}{dx} = \mathrm{Tr}\left(A^{-1} \frac{dA}{dx}\right) \\
&= \frac{1}{\det A} \frac{d}{dx} \det A.
\end{aligned}$$

Finally we reintroduce the explicit t-dependence to write the N-soliton solution as

$$u(x, t) = -2 \frac{\partial^2}{\partial x^2} \ln[\det A(x)] \quad \text{where} \quad A_{nm}(x) = \delta_{nm} + \frac{\beta_n e^{-(\chi_n + \chi_m)x}}{\chi_n + \chi_m}. \tag{2.3.15}$$

2.3.3 Two-soliton asymptotics

Let us analyse a two-soliton solution with $\chi_1 > \chi_2$ in more detail. Set

$$\tau_k = \chi_k x - 4\chi_k^3 t, \quad k = 1, 2$$

and consider the determinant

$$\det A = \left[1 + \frac{\beta_1(0)}{2\chi_1} e^{-2\tau_1}\right]\left[1 + \frac{\beta_2(0)}{2\chi_2} e^{-2\tau_2}\right] - \frac{\beta_1(0)\beta_2(0)}{(\chi_1 + \chi_2)^2} e^{-2(\tau_1 + \tau_2)}.$$

We first analyse the case $t \to -\infty$. In the limit $x \to -\infty$ we have $\det A \sim e^{-2(\tau_1 + \tau_2)}$ so

$$\log(\det A) \sim \mathrm{const} - 2(\tau_1 + \tau_2)$$

and $u \sim 0$ which we already knew. Now move along the x-axis and consider the leading term in $\det A$ when $\tau_1 = 0$ and then when $\tau_2 = 0$. We first reach the point $\tau_1 = 0$ or

$$x = 4\chi_1^2 t.$$

In the neighbourhood of this point $\tau_2 = 4t\chi_2(\chi_1^2 - \chi_2^2) \ll 0$ and

$$\det A \sim \frac{\beta_2(0)}{2\chi_2} e^{-2\tau_2} \left[1 + \frac{\beta_1(0)}{2\chi_1} \left(\frac{\chi_1 - \chi_2}{\chi_1 + \chi_2} \right)^2 e^{-2\tau_1} \right].$$

Differentiating the logarithm of $\det A$ yields

$$u \sim -2\frac{\partial^2}{\partial x^2} \left[1 + \frac{\beta_1(0)}{2\chi_1} \left(\frac{\chi_1 - \chi_2}{\chi_1 + \chi_2} \right)^2 e^{-2\chi_1(x - 4\chi_1^2 t)} \right]$$

which looks like a one-soliton solution with a phase

$$(\phi_1)_- = \frac{1}{2\chi_1} \log \left[\frac{\beta_1(0)}{2\chi_1} \left(\frac{\chi_1 - \chi_2}{\chi_1 + \chi_2} \right)^2 \right].$$

We now move along the x-axis until we reach $\tau_2 = 0$. Repeating the above analysis shows that now $\tau_1 = 4\chi_1(\chi_2^2 - \chi_1^2)t \gg 0$ and around the point $x = 4\chi_2^2 t$ we have

$$\det A \sim 1 + \frac{\beta_2(0)}{2\chi_2} e^{-2\tau_2}.$$

Therefore the function u looks like a one-soliton solution with a phase

$$(\phi_2)_- = \frac{1}{2\chi_2} \log \left[\frac{\beta_2(0)}{2\chi_2} \right].$$

As t approaches 0 the two solitons coalesce and the exact behaviour depends on the ratio χ_1/χ_2.

We perform an analogous analysis as $t \to \infty$. If $x \to \infty$ then $\det A \sim 1$ and $u \sim 0$. We move along the x-axis to the left until we reach $\tau_1 = 0$ where $\tau_2 \gg 0$ and the profile of u is given by a one-soliton with the phase

$$(\phi_1)_+ = \frac{1}{2\chi_1} \log \left[\frac{\beta_1(0)}{2\chi_1} \right].$$

Then we reach the point $\tau_2 = 0$, $\tau_1 \ll 0$ where there is a single soliton with the phase

$$(\phi_2)_+ = \frac{1}{2\chi_2} \log \left[\frac{\beta_2(0)}{2\chi_2} \left(\frac{\chi_1 - \chi_2}{\chi_1 + \chi_2} \right)^2 \right].$$

Thus the larger soliton has overtaken the smaller one. This asymptotic analysis shows that the solitons have preserved their shape but their phases have changed

$$\Delta\phi_1 = (\phi_1)_+ - (\phi_1)_- = -\frac{1}{\chi_1}\log\frac{\chi_1 - \chi_2}{\chi_1 + \chi_2},$$

$$\Delta\phi_2 = (\phi_2)_+ - (\phi_2)_- = \frac{1}{\chi_2}\log\frac{\chi_1 - \chi_2}{\chi_1 + \chi_2}.$$

The only result of the interaction can be measured by

$$-\log\frac{\chi_1 - \chi_2}{\chi_1 + \chi_2}$$

which is large if the difference between the velocities χ_1 and χ_2 is small.

The figures show the two-soliton solution (the graphs show $-u$ as a function of x) at $t = -1, t = 0$, and $t = 1$ (for the chosen parameters $t = -1$ is considered to be a large negative time when the two solitons are separated). It should be interpreted as a passing collision of fast and slow solitons. The larger, faster soliton has amplitude 8, and the slower, smaller soliton has amplitude 2. Its velocity is one half of that of the fast soliton. The solitons are separated at $t = -1$. At $t = 0$ the collision takes place. The wave amplitude becomes smaller than the sum of the two waves. At $t = 1$ the larger soliton has overtaken the smaller one. The amplitudes and shapes have not changed.

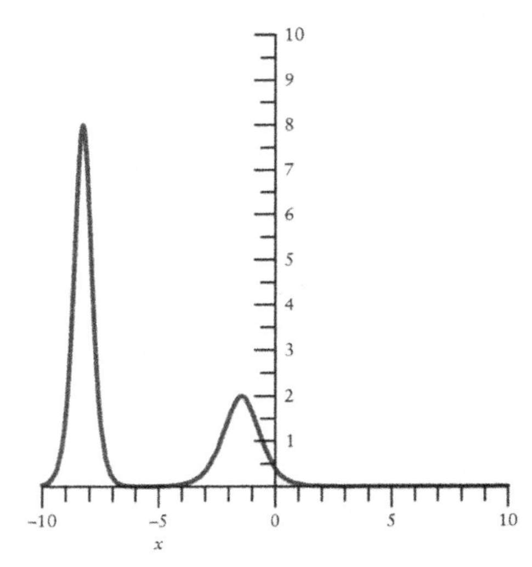

Two-solition solution at $t = -1$.

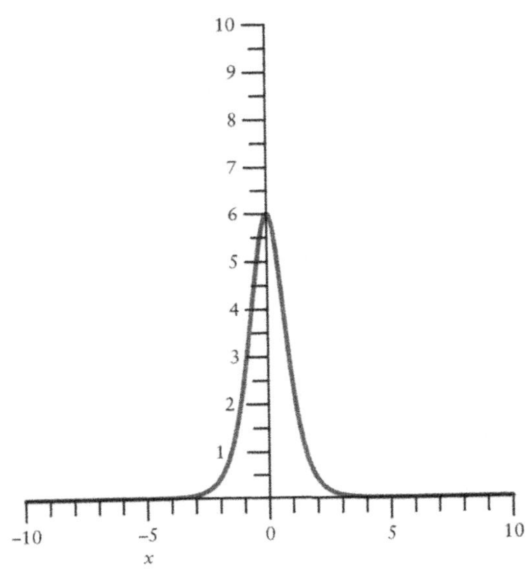

Two-soliton solution at $t = 0$. The total amplitude is smaller than the sum of the two amplitudes.

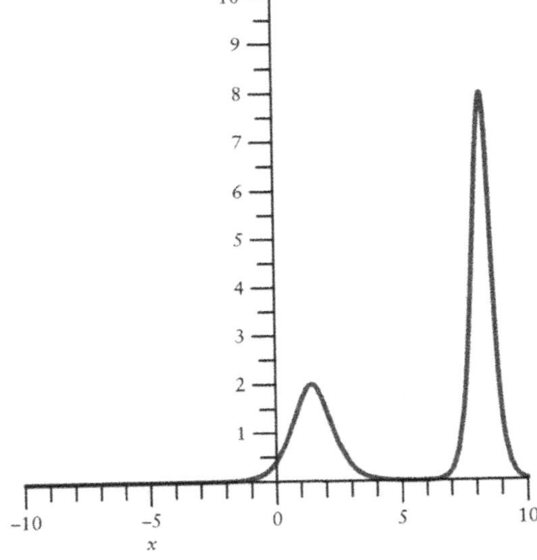

Two-soliton solution at $t = 1$. Amplitudes and shapes preserved by the collision.

This picture generalizes to $N > 2$. The general solution (2.3.15) asymptotically represents N separate solitons ordered accordingly to their speed. The tallest (and therefore fastest) soliton is at the front, followed by the second tallest, etc.

At $t = 0$ the 'interaction' takes place and then the individual solitons re-emerge in the opposite order as $t \to \infty$. The total phase shift is the sum of pairwise phase shifts [122].

The number of discrete eigenvalues N for the Schrödinger operator is equal to the number of solitons at $t \to \pm\infty$. This number is of course encoded in the initial conditions. To see this consider

$$u(x, 0) = u_0(x) = -\frac{N(N+1)}{\cosh^2(x)}, \qquad N \in \mathbb{Z}^+.$$

Substituting $\xi = \tanh(x) \in (-1, 1)$ in the Schrödinger equation

$$-\frac{d^2 f}{dx^2} + u_0(x) f = k^2 f$$

yields the associated Legendre equation

$$\frac{d}{d\xi}\left[(1-\xi^2)\frac{df}{d\xi}\right] + \left[N(N+1) + \frac{k^2}{1-\xi^2}\right] f = 0.$$

Analysis of the power series solution shows that square-integrable solutions exist if $k^2 = -\chi^2$ and $\chi = 1, 2, \ldots, N$. Therefore $F(x)$ in the GLM equation is given by

$$F(x) = \sum_{n=1}^{N} \beta_n e^{-\chi_n x},$$

and the earlier calculation applies leading to a particular case of the N-soliton solution (2.3.15). See the more complete discussion of this point in [42].

Exercises

1. Verify that the equation

$$\frac{1}{v}\Psi_t + \Psi_x + \beta\Psi_{xxx} + \alpha\Psi\Psi_x = 0.$$

where $\Psi = \Psi(x, t)$ and (v, β, α) are non-zero constants is equivalent to the KdV equation

$$u_t - 6uu_x + u_{xxx} = 0, \text{ where } u = u(x, t)$$

after a suitable change of dependent and independent variables.

2. Assume that a solution of the KdV equation is of the form

$$u(x, t) = f(\xi), \quad \text{where} \quad \xi = x - ct$$

for some constant c. Show that the function $f(\xi)$ satisfies the ODE

$$\frac{1}{2}(f')^2 = f^3 + \frac{1}{2}cf^2 + \alpha f + \beta$$

where (α, β) are arbitrary constants. Assume that f and its first two derivatives tend to zero as $|\xi| \to \infty$ and solve the ODE to construct the one-soliton solution to the KdV equation.

3. Let $v = v(x,t)$ satisfy the *modified KdV equation*

$$v_t - 6v^2 v_x + v_{xxx} = 0.$$

Show that the function $u(x,t)$ given by the Miura transformation

$$u = v^2 + v_x \tag{2.3.16}$$

satisfies the KdV equation. Is it true that any solution u to the KdV equation gives rise, via (2.3.16), to a solution of the modified KdV equation?

4. Show that the KdV equation is equivalent to

$$L_t = [L, A]$$

where the Lax operators are

$$L = -\frac{d^2}{dx^2} + u \quad \text{and} \quad A = 4\frac{d^3}{dx^3} - 3\left(u\frac{d}{dx} + \frac{d}{dx}u\right), \quad \text{where } u = u(x,t).$$

5. Let (L, A) be the KdV Lax pair. Set $u(x,t) = U(X,T)$. Substitute

$$\frac{\partial}{\partial x} = \varepsilon\frac{\partial}{\partial X} \quad \text{and} \quad \frac{\partial}{\partial t} = \varepsilon\frac{\partial}{\partial T},$$

and consider the operators acting on functions of the form

$$\psi(x,t) = \exp\left[\varepsilon^{-1}\Psi(X,T)\right].$$

Show that in the dispersionless limit $\varepsilon \longrightarrow 0$ commutators of the differential operators are replaced by Poisson brackets according to the relation

$$\frac{\partial^k}{\partial x^k}\psi \longrightarrow (\Psi_X)^k\psi, \quad [L,A] \longrightarrow \frac{\partial \mathrm{L}}{\partial p}\frac{\partial \mathrm{A}}{\partial X} - \frac{\partial \mathrm{L}}{\partial X}\frac{\partial \mathrm{A}}{\partial p} = -\{\mathrm{L}, \mathrm{A}\},$$

where $p = \Psi_X$ and

$$\mathrm{L} = -p^2 + U(X,T) \quad \text{and} \quad \mathrm{A} = 4p^3 - 6U(X,T)p.$$

Deduce that dispersionless limit of the Lax representation is

$$\mathrm{L}_T + \{\mathrm{L}, \mathrm{A}\} = 0,$$

and that $U = U(X,T)$ satisfies the dispersionless KdV equation

$$U_T = 6UU_X.$$

[The functions (L, A) are called the *symbols* of operators (L, A). This method of taking the dispersionless limit is analogous to the WKB approximation in QM.] Obtain the same limit by making the substitutions directly in the KdV equation.

6. Use the chain rule to verify that the implicit solution to the dispersionless KdV is given by

$$U(X, T) = F[X + 6TU(X, T)],$$

where F is an arbitrary differentiable function of one variable. (This solution is obtained by the method of characteristics. Read about it in Appendix C.)
Assume that

$$U(X, 0) = -\frac{2}{\cosh^2(X)}$$

and show that U_X is unbounded, that is, that for any $M > 0$ there exists $T > 0$ so that $U_X(X_0, T) > M$ for some X_0. Deduce that U_X becomes infinite after finite time. This is called a *gradient catastrophe*, or a *shock*. Draw a graph of $U(X, T)$ illustrating this situation. Compare it with the one-soliton solution to the KdV equation with the same initial condition.

7. Assume that the scattering data consists of two energy levels $E_1 = -\chi_1^2$ and $E_2 = -\chi_2^2$ where $\chi_1 > \chi_2$ and a vanishing reflection coefficient. Solve the GLM equation to find the two-soliton solution.
[Follow the derivation of the one-soliton in this chapter but try not to look at the N-soliton unless you really get stuck.]

8. Let $L\psi = k^2\psi$ where $L = -\partial_x^2 + u$. Consider ψ of the form

$$\psi(x) = e^{ikx} + \int_x^\infty K(x, z)e^{ikz}dz$$

where $K(x, z), \partial_z K(x, z) \to 0$ as $z \to \infty$ for any fixed x. Use integration by parts to show that

$$\psi = e^{ikx}\left(1 + \frac{i\hat{K}}{k} - \frac{\hat{K}_z}{k^2}\right) - \frac{1}{k^2}\int_x^\infty K_{zz}e^{ikz}dz,$$

where $\hat{K} = K(x, x)$ and $\hat{K}_z = (\partial_z K)|_{z=x}$. Deduce that the Schrödinger equation is satisfied if

$$u(x) = -2(\hat{K}_x + \hat{K}_z) \quad \text{and} \quad K_{xx} - K_{zz} - uK = 0 \quad \text{for} \quad z > x.$$

3 Hamiltonian formalism and zero-curvature representation

3.1 First integrals

We shall make contact with the Definition 1.2.1 of finite-dimensional integrable systems and show that KdV has infinitely many first integrals. Rewrite the expression (2.2.6)

$$\phi(x, k) = \begin{cases} e^{-ikx}, & \text{for } x \to -\infty \\ a(k, t)e^{-ikx} + b(k, t)e^{ikx}, & \text{for } x \to \infty, \end{cases}$$

when the time dependence of the scattering data has been determined using the KdV equation. The formula (2.2.14) gives

$$\frac{\partial}{\partial t} a(k, t) = 0, \quad \forall k$$

so the scattering data gives infinitely many first integrals provided that they are non-trivial and independent. We aim to express these first integrals in the form

$$I[u] = \int_{\mathbb{R}} P(u, u_x, u_{xx}, \ldots)dx$$

where P is a polynomial in u and its derivatives.

For large $|k|$ we set

$$\phi(x, t, k) = e^{-ikx + \int_{-\infty}^{x} S(y,t,k)dy}.$$

For large x the formula (2.2.6) gives

$$e^{ikx}\phi \cong a(k) + b(k, t)e^{2ikx}.$$

If we assume that k is in the upper half plane $\mathrm{Im}(k) > 0$ the second term on the RHS goes to 0 as $x \to \infty$. Thus

$$\begin{aligned} a(k) &= \lim_{x\to\infty} e^{ikx}\phi(x, t, k) = \lim_{x\to\infty} e^{\int_{-\infty}^{x} S(y,t,k)dy} \\ &= e^{\int_{-\infty}^{\infty} S(y,t,k)dy}, \end{aligned} \tag{3.1.1}$$

where the above formula also holds in the limit $\mathrm{Im}(k) \to 0$ because of the real analyticity. Now we shall use the Schrödinger equation with t regarded as a parameter

$$-\frac{d^2\phi}{dx^2} + u\phi = k^2\phi$$

to find an equation for S. Substituting

$$\frac{d\phi}{dx} = [-ik + S(x,k)]\phi, \quad \frac{d^2\phi}{dx^2} = \frac{dS}{dx}\phi + [-ik + S(x,k)]^2\phi$$

gives the Riccati type equation

$$\frac{dS}{dx} - 2ikS + S^2 = u \tag{3.1.2}$$

(we stress that both S and u depend on x as well as t). Look for solutions in the form of the asymptotic expansion

$$S = \sum_{n=1}^{\infty} \frac{S_n(x,t)}{(2ik)^n}.$$

Substituting this to (3.1.2) yields a recursion relation

$$S_1(x,t) = -u(x,t), \quad S_{n+1} = \frac{dS_n}{dx} + \sum_{m=1}^{n-1} S_m S_{n-m} \tag{3.1.3}$$

which can be solved for the first few terms

$$S_2 = -\frac{\partial u}{\partial x}, \quad S_3 = -\frac{\partial^2 u}{\partial x^2} + u^2, \quad S_4 = -\frac{\partial^3 u}{\partial x^3} + 2\frac{\partial}{\partial x}u^2, \quad \text{and}$$

$$S_5 = -\frac{\partial^4 u}{\partial x^4} + 2\frac{\partial^2}{\partial x^2}u^2 + \left(\frac{\partial u}{\partial x}\right)^2 + 2\frac{\partial^2 u}{\partial x^2}u - 2u^3.$$

Now using the time independence (2.2.14) of $a(k)$ for all k and combining it with (3.1.1) implies that

$$\int_{\mathbb{R}} S_n(x,t)dx$$

are first integrals of the KdV equation. Not all of these integrals are non-trivial. For example, S_2 and S_4 given above are total x derivatives so they integrate to 0 (using the boundary conditions for u). The same is true for all even terms S_{2n}. To see this set

$$S = S_{\mathrm{R}} + iS_{\mathrm{I}}$$

where $S_{\rm R}$ and $S_{\rm I}$ are real-valued functions and substitute this to (3.1.2). Taking the imaginary part gives

$$\frac{dS_{\rm I}}{dx} + 2S_{\rm R}S_{\rm I} - 2kS_{\rm R} = 0$$

which integrates to

$$S_{\rm R} = -\frac{1}{2}\frac{d}{dx}\log(S_{\rm I} - k).$$

The even terms

$$\frac{S_{2n}(x)}{(2ik)^{2n}}, \qquad n = 1, 2, \ldots$$

in the expansion of a are real. Comparing this with the expansion of $S_{\rm R}$ in k shows that S_{2n} are all total derivatives and therefore

$$\int_{\mathbb{R}} S_{2n}dx = 0.$$

Let us now concentrate on the remaining non-trivial first integrals. Set

$$I_{n-1}[u] = \frac{1}{2}\int_{\mathbb{R}} S_{2n+1}(x,t)dx, \qquad n = 0, 1, 2, \ldots . \tag{3.1.4}$$

Our analysis shows

$$\frac{dI_n}{dt} = 0.$$

The first of these is just the integral of u itself. The next two are known as momentum and energy, respectively

$$I_0 = \frac{1}{2}\int_{\mathbb{R}} u^2 dx \quad \text{and} \quad I_1 = -\frac{1}{2}\int_{\mathbb{R}} (u_x^2 + 2u^3)dx,$$

where in the last integral we have isolated the total derivative in

$$S_5 = -\frac{\partial^4}{\partial x^4}u + 2\frac{\partial^2}{\partial x^2}u^2 + 2\frac{\partial}{\partial x}\left(u\frac{\partial u}{\partial x}\right) - \left(\frac{\partial u}{\partial x}\right)^2 - 2u^3$$

and eliminated it using integration by parts and boundary conditions. These two first integrals are associated, via Noether's theorem, with the translational invariance of KdV: if $u(x,t)$ is a solution then $u(x+x_0,t)$ and $u(x,t+t_0)$ are also solutions. The systematic way of constructing such symmetries will be presented in Chapter 4.

3.2 Hamiltonian formalism

We can now cast the KdV in the Hamiltonian form with the Hamiltonian functional given by the energy integral $H[u] = -I_1[u]$. First calculate

$$\frac{\delta I_1[u]}{\delta u(x)} = -3u^2 + u_{xx}, \quad \frac{\partial}{\partial x}\frac{\delta I_1[u]}{\delta u(x)} = -6uu_x + u_{xxx}.$$

Recall that the Hamilton canonical equations for PDEs take the form (1.3.11)

$$\frac{\partial u}{\partial t} = \frac{\partial}{\partial x}\frac{\delta H[u]}{\delta u(x)}.$$

Therefore

$$\frac{\partial u}{\partial t} = -\frac{\partial}{\partial x}\frac{\delta I_1[u]}{\delta u(x)}, \tag{3.2.5}$$

is the KdV equation. With some more work (see [122]) it can be shown that

$$\{I_m, I_n\} = 0$$

where the Poisson bracket is given by (1.3.10) so that KdV is indeed integrable in the Arnold–Liouville sense. For example

$$\begin{aligned}
\{I_n, I_1\} &= \int_{\mathbb{R}} \frac{\delta I_n}{\delta u(x)}\frac{\partial}{\partial x}\frac{\delta I_1}{\delta u(x)}dx = -\int_{\mathbb{R}} \frac{\delta I_n}{\delta u(x)} u_t\, dx \\
&= -\frac{1}{2}\int_{\mathbb{R}} \sum_{k=0}^{2n} (-1)^k \left[\left(\frac{\partial}{\partial x}\right)^k \frac{\partial S_{2n+1}}{\partial u^{(k)}}\right] u_t\, dx \\
&= -\frac{1}{2}\int_{\mathbb{R}} \sum_{k=0}^{2n} \frac{\partial S_{2n+1}}{\partial u^{(k)}}\frac{\partial}{\partial t}u^{(k)}\, dx \\
&= -\frac{d}{dt}I_n[u] = 0
\end{aligned}$$

where we used integration by parts and the boundary conditions.

3.2.1 Bi-Hamiltonian systems

Most systems integrable by the IST are Hamiltonian in two distinct ways. This means that for a given evolution equation $u_t = F(u, u_x, \ldots)$ there exist two Poisson structures $\mathcal{D}$ and $\mathcal{E}$ and two functionals $H_0[u]$ and $H_1[u]$ such that

$$\frac{\partial u}{\partial t} = \mathcal{D}\frac{\delta H_1}{\delta u(x)} = \mathcal{E}\frac{\delta H_0}{\delta u(x)}. \tag{3.2.6}$$

One of these Poisson structures can be put in a form $\mathcal{D} = \partial/\partial x$ and corresponds to the standard Poisson bracket (1.3.10) (however the transformation between these two forms can be quite non-trivial), but the second structure $\mathcal{E}$ gives a new Poisson bracket.

In the finite-dimensional context discussed in Section 1.2 this would correspond to having two skew-symmetric matrices ω and Ω which satisfy the Jacobi identity and such that ω is non-degenerate.

The Darboux theorem implies the existence of a local coordinate system (p, q) in which one of these, say ω, is a constant skew-symmetric matrix. The matrix components of the second structure Ω will however be non-constant functions of (p, q). Using (1.3.9) we write the bi–Hamiltonian condition as

$$\omega^{ab}\frac{\partial H_1}{\partial \xi^a} = \Omega^{ab}\frac{\partial H_0}{\partial \xi^a},$$

where $\xi^a, a = 1, \ldots, 2n$ are local coordinates on the phase space M, and H_0, H_1 are two distinct functions on M. The matrix valued function

$$R_a{}^c = \Omega^{bc}(\omega^{-1})_{ab}$$

is called the *recursion operator*. It should be thought of as an endomorphism $R = \Omega \circ \omega^{-1}$ acting on the tangent space T_pM, where $p \in M$. This endomorphism smoothly depends on a point p. The existence of such recursion operator is, under certain technical assumptions [109], equivalent to Arnold–Liouville integrability in the sense of Theorem 1.2.2. This is because given one first integral H_0 the remaining $(n-1)$ integrals $H_1, \ldots, H_{n-1}$ can be constructed recursively by

$$\omega^{ab}\frac{\partial H_i}{\partial \xi^a} = R^i\left(\omega^{ab}\frac{\partial H_0}{\partial \xi^a}\right), \qquad i = 1, 2, \ldots, n-1.$$

The extension of this formalism to the infinite-dimensional setting provides a practical way of constructing first integrals. In the case of KdV the first Hamiltonian formulation (3.2.5) has $\mathcal{D} = \partial/\partial x$ and

$$H_1[u] = \int_{\mathbb{R}} \left(\frac{1}{2}{u_x}^2 + u^3\right) dx.$$

The second formulation can be obtained taking

$$H_0[u] = \frac{1}{2}\int_{\mathbb{R}} u^2 dx \quad \text{and} \quad \mathcal{E} = -\partial_x^3 + 4u\partial_x + 2u_x.$$

In general it is required that a pencil of Poisson structures $\mathcal{D} + c\mathcal{E}$ is also a Poisson structure (i.e. satisfies the Jacobi identity) for any constant $c \in \mathbb{R}$. If this condition is satisfied, the bi-Hamiltonian formulation gives an effective way to construct first integrals. The following result is proved in the book of Olver [124]

Theorem 3.2.1 *Let (3.2.6) be a bi-Hamiltonian system, such that the Poisson structure $\mathcal{D}$ is non-degenerate,*[1] *and let*

$$R = \mathcal{E} \circ \mathcal{D}^{-1}$$

be the corresponding recursion operator. Assume that

$$R^n \left[\mathcal{D} \frac{\delta H_0}{\delta u(x)} \right]$$

lies in the image of $\mathcal{D}$ for each $n = 1, 2, \ldots$. Then there exists conserved functionals

$$H_1[u], H_2[u], \ldots$$

which are in involution, that is,

$$\{H_m, H_n\} := \int_{\mathbb{R}} \frac{\delta H_m}{\delta u(x)} \mathcal{D} \frac{\delta H_n}{\delta u(x)} dx = 0.$$

The conserved functionals $H_n[u]$ are constructed recursively from H_0 by

$$\mathcal{D} \frac{\delta H_n}{\delta u(x)} = R^n \left[\mathcal{D} \frac{\delta H_0}{\delta u(x)} \right], \qquad n = 1, 2, \ldots . \tag{3.2.7}$$

In the case of the KdV equation the recursion operator is

$$R = -\partial_x^2 + 4u + 2u_x \partial_x{}^{-1}, \tag{3.2.8}$$

where $\partial_x{}^{-1}$ is formally defined as integration with respect to x, and formula (3.2.7) gives an alternative way of constructing the first integrals (3.1.4).

3.3 Zero-curvature representation

We shall discuss a more geometric form of the Lax representation where integrable systems arise as compatibility conditions of overdetermined systems of matrix PDEs. Let $U(\lambda)$ and $V(\lambda)$ be matrix-valued functions of (ρ, τ) depending on the auxiliary variable λ called the spectral parameter. Consider a system of linear PDEs

$$\frac{\partial}{\partial \rho} v = U(\lambda) v \quad \text{and} \quad \frac{\partial}{\partial \tau} v = V(\lambda) v \tag{3.3.9}$$

where v is a column vector whose components depend on (ρ, τ, λ). This is an overdetermined system as there are twice as many equations as unknowns (see Appendix C for the general discussion of overdetermined systems). The

[1] A differential operator $\mathcal{D}$ is degenerate if there exists a non-zero differential operator $\hat{\mathcal{D}}$ such that the operator $\hat{\mathcal{D}} \circ \mathcal{D}$ is identically zero.

compatibility conditions can be obtained by cross-differentiating and commuting the partial derivatives

$$\frac{\partial}{\partial \tau}\frac{\partial}{\partial \rho}v - \frac{\partial}{\partial \rho}\frac{\partial}{\partial \tau}v = 0$$

which gives

$$\frac{\partial}{\partial \tau}[U(\lambda)v] - \frac{\partial}{\partial \rho}[V(\lambda)v] = \left[\frac{\partial}{\partial \tau}U(\lambda) - \frac{\partial}{\partial \rho}V(\lambda) + [U(\lambda), V(\lambda)]\right] v = 0.$$

This has to hold for all characteristic initial data so the linear system (3.3.9) is consistent iff the non-linear equation

$$\frac{\partial}{\partial \tau}U(\lambda) - \frac{\partial}{\partial \rho}V(\lambda) + [U(\lambda), V(\lambda)] = 0 \qquad (3.3.10)$$

holds. The whole scheme is known as the zero-curvature representation.[2] Most non-linear integrable equations admit a zero-curvature representation analogous to (3.3.10).

- **Example.** If

$$U = \frac{i}{2}\begin{pmatrix} 2\lambda & \phi_\rho \\ \phi_\rho & -2\lambda \end{pmatrix} \quad \text{and} \quad V = \frac{1}{4i\lambda}\begin{pmatrix} \cos\phi & -i\sin\phi \\ i\sin\phi & -\cos\phi \end{pmatrix} \qquad (3.3.11)$$

where $\phi = \phi(\rho, \tau)$ then (3.3.10) is equivalent to the Sine-Gordon equation

$$\phi_{\rho\tau} = \sin\phi.$$

- **Example.** Consider the zero-curvature representation with

$$U = i\lambda\begin{pmatrix} 1 & 0 \\ 0 & -1 \end{pmatrix} + i\begin{pmatrix} 0 & \overline{\phi} \\ \phi & 0 \end{pmatrix} \quad \text{and} \qquad (3.3.12)$$

$$V = 2i\lambda^2\begin{pmatrix} 1 & 0 \\ 0 & -1 \end{pmatrix} + 2i\lambda\begin{pmatrix} 0 & \overline{\phi} \\ \phi & 0 \end{pmatrix} + \begin{pmatrix} 0 & \overline{\phi}_\rho \\ -\phi_\rho & 0 \end{pmatrix} - i\begin{pmatrix} |\phi|^2 & 0 \\ 0 & -|\phi|^2 \end{pmatrix}.$$

The condition (3.3.10) holds if the complex valued function $\phi = \phi(\tau, \rho)$ satisfies the non-linear Schrödinger equation (NLS)

$$i\phi_\tau + \phi_{\rho\rho} + 2|\phi|^2\phi = 0.$$

This is another famous soliton equation which can be solved by IST.

[2] The terminology, due to Zaharov and Shabat [191], comes from differential geometry where (3.3.10) means that the curvature of a connection $U d\rho + V d\tau$ is zero. In Chapter 7 we shall make a full use of this interpretation.

There is a freedom in the matrices $U(\lambda)$ and $V(\lambda)$ known as *gauge invariance*. Let $g = g(\tau, \rho)$ be an arbitrary invertible matrix. The transformations

$$\tilde{U} = gUg^{-1} + \frac{\partial g}{\partial \rho}g^{-1} \quad \text{and} \quad \tilde{V} = gVg^{-1} + \frac{\partial g}{\partial \tau}g^{-1} \tag{3.3.13}$$

map solutions to the zero-curvature equation into new solutions: if the matrices (U, V) satisfy (3.3.10) then so do the matrices $(\tilde{U}, \tilde{V})$. To see this assume that $v(\rho, \tau, \lambda)$ is a solution to the linear system (3.3.9), and demand that $\tilde{v} = g(\rho, \tau)v$ be another solution for some $(\tilde{U}, \tilde{V})$. This leads to the gauge transformation (3.3.13).

One can develop a version of the IST which recovers $U(\lambda)$ and $V(\lambda)$ from a linear-scattering problem (3.3.9). The representation (3.3.10) can also be an effective direct method of finding solutions if we know n linearly independent solutions $v_1, \ldots, v_n$ to the linear system (3.3.9) in the first place. Let $\Phi(\rho, \tau, \lambda)$ be a fundamental matrix solution to (3.3.9). The columns of Φ are the n linearly independent solutions $v_1, \ldots, v_n$. Then (3.3.9) holds with v replaced by Φ and we can write

$$U(\lambda) = \frac{\partial \Phi}{\partial \rho}\Phi^{-1} \quad \text{and} \quad V(\lambda) = \frac{\partial \Phi}{\partial \tau}\Phi^{-1}.$$

In practice one assumes a simple λ dependence in Φ, characterized by a finite number of poles with given multiplicities. One general scheme of solving (3.3.10), known as the dressing method, is based on the Riemann–Hilbert problem which we shall review next.

3.3.1 Riemann–Hilbert problem

Let $\lambda \in \overline{\mathbb{C}} = \mathbb{C} \cup \{\infty\}$ and let Γ be a closed contour in the extended complex plane. In particular we can consider Γ to be a real line $-\infty < \lambda < \infty$ regarded as a circle in $\overline{\mathbb{C}}$ passing through ∞. Let $G = G(\lambda)$ be a matrix-valued function on the contour Γ. The Riemann–Hilbert problem is to construct two matrix-valued functions $G_+(\lambda)$ and $G_-(\lambda)$ holomorphic, respectively, inside and outside the contour such that on Γ

$$G(\lambda) = G_+(\lambda)G_-(\lambda). \tag{3.3.14}$$

In the case when Γ is the real axis G_+ is required to be holomorphic in the upper half-plane and G_- is required to be holomorphic in the lower half-plane. If (G_+, G_-) is a solution of the Riemann–Hilbert problem, then

$$\tilde{G}_+ = G_+ g^{-1} \quad \text{and} \quad \tilde{G}_- = g\, G_-$$

will also be a solution for any constant invertible matrix g. This ambiguity can be avoided by fixing a values of G_+ or G_- at some point in their domain, for example by setting $G_-(\infty) = I$. If the matrices $G_\pm$ are everywhere invertible then this normalization guarantees that the solution to (3.3.14) is unique.

Solving a Riemann–Hilbert problem comes down to an integral equation. Choose a normalization $G_+(\lambda_0) = I$ and set $G_-(\lambda_0) = g$ for some $\lambda_0 \in \mathbb{C}$. Assume that the Riemann–Hilbert problem has a solution of the form

$$(G_+)^{-1} = h + \oint_\Gamma \frac{\Phi(\xi)}{\xi - \lambda} d\xi$$

for λ inside the contour Γ, and

$$G_- = h + \oint_\Gamma \frac{\Phi(\xi)}{\xi - \lambda} d\xi$$

for λ outside Γ, where h is determined by the normalization condition to be

$$h = g - \oint_\Gamma \frac{\Phi(\xi)}{\xi - \lambda_0} d\xi.$$

The Plemelj formula [3] can be used to determine $(G_+)^{-1}$ and G_- on the contour: If $\lambda \in \Gamma$ then

$$(G_+)^{-1}(\lambda) = h + \oint_\Gamma \frac{\Phi(\xi)}{\xi - \lambda} d\xi + \pi i \Phi(\lambda) \text{ and}$$

$$G_-(\lambda) = h + \oint_\Gamma \frac{\Phi(\xi)}{\xi - \lambda} d\xi - \pi i \Phi(\lambda),$$

where the integrals are assumed to be defined by the principal value. Substituting these expressions to (3.3.14) yields an integral equation for $\Phi = \Phi(\lambda)$. If the normalization is canonical, so that $h = g = 1$, the equation is

$$\frac{1}{\pi i}\left[\int_\Gamma \frac{\Phi(\xi)}{\xi - \lambda} d\xi + I\right] + \Phi(\lambda)(G + I)(G - I)^{-1} = 0.$$

The simplest case is the scalar Riemann–Hilbert problem where G, G_+, and G_- are ordinary functions. In this case the solution can be written down explicitly as

$$G_+ = \exp\left\{-\left[\frac{1}{2\pi i}\int_{-\infty}^{\infty} \frac{\log G(\xi)}{\xi - \lambda} d\xi\right]\right\}, \quad \operatorname{Im}(\lambda) > 0$$

$$G_- = \exp\left[\frac{1}{2\pi i}\int_{-\infty}^{\infty} \frac{\log G(\xi)}{\xi - \lambda} d\xi\right], \quad \operatorname{Im}(\lambda) < 0.$$

This is verified by taking a logarithm of (3.3.14)

$$\log G = \log(G_-) - \log(G_+)^{-1}$$

and applying the Cauchy integral formulae. (Compare the cohomological interpretation given by formula (B8).)

3.3.2 Dressing method

We shall assume that the matrices (U, V) in the zero-curvature representation (3.3.10) have rational dependence on the spectral parameter λ. The complex analytic data for each of these matrices consist of a set of poles (including a pole at $\lambda = \infty$) with the corresponding multiplicities. Define the divisors to be the sets

$$S_U = \{\alpha_i, n_i, n_\infty\} \quad \text{and} \quad S_V = \{\beta_j, m_j, m_\infty\}, \qquad i = 1, \dots, n, \qquad j = 1, \dots, m$$

so that

$$U(\rho, \tau, \lambda) = \sum_{i=1}^{n} \sum_{r=1}^{n_i} \frac{U_{i,r}(\rho, \tau)}{(\lambda - \alpha_i)^r} + \sum_{k=0}^{n_\infty} \lambda^k U_k(\rho, \tau)$$

$$V(\rho, \tau, \lambda) = \sum_{j=1}^{m} \sum_{r=1}^{m_i} \frac{V_{j,r}(\rho, \tau)}{(\lambda - \beta_j)^r} + \sum_{k=0}^{m_\infty} \lambda^k V_k(\rho, \tau). \tag{3.3.15}$$

The zero-curvature condition (3.3.10) is a system of non-linear PDEs for the coefficients

$$U_{i,r}, \quad U_k, \quad V_{j,r}, \quad \text{and} \quad V_k$$

of U and V. Consider a trivial solution to (3.3.10)

$$U = U_0(\rho, \lambda), \quad \text{and} \quad V = V_0(\tau, \lambda)$$

where U_0 and V_0 are any two commuting matrices with divisors S_U and S_V, respectively.

Let Γ be a contour in the extended complex plane which does not contain any points from $S_U \cup S_V$, and let $G(\lambda)$ be a smooth matrix-valued function defined on Γ. The dressing method [191] is a way of constructing a non-trivial solution with analytic structure specified by divisors S_U and S_V out of the data

$$(U_0, \ V_0, \ \Gamma, \ G).$$

It consists of the following steps:

1. Find a fundamental matrix solution to the linear system of equations

$$\frac{\partial}{\partial \rho}\Psi_0 = U_0(\lambda)\Psi_0 \quad \text{and} \quad \frac{\partial}{\partial \tau}\Psi_0 = V_0(\lambda)\Psi_0. \tag{3.3.16}$$

   This overdetermined system is compatible as U_0 and V_0 satisfy (3.3.10).
2. Define a family of smooth functions $G(\rho, \tau, \lambda)$ parameterized by (ρ, τ) on Γ

$$G(\rho, \tau, \lambda) = \Psi_0(\rho, \tau, \lambda)G(\lambda)\Psi_0{}^{-1}(\rho, \tau, \lambda). \tag{3.3.17}$$

This family admits a factorization

$$G(\rho,\tau,\lambda) = G_+(\rho,\tau,\lambda)G_-(\rho,\tau,\lambda) \tag{3.3.18}$$

where $G_+(\rho,\tau,\lambda)$ and $G_-(\rho,\tau,\lambda)$ are solutions to the Riemann–Hilbert problem described in the last subsection, and are holomorphic, respectively, inside and outside the contour Γ.

3. Differentiate (3.3.18) with respect to ρ and use (3.3.16) and (3.3.17). This yields

$$\frac{\partial G_+}{\partial\rho}G_- + G_+\frac{\partial G_-}{\partial\rho} = U_0G_+G_- - G_+G_-U_0.$$

Therefore we can define

$$U(\rho,\tau,\lambda) := \left(\frac{\partial G_-}{\partial\rho} + G_-U_0\right)G_-{}^{-1} = -G_+{}^{-1}\left(\frac{\partial G_+}{\partial\rho} - U_0G_+\right)$$

which is holomorphic in $\overline{\mathbb{C}}/S_U$. The Liouville theorem (Theorem B.0.4) applied to the extended complex plane implies that $U(\rho,\tau,\lambda)$ is rational in λ and has the same pole structure as U_0.

Analogous argument leads to

$$V(\rho,\tau,\lambda) := \left(\frac{\partial G_-}{\partial\tau} + G_-V_0\right)G_-{}^{-1} = -G_+{}^{-1}\left(\frac{\partial G_+}{\partial\tau} - V_0G_+\right)$$

which has the same pole structure as V_0.

4. Define two matrix-valued functions

$$\Psi_+ = G_+{}^{-1}\Psi_0 \quad\text{and}\quad \Psi_- = G_-{}^{-1}\Psi_0.$$

Equations (3.3.16) and the definitions of (U, V) imply that these matrices both satisfy the overdetermined system

$$\frac{\partial}{\partial\rho}\Psi_\pm = U(\lambda)\Psi_\pm \quad\text{and}\quad \frac{\partial}{\partial\tau}\Psi_\pm = V(\lambda)\Psi_\pm.$$

We can therefore deduce that $U(\rho,\tau,\lambda)$ and $V(\rho,\tau,\lambda)$ are of the form (3.3.15) and satisfy the zero-curvature relation (3.3.10).

This procedure is called 'dressing' as the bare, trivial solution (U_0, V_0) has been dressed by an application of a Riemann–Hilbert problem to a non-trivial (U, V). Now, given another matrix-valued function $G = G'(\lambda)$ on the contour we could repeat the whole procedure and apply it to (U, V) instead of (U_0, V_0). This would lead to another solution (U', V') with the same pole structure. Thus dressing transformations act on the space of solutions to (3.3.10) and form a group. If $G = G_+G_-$ and $G' = G'_+G'_-$ then

$$(G \circ G') = G_+G'_+G'_-G_-.$$

The solution to the Riemann–Hilbert problem (3.3.18) is not unique. If $G_{\pm}$ give a factorization of $G(\rho, \tau, \lambda)$ then so do

$$\tilde{G}_+ = G_+ g^{-1} \quad \text{and} \quad \tilde{G}_- = gG_-$$

where $g = g(\rho, \tau)$ is a matrix-valued function. The corresponding solutions $(\tilde{U}, \tilde{V})$ are related to (U, V) by the gauge transformation (3.3.13). Fixing the gauge is therefore equivalent to fixing the value of G_+ or G_- at one point of the extended complex plane, say $\lambda = \infty$. This leads to a unique solution of the Riemann–Hilbert problem with $G_{\pm}(\infty) = G(\infty) = I$.

The dressing method leads to the general form of U and V with prescribed singularities, but more work is required to make contact with specific integrable models when additional algebraic constraints need to be imposed on U and V. For example, in the Sine-Gordon case (3.3.11) the matrices are anti-Hermitian. The anti-Hermiticity condition gives certain constraints on the contour Γ and the function G. Only if these constraints hold, the matrices resulting from the dressing procedure will be given (in some gauge) in terms of the solution to the Sine-Gordon equation. The problem of gauge-invariant characterization of various integrable equations will be studied in Chapter 7.

3.3.3 From Lax representation to zero curvature

The zero-curvature representation (3.3.10) is more general than the scalar Lax representation but there is a connection between the two. The first similarity is that the Lax equation (2.2.11) also arises as a compatibility condition for two overdetermined PDEs. To see this take f to be an eigenfunction of L with a simple eigenvalue $E = \lambda$ and consider the relation (2.2.12) which follows from the Lax equations. If $E = \lambda$ is a simple eigenvalue then

$$\frac{\partial f}{\partial t} + Af = C(t) f$$

for some function C which depends on t but not on x. Therefore one can use an integrating factor to find a function $\hat{f} = \hat{f}(x, t, \lambda)$ such that

$$L\hat{f} = \lambda \hat{f} \quad \text{and} \quad \frac{\partial \hat{f}}{\partial t} + A\hat{f} = 0, \tag{3.3.19}$$

where L is the Schrödinger operator and A is some differential operator (e.g. given by (2.2.13)). Therefore the Lax relation

$$\dot{L} = [L, A]$$

is the compatibility of an overdetermined system (3.3.19).

Consider a general scalar Lax pair

$$L = \frac{\partial^n}{\partial x^n} + u_{n-1}(x,t)\frac{\partial^{n-1}}{\partial x^{n-1}} + \dots + u_1(x,t)\frac{\partial}{\partial x} + u_0(x,t)$$

$$A = \frac{\partial^m}{\partial x^m} + v_{m-1}(x,t)\frac{\partial^{m-1}}{\partial x^{m-1}} + \dots + v_1(x,t)\frac{\partial}{\partial x} + v_0(x,t)$$

given by differential operators with coefficients depending on (x, t). The Lax equations

$$\dot{L} = [L, A]$$

(in general there will be more than one) are non-linear PDEs for the coefficients

$$(u_0, \dots, u_{n-1}, v_0, \dots, v_{m-1}).$$

The linear nth-order scalar PDE

$$L\hat{f} = \lambda \hat{f} \tag{3.3.20}$$

is equivalent to the first-order matrix PDE

$$\frac{\partial F}{\partial x} = U_L F$$

where $U_L = U_L(x, t, \lambda)$ is an n by n matrix

$$U_L = \begin{pmatrix} 0 & 1 & 0 & \dots & 0 & 0 \\ 0 & 0 & 1 & \dots & 0 & 0 \\ \vdots & \vdots & \vdots & \dots & \vdots & \vdots \\ \vdots & \vdots & \vdots & \dots & \vdots & \vdots \\ 0 & 0 & 0 & \dots & 0 & 1 \\ \lambda - u_0 & -u_1 & -u_2 & \dots & -u_{n-2} & -u_{n-1} \end{pmatrix}$$

and F is a column vector

$$F = (f_0, f_1, \dots, f_{n-1})^{\mathrm{T}}, \qquad \text{where} \quad f_k = \frac{\partial^k \hat{f}}{\partial x^k}.$$

Now consider the second equation in (3.3.19)

$$\frac{\partial \hat{f}}{\partial t} + A\hat{f} = 0$$

which is compatible with (3.3.20) if the Lax equations hold. We differentiate this equation with respect to x and use (3.3.20) to express $\partial_x^n \hat{f}$ in terms of λ and lower order derivatives. Repeating this process $(n-1)$ times gives an action of A on components of the vector F. We write it as

$$\frac{\partial F}{\partial t} = V_A F$$

using the method described above. This leads to a pair of first-order linear matrix equations with the zero-curvature compatibility conditions

$$\frac{\partial U_L}{\partial t} - \frac{\partial V_A}{\partial x} + [U_L, V_A] = 0.$$

These conditions hold if the operators (L, A) satisfy the Lax relations $\dot{L} = [L, A]$.

- **Example.** Let us apply this procedure to the KdV Lax pair (2.2.13). Set

$$f_0 = \hat{f}(x, t, \lambda) \quad \text{and} \quad f_1 = \partial_x \hat{f}(x, t, \lambda).$$

The eigenvalue problem $L\hat{f} = \lambda \hat{f}$ gives

$$(f_0)_x = f_1 \quad \text{and} \quad (f_1)_x = (u - \lambda) f_0.$$

The equation $\partial_t \hat{f} + A\hat{f} = 0$ gives

$$\begin{aligned}(f_0)_t &= -4(f_0)_{xxx} + 6uf_1 + 3u_x f_0\\ &= -u_x(f_0) + (2u + 4\lambda) f_1.\end{aligned}$$

We differentiate this equation with respect to x and eliminate the second derivatives of $\hat{f}$ to get

$$(f_1)_t = [(2u + 4\lambda)(u - \lambda) - u_{xx}] f_0 + u_x f_1.$$

We now collect the equations in the matrix form $\partial_x F = U_L F$ and $\partial_t F = V_A F$ where $F = (f_0, f_1)^{\mathrm{T}}$ and

$$U_L = \begin{pmatrix} 0 & 1 \\ u - \lambda & 0 \end{pmatrix} \quad \text{and} \quad V_A = \begin{pmatrix} -u_x & 2u + 4\lambda \\ 2u^2 - u_{xx} + 2u\lambda - 4\lambda^2 & u_x \end{pmatrix}. \tag{3.3.21}$$

We have therefore obtained a zero-curvature representation for KdV.

3.4 Hierarchies and finite-gap solutions

We shall end our discussion of the KdV equation with a description of the KdV hierarchy. Recall that KdV is a Hamiltonian system (3.2.5) with the Hamiltonian given by the first integral $-I_1[u]$. Now choose a (constant multiple of) different first integral $I_n[u]$ as a Hamiltonian and consider the equation

$$\frac{\partial u}{\partial t_n} = (-1)^n \frac{\partial}{\partial x} \frac{\delta I_n[u]}{\delta u(x)} \tag{3.4.22}$$

for a function $u = u(x, t_n)$. This leads to an infinite set of equations known as higher KdVs. The first three equations are

$$
\begin{aligned}
u_{t_0} &= u_x, \\
u_{t_1} &= 6uu_x - u_{xxx}, \quad \text{and} \\
u_{t_2} &= 10uu_{xxx} - 20u_x u_{xx} - 30u^2 u_x - u_{xxxxx}.
\end{aligned}
$$

Each of these equations can be solved by the inverse scattering method we have discussed, and the functionals I_k, $k = -1, 0, \ldots$, are first integrals regardless of which one of them is chosen as a Hamiltonian. In the associated Lax representation L stays unchanged, but A is replaced by a differential operator of degree $(2n + 1)$. One can regard the higher KdVs as a system of overdetermined PDEs for

$$
u = u(t_0 = x, t_1 = t, t_2, t_3, \ldots),
$$

where we have identified t_0 with x using the first equation in (3.4.22).

This system is called a hierarchy and the coordinates $(t_2, t_3, \ldots)$ are known as higher times. The equations of the hierarchy are consistent as the flows generated by time translations commute:

$$
\begin{aligned}
\frac{\partial}{\partial t_m}\frac{\partial}{\partial t_n}u - \frac{\partial}{\partial t_n}\frac{\partial}{\partial t_m}u &= (-1)^n \frac{\partial}{\partial t_m}\frac{\partial}{\partial x}\frac{\delta I_n[u]}{\delta u(x)} - (-1)^m \frac{\partial}{\partial t_n}\frac{\partial}{\partial x}\frac{\delta I_m[u]}{\delta u(x)} \\
&= \{u, \partial_m I_n - \partial_n I_m + \{I_m, I_n\}\} = 0
\end{aligned}
$$

where we used the Jacobi identity and the fact that $I_n[u]$ Poisson commute.

The concept of the hierarchy leads to a beautiful method of finding solutions to KdV with periodic initial data, that is,

$$
u(x, 0) = u(x + X_0, 0)
$$

for some period X_0. The method is based on the concept of stationary (i.e. time-independent) solutions, albeit applied to a combination of the higher times.

Consider the first $(n + 1)$ higher KdVs and take $(n + 1)$ real constants $c_0, \ldots, c_n$. Therefore

$$
\sum_{k=0}^{n} c_k \frac{\partial u}{\partial t_k} = \sum_{k=0}^{n} (-1)^k c_k \frac{\partial}{\partial x}\frac{\delta I_k[u]}{\delta u(x)}.
$$

The stationary solutions correspond to u being independent of the combination of higher times on the LHS. This leads to an ODE

$$
\sum_{k=0}^{n} (-1)^k c_k \frac{\delta I_k[u]}{\delta u(x)} = c_{n+1}, \qquad c_{n+1} = \text{const.} \tag{3.4.23}
$$

The recursion relations (3.1.3) for S_k's imply that this ODE is of order $2n$. Its general solution depends on $2n$ constants of integration as well as $(n+1)$ parameters c as we can always divide (3.4.23) by $c_n \neq 0$. Altogether one has $3n+1$ parameters. The beauty of this method is that the ODE is integrable in the sense of the Arnold–Liouville theorem and its solutions can be constructed by hyper-elliptic functions. The corresponding solutions to KdV are known as finite-gap solutions. Their description in terms of a spectral data is rather involved and uses Riemann surfaces and algebraic geometry (see Chapter 2 of [122]).

We shall now present the construction of the first integrals to equation (3.4.23) (we stress that (3.4.23) is an ODE in x so the first integrals are functions of u and its derivatives which do not depend on x when (3.4.23) holds). The higher KdV equations (3.4.22) admit a zero-curvature representation

$$\frac{\partial}{\partial t_n}U - \frac{\partial}{\partial x}V_n + [U, V_n] = 0$$

where

$$U = \begin{pmatrix} 0 & 1 \\ u-\lambda & 0 \end{pmatrix}$$

is the matrix obtained for KdV in section (3.3.3) and $V_n = V_n(x, t, \lambda)$ are traceless 2×2 matrices analogous to V_Λ which can be obtained using (3.4.22) and the recursion relations for (3.1.2). The components of V_n depend on (x, t) and are polynomials in λ of degree $(n+1)$. Now set

$$\Lambda = c_0 V_0 + \cdots + c_n V_n$$

where c_k are constants and consider solutions to

$$\frac{\partial}{\partial T}U - \frac{\partial}{\partial x}\Lambda + [U, \Lambda] = 0,$$

such that

$$\frac{\partial}{\partial T}U = 0, \qquad \text{where} \qquad \frac{\partial}{\partial T} = c_0\frac{\partial}{\partial t_0} + \cdots + c_n\frac{\partial}{\partial t_n}.$$

This gives rise to the ODE

$$\frac{d}{dx}\Lambda = [U, \Lambda]$$

which is the Lax representation of (3.4.23). This representation reveals the existence of many first integrals for (3.4.23) . We have

$$\frac{d}{dx}\mathrm{Tr}(\Lambda^p) = \mathrm{Tr}(p[U, \Lambda]\Lambda^{p-1}) = p\mathrm{Tr}(-\Lambda U\Lambda^{p-1} + U\Lambda^p) = 0, \quad p = 2, 3, \ldots$$

by the cyclic property of trace. Therefore all the coefficients of the polynomials $\mathrm{Tr}[\Lambda(\lambda)^p]$ for all p are conserved (which implies that the whole spectrum of $\Lambda(\lambda)$ is constant in x). It turns out [122] that one can find n-independent non-trivial integrals in this set which are in involution thus guaranteeing the integrability of (3.4.23) is the sense of the Arnold–Liouville theorem 1.2.2.

The resulting solutions to KdV are known as 'finite-gap' potentials. Let us justify this terminology. The spectrum of Λ does not depend on x and so the coefficients of the characteristic polynomial

$$\det[\mathbf{1}\mu - \Lambda(\lambda)] = 0$$

also do not depend on x. Using the fact that $\Lambda(\lambda)$ is trace free we can rewrite this polynomial as

$$\mu^2 + R(\lambda) = 0 \tag{3.4.24}$$

where

$$\begin{aligned} R(\lambda) &= \lambda^{2n+1} + a_1\lambda^{2n} + \cdots + a_{2n}\lambda + a_{2n+1} \\ &= (\lambda - \lambda_0)\cdots(\lambda - \lambda_{2n}). \end{aligned}$$

Therefore the coefficients $a_1, \ldots, a_{2n+1}$ (or equivalently $\lambda_0, \ldots, \lambda_{2n}$) do not depend on x. However $(n+1)$ of those coefficients can be expressed in terms of the constants c_k and thus the corresponding first integrals are trivial. This leaves us with n first integrals for an ODE (3.4.23) of order $2n$.

It is possible to show [122] that

- All solutions to the KdV equation with periodic initial data arise from (3.4.23).
- For each λ the corresponding eigenfunctions of the Schrödinger operator $L\psi = \lambda\psi$ can be expanded in a basis $\psi_\pm$ such that

$$\psi_\pm(x + X_0) = e^{\pm ipX_0}\psi_\pm(x)$$

for some $p = p(\lambda)$ ($\psi_\pm$ are called Bloch functions). The set of real λ for which $p(\lambda) \in \mathbb{R}$ is called the permissible zone. The roots of the polynomial $R(\lambda)$ are the end-points of the permissible zones

$$(\lambda_0, \lambda_1), \quad (\lambda_2, \lambda_3), \quad \ldots, \quad (\lambda_{2n-2}, \lambda_{2n-1}), \quad (\lambda_{2n}, \infty).$$

The equation (3.4.24) defines a Riemann surface Γ of genus n. The number of forbidden zones (gaps) is therefore finite for the periodic solutions as the Riemann surface (3.4.24) has finite genus. This justifies the name 'finite-gap' potentials.

- **Example.** If $n = 0$ the Riemann surface Γ has the topology of the sphere and the corresponding solution to KdV is a constant. If $n = 1$ Γ is called an elliptic curve (it has topology of a two-torus) and the ODE (3.4.23) is solvable by elliptic functions: the stationary condition

$$c_0 \frac{\partial u}{\partial t_0} + c_1 \frac{\partial u}{\partial t_1} = 0$$

yields

$$c_0 u_x + c_1 (6uu_x - u_{xxx}) = 0$$

where we used the first two equations of the hierarchy. We can set $c_1 = 1$ redefining the other constant. This ODE can be integrated and the general solution is a Weierstrass elliptic function

$$\int \frac{du}{\sqrt{2u^3 + c_0 u^2 + bu + d}} = x - x_0$$

for some constants (b, d). The stationary condition implies that $u = u(x - c_0 t)$ where we have identified $t_0 = x$ and $t_1 = t$. Thus $x_0 = c_0 t$. These solutions are called cnoidal waves because the corresponding elliptic function is often denoted 'cn'.

If $n > 1$ the Riemann surface Γ is a hyper-elliptic curve and the corresponding KdV potential is given in terms of Riemann's theta function [122].

Exercises

1. Consider the Riccati equation

$$\frac{dS}{dx} - 2ikS + S^2 = u,$$

for the first integrals of KdV. Assume that

$$S = \sum_{n=1}^{\infty} \frac{S_n(x)}{(2ik)^n}$$

and find the recursion relations

$$S_1(x, t) = -u(x, t) \quad \text{and} \quad S_{n+1} = \frac{dS_n}{dx} + \sum_{m=1}^{n-1} S_m S_{n-m}.$$

Solve the first few relations to show that

$$S_2 = -\frac{\partial u}{\partial x}, \quad S_3 = -\frac{\partial^2 u}{\partial x^2} + u^2, \quad \text{and} \quad S_4 = -\frac{\partial^3 u}{\partial x^3} + 2\frac{\partial}{\partial x} u^2$$

and find S_5. Use the KdV equation to verify directly that

$$\frac{d}{dt}\int_{\mathbb{R}} S_3 dx = 0 \quad \text{and} \quad \frac{d}{dt}\int_{\mathbb{R}} S_5 dx = 0.$$

2. Apply the recursion operator (3.2.8) twice to $H_0[u]$ to show that

$$\int_{\mathbb{R}} \left(\frac{u_{xx}^2}{2} - \frac{5}{2}u^2 u_{xx} + \frac{5}{2}u^4 \right) dx$$

is a first integral of the KdV equation.

3. Consider a one-parameter family of self-adjoint operators $L(t)$ in some complex inner product space such that

$$L(t) = U(t)L(0)U(t)^{-1}$$

where $U(t)$ is a unitary operator, that is, $U(t)U(t)^* = 1$ where U^* is the adjoint of U.

Show that $L(t)$ and $L(0)$ have the same eigenvalues. Show that there exist an anti-self-adjoint operator A such that $U_t = -AU$ and

$$L_t = [L, A].$$

4. Let $L = -\partial_x^2 + u(x, t)$ be a Schrödinger operator and let

$$A = a_n \partial_x^n + \cdots + a_1 \partial_x + a_0,$$

where $a_k = a_k(x, t)$ are functions, be another operator such that

$$L_t = [L, A]. \tag{3.4.25}$$

Show that the eigenvalues of L are independent on t.

Let f be an eigenfunction of L corresponding to an eigenvalue λ which is non-degenerate. Show that there exists a function $\hat{f} = \hat{f}(x, t, \lambda)$ such that

$$L\hat{f} = \lambda \hat{f} \quad \text{and} \quad \hat{f}_t + A\hat{f} = 0. \tag{3.4.26}$$

Now assume that $n = 3$ and $a_3 = 1, a_2 = 0$. Show that the Lax representation (3.4.25) is equivalent to a zero-curvature representation

$$\partial_t U_L - \partial_x V_A + [U_L, V_A] = 0$$

where U_L and V_A are some 2×2 matrices which should be determined.

[Hint: Consider (3.4.26) as a first-order system on a pair of functions $(\hat{f}, \partial_x \hat{f})$.]

5. Let $L(t)$ and $A(t)$ be complex valued $n \times n$ matrices such that

$$\dot{L} = [L, A].$$

Deduce that $\text{Tr}(L^p)$, $p \in \mathbb{Z}$ does not depend on t.

Assume that

$$L = (\Phi_1 + i\Phi_2) + 2\Phi_3\lambda - (\Phi_1 - i\Phi_2)\lambda^2 \quad \text{and}$$
$$A = -i\Phi_3 + i(\Phi_1 - i\Phi_2)\lambda$$

where λ is a parameter and find the system of ODEs satisfied by the matrices $\Phi_j(t)$, $j = 1, 2, 3$.

[Hint: The Lax relations should hold for any value of the parameter λ.] Now take $\Phi_j(t) = -i\sigma_j w_j(t)$ (no summation) where σ_j are the matrices

$$\sigma_1 = \frac{1}{2}\begin{pmatrix} 0 & 1 \\ 1 & 0 \end{pmatrix}, \quad \sigma_2 = \frac{1}{2}\begin{pmatrix} 0 & -i \\ i & 0 \end{pmatrix}, \quad \text{and} \quad \sigma_3 = \frac{1}{2}\begin{pmatrix} 1 & 0 \\ 0 & -1 \end{pmatrix}$$

which satisfy $[\sigma_j, \sigma_k] = i\sum_{l=1}^{3} \varepsilon_{jkl}\sigma_l$. Show that the system reduces to the Euler equations

$$\dot{w}_1 = w_2 w_3, \qquad \dot{w}_2 = w_1 w_3, \quad \text{and} \quad \dot{w}_3 = w_1 w_2.$$

Use $\mathrm{Tr}(L^p)$ to construct first integrals of this system.

6. Let $g = g(\tau, \rho)$ be an arbitrary invertible matrix. Show that the transformations

$$\widetilde{U} = gUg^{-1} + \frac{\partial g}{\partial \rho}g^{-1} \quad \text{and} \quad \widetilde{V} = gVg^{-1} + \frac{\partial g}{\partial \tau}g^{-1}$$

map solutions to the zero-curvature equation into new solutions: if the matrices (U, V) satisfy

$$\frac{\partial}{\partial \tau}U(\lambda) - \frac{\partial}{\partial \rho}V(\lambda) + [U(\lambda), V(\lambda)] = 0$$

then so do $\widetilde{U}(\lambda)$ and $\widetilde{V}(\lambda)$. What is the relationship between the solutions of the associated linear problems?

7. Consider solutions to the KdV hierarchy which are stationary with respect to

$$c_0\frac{\partial}{\partial t_0} + c_1\frac{\partial}{\partial t_1}$$

where the kth KdV flow is generated by the Hamiltonian $(-1)^k I_k[u]$ and $I_k[u]$ are the first integrals constructed in lectures.

Show that the resulting solution to KdV is

$$F(u) = c_1 x - c_0 t,$$

where $F(u)$ is given by an integral which should be determined and $t_0 = x$, $t_1 = t$.

Find the zero-curvature representation for the ODE characterizing the stationary solutions.

8. Consider the zero-curvature representation with

$$U = i\lambda \begin{pmatrix} 1 & 0 \\ 0 & -1 \end{pmatrix} + i \begin{pmatrix} 0 & \overline{\phi} \\ \phi & 0 \end{pmatrix} \quad \text{and}$$

$$V = 2i\lambda^2 \begin{pmatrix} 1 & 0 \\ 0 & -1 \end{pmatrix} + 2i\lambda \begin{pmatrix} 0 & \overline{\phi} \\ \phi & 0 \end{pmatrix} + \begin{pmatrix} 0 & \overline{\phi}_\rho \\ -\phi_\rho & 0 \end{pmatrix} - i \begin{pmatrix} |\phi|^2 & 0 \\ 0 & -|\phi|^2 \end{pmatrix}$$

and show that complex-valued function $\phi = \phi(\tau, \rho)$ satisfies the non-linear Schrödinger equation

$$i\phi_\tau + \phi_{\rho\rho} + 2|\phi|^2\phi = 0.$$

9. Consider the zero-curvature representation with

$$U = \frac{u(\rho, \tau)}{1 - \lambda} \quad \text{and} \quad V = \frac{v(\rho, \tau)}{1 + \lambda},$$

where $u(\rho, \tau)$ and $v(\rho, \tau)$ are matrices. Deduce the existence of a unitary matrix $g(\rho, \tau)$ such that

$$u = g^{-1}\partial_\rho g \quad \text{and} \quad v = g^{-1}\partial_\tau g$$

and thus show that the solutions to the principal chiral model

$$\frac{\partial}{\partial \tau}\left(g^{-1}\frac{\partial g}{\partial \rho}\right) + \frac{\partial}{\partial \rho}\left(g^{-1}\frac{\partial g}{\partial \tau}\right)$$

are given by

$$g = g(\rho, \tau) = \Psi^{-1}(\rho, \tau, \lambda = 0),$$

where Ψ satisfies the linear system (3.3.9).

4 Lie symmetries and reductions

4.1 Lie groups and Lie algebras

Phrases like 'the unifying role of symmetry in ...' feature prominently in the popular science literature. Depending on the subject, the symmetry may be 'cosmic', 'Platonic', 'perfect', 'broken', or even 'super'.[1]

The mathematical framework used to define and describe symmetries is group theory. Recall that a group is a set G with a map

$$G \times G \to G, \quad (g_1, g_2) \to g_1 g_2$$

called the group multiplication which satisfies the following properties:

- Associativity

$$(g_1 g_2) g_3 = g_1 (g_2 g_3) \quad \forall g_1, g_2, g_3 \in G.$$

- There exist an identity element $e \in G$ such that

$$eg = ge = g, \quad \forall g \in G.$$

- For any $g \in G$ there exists an inverse element $g^{-1} \in G$ such that

$$gg^{-1} = g^{-1} g = e.$$

A group G acts on a set X if there exists a map $G \times X \to X$, $(g, p) \to g(p)$ such that

$$e(p) = p, \quad g_2 [g_1(p)] = (g_2 g_1)(p)$$

for all $p \in X$, and $g_1, g_2 \in G$. The set $\text{Orb}(p) = \{g(p), \ g \in G\} \subset X$ is called the orbit of p. Groups acting on sets are often called groups of transformations.

In this chapter we shall explore the groups which act on solutions to DEs. Such group actions occur both for integrable and non-integrable systems so

[1] Supersymmetry is a symmetry between elementary particles known as bosons and fermions. It is a symmetry of equations underlying the current physical theories. Supersymmetry predicts that each elementary particle has its supersymmetric partner. No one has yet observed supersymmetry. Perhaps it will be found in the LHC. See a footnote on page 26.

the methods we shall study are quite universal.[2] In fact all the techniques of integration of DEs (like separation of variables, integrating factors, homogeneous equations, etc.) students have encountered in their education are special cases of the symmetry approach. See [124] for a very complete treatment of this subject and [84] for an elementary introduction at an undergraduate level.

The symmetry programme goes back to a nineteenth century Norwegian mathematician Sophus Lie who developed a theory of continuous transformations now known as Lie groups. One of the most important of Lie's discoveries was that a continuous group G of transformations is easy to describe by infinitesimal transformations characterizing group elements close (in the sense of Taylor's theorem) to the identity element. These infinitesimal transformations are elements of the Lie algebra $\mathfrak{g}$. For example, a general element of the rotation group $G = SO(2)$

$$g(\varepsilon) = \begin{pmatrix} \cos\varepsilon & -\sin\varepsilon \\ \sin\varepsilon & \cos\varepsilon \end{pmatrix}$$

depends on one parameter ε. The group $SO(2)$ is a Lie group as g, its inverse and the group multiplication depend on ε in a differentiable way. This Lie group is one-dimensional as one parameter – the angle of rotation – is sufficient to describe any rotation around the origin in $\mathbb{R}^2$. A rotation in $\mathbb{R}^3$ depends on three such parameters – the Euler angles used in classical dynamics – so $SO(3)$ is a three-dimensional Lie group. Now consider the Taylor series

$$g(\varepsilon) = \begin{pmatrix} 1 & 0 \\ 0 & 1 \end{pmatrix} + \varepsilon \begin{pmatrix} 0 & -1 \\ 1 & 0 \end{pmatrix} + O(\varepsilon^2).$$

The antisymmetric matrix

$$A = \begin{pmatrix} 0 & -1 \\ 1 & 0 \end{pmatrix}$$

represents an infinitesimal rotation as $A\mathbf{x} = (-y, x)^{\mathrm{T}}$ are components of the vector tangent to the orbit of $\mathbf{x}$ at $\mathbf{x}$. The one-dimensional vector space spanned by A is called a Lie algebra of $SO(2)$.

The following definition is not quite correct (Lie groups should be defined as manifolds – see the Definition A.1.1 in Appendix A) but it is sufficient for our purposes.

Definition 4.1.1 *An m-dimensional Lie group is a group whose elements depend smoothly of m parameters such that the maps $(g_1, g_2) \to g_1 g_2$ and $g \to g^{-1}$ are smooth (infinitely differentiable) functions of these parameters.*

[2] It is however the case that integrable systems admit 'large' groups of symmetries and non-integrable systems usually do not.

The infinitesimal description of Lie groups is given by Lie algebras.

Definition 4.1.2 *A Lie algebra is a vector space* $\mathfrak{g}$ *with an antisymmetric bilinear operation called a Lie bracket* $[\,,\,]_{\mathfrak{g}} : \mathfrak{g} \times \mathfrak{g} \to \mathfrak{g}$ *which satisfies the Jacobi identity*

$$[A, [B, C]] + [C, [A, B]] + [B, [C, A]] = 0, \quad \forall A, B, C \in \mathfrak{g}.$$

If the vectors $A_1, \ldots, A_{\dim \mathfrak{g}}$ span $\mathfrak{g}$, the algebra structure is determined by the structure constants $f^{\gamma}_{\alpha\beta}$ such that

$$[A_\alpha, A_\beta]_{\mathfrak{g}} = \sum_{\gamma} f^{\gamma}_{\alpha\beta} A_\gamma, \quad \alpha, \beta, \gamma = 1, \ldots, \dim \mathfrak{g}.$$

The Lie bracket is related to non-commutativity of the group operation as the following argument demonstrates. Let $a, b \in G$. Set

$$a = I + \varepsilon A + O(\varepsilon^2) \quad \text{and} \quad b = I + \varepsilon B + O(\varepsilon^3)$$

for some A, B and calculate

$$\begin{aligned} aba^{-1}b^{-1} &= (I + \varepsilon A + \cdots)(I + \varepsilon B + \cdots)(I - \varepsilon A + \cdots)(I - \varepsilon B + \cdots) \\ &= I + \varepsilon^2 [A, B] + O(\varepsilon^3) \end{aligned}$$

where $\cdots$ denote terms of higher order in ε and we used the fact $(1 + \varepsilon A)^{-1} = 1 - \varepsilon A + \cdots$ which follows from the Taylor series. Some care needs to be taken with the above argument as we have neglected the second-order terms in the group elements but not in the answer. The readers should convince themselves that these terms indeed cancel out.

- **Example.** Consider the group of special orthogonal transformations $SO(n)$ which consist of $n \times n$ matrices a such that

  $$aa^{\mathrm{T}} = I, \quad \det a = 1.$$

  These conditions imply that only $n(n-1)/2$ matrix components are independent and $SO(n)$ is a Lie group of dimension $n(n-1)/2$. Setting $a = I + \varepsilon A + O(\varepsilon^2)$ shows that infinitesimal version of the orthogonal condition is antisymmetry

  $$A + A^{\mathrm{T}} = 0.$$

  Given two antisymmetric matrices their commutator is also antisymmetric as

  $$[A, B]^{\mathrm{T}} = B^{\mathrm{T}} A^{\mathrm{T}} - A^{\mathrm{T}} B^{\mathrm{T}} = -[A, B].$$

  Therefore the vector space of antisymmetric matrices is a Lie algebra with the Lie bracket defined to be the matrix commutator. This Lie algebra, called $\mathfrak{so}(n)$, is a vector space of dimension $n(n-1)/2$. This is equal to the dimension (the number of parameters) of the corresponding Lie group $SO(n)$.

- **Example.** An example of a three-dimensional Lie group is given by the group of 3×3 upper triangular matrices with diagonal entries equal to 1

$$g(m_1, m_2, m_3) = \begin{pmatrix} 1 & m_1 & m_3 \\ 0 & 1 & m_2 \\ 0 & 0 & 1 \end{pmatrix}. \tag{4.1.1}$$

Note that $g = \mathbf{1} + \sum_\alpha m_\alpha T_\alpha$, where the matrices T_α are

$$T_1 = \begin{pmatrix} 0 & 1 & 0 \\ 0 & 0 & 0 \\ 0 & 0 & 0 \end{pmatrix}, \quad T_2 = \begin{pmatrix} 0 & 0 & 0 \\ 0 & 0 & 1 \\ 0 & 0 & 0 \end{pmatrix}, \quad \text{and} \quad T_3 = \begin{pmatrix} 0 & 0 & 1 \\ 0 & 0 & 0 \\ 0 & 0 & 0 \end{pmatrix}. \tag{4.1.2}$$

This Lie group is called **Nil**, as the matrices T_α are all nilpotent. These matrices span the Lie algebra of the group **Nil** and have the commutation relations

$$[T_1, T_2] = T_3, \quad [T_1, T_3] = 0, \quad \text{and} \quad [T_2, T_3] = 0. \tag{4.1.3}$$

This gives the structure constants $f_{12}^3 = -f_{21}^3 = 1$ and all other constants vanish.

A three-dimensional Lie algebra with these structure constants is called the Heisenberg algebra because of its connection with QM – think of T_1 and T_2 as position and momentum operators, respectively, and T_3 as $i\hbar$ times the identity operator.

In the above example the Lie algebra of a Lie group was represented by matrices. If the group acts on a subset X of $\mathbb{R}^n$, its Lie algebra is represented by vector fields[3] on X. This approach underlies the application of Lie groups to DEs so we shall study it next.

4.2 Vector fields and one-parameter groups of transformations

Let X be an open set in $\mathbb{R}^n$ with local coordinates $x^1, \ldots, x^n$ and let $\gamma : [0,1] \longrightarrow X$ be a parameterized curve, so that $\gamma(\varepsilon) = (x^1(\varepsilon), \ldots, x^n(\varepsilon))$. The tangent vector $V|_p$ to this curve at a point $p \in X$ has components

$$V^a = \dot{x}^a|_p, \quad a = 1, \ldots, n, \quad \text{where} \quad \cdot = \frac{d}{d\varepsilon}.$$

[3] The structure constants $f_{\alpha\beta}^{\gamma}$ do not depend on which of these representations are used.

The collection of all tangent vectors to all possible curves through p is an n-dimensional vector space called the tangent space T_pX. The collection of all tangent spaces as x varies in X is called a tangent bundle $TX = \bigcup_{p\in X} T_pX$. The tangent bundle is a manifold of dimension $2n$ (see Appendix A).

A vector field V on X assigns a tangent vector $V|_p \in T_pX$ to each point in X. Let $f : X \longrightarrow \mathbb{R}$ be a function on X. The rate of change of f along the curve is measured by a derivative

$$\begin{aligned}\frac{d}{d\varepsilon} f\,[x(\varepsilon)]\,|_{\varepsilon=0} &= V^a \frac{\partial f}{\partial x^a}\\ &= V(f)\end{aligned}$$

where

$$V = V^1 \frac{\partial}{\partial x^1} + \cdots + V^n \frac{\partial}{\partial x^1}.$$

Thus vector fields can be thought of as first-order differential operators. The derivations $\{\frac{\partial}{\partial x^1}, \ldots, \frac{\partial}{\partial x^n}\}$ at the point p denote the elements of the basis of T_pX.

An integral curve γ of a vector field V is defined by $\dot{\gamma}(\varepsilon) = V|_{\gamma(\varepsilon)}$or equivalently

$$\frac{dx^a}{d\varepsilon} = V^a(x). \tag{4.2.4}$$

This system of ODEs has a unique solution for each initial data, and the integral curve passing through p with coordinates x^a is called a flow $\tilde{x}^a(\varepsilon, x^b)$. The vector field V is called a generator of the flow, as

$$\tilde{x}^a(\varepsilon, x) = x^a + \varepsilon V^a(x) + O(\varepsilon^2).$$

Determining the flow of a given vector field comes down to solving a system of ODEs (4.2.4).

- **Example.** Integral curves of the vector field

  $$V = x\frac{\partial}{\partial x} + \frac{\partial}{\partial y}$$

  on $\mathbb{R}^2$ are found by solving a pair of ODEs $\dot{x} = x,\ \dot{y} = 1$. Thus

  $$(x(\varepsilon), y(\varepsilon)) = (x(0)e^{\varepsilon}, y(0) + \varepsilon).$$

  There is one integral curve passing through each point in $\mathbb{R}^2$.

The flow is an example of one-parameter group of transformations, as

$$\tilde{x}(\varepsilon_2, \tilde{x}(\varepsilon_1, x)) = \tilde{x}(\varepsilon_1 + \varepsilon_2, x), \quad \tilde{x}(0, x) = x.$$

An invariant of a flow is a function $f(x^a)$ such that $f(x^a) = f(\tilde{x}^a)$ or equivalently

$$V(f) = 0$$

where V is the generating vector field.

- **Example.** The one-parameter group $SO(2)$ of rotations of the plane

$$(\tilde{x}, \tilde{y}) = (x\cos\varepsilon - y\sin\varepsilon, x\sin\varepsilon + y\cos\varepsilon)$$

is generated by

$$V = \left(\frac{\partial\tilde{y}}{\partial\varepsilon}|_{\varepsilon=0}\right)\frac{\partial}{\partial y} + \left(\frac{\partial\tilde{x}}{\partial\varepsilon}|_{\varepsilon=0}\right)\frac{\partial}{\partial x}$$
$$= x\frac{\partial}{\partial y} - y\frac{\partial}{\partial x}.$$

The function $r = \sqrt{x^2 + y^2}$ is an invariant of V.

The Lie bracket of two vector fields V, W is a vector field $[V, W]$ defined by its action on functions as

$$[V, W](f) := V[W(f)] - W[V(f)]. \tag{4.2.5}$$

The components of the Lie bracket are

$$[V, W]^a = V^b\frac{\partial W^a}{\partial x^b} - W^b\frac{\partial V^a}{\partial x^b}.$$

From its definition the Lie bracket is bilinear, antisymmetric and satisfies the Jacobi identity

$$[V, [W, U]] + [U, [V, W]] + [W, [U, V]] = 0. \tag{4.2.6}$$

A geometric interpretation of the Lie bracket is as the infinitesimal commutator of two flows. To see this consider $\tilde{x}_1(\varepsilon_1, x)$ and $\tilde{x}_2(\varepsilon_2, x)$ which are the flows of vector fields V_1 and V_2, respectively. For any $f : X \to \mathbb{R}$ define

$$F(\varepsilon_1, \varepsilon_2, x) := f\{\tilde{x}_1(\varepsilon_1, [\tilde{x}_2(\varepsilon_2, x)])\} - f\{\tilde{x}_2(\varepsilon_2, [\tilde{x}_1(\varepsilon_1, x)])\}.$$

Then

$$\frac{\partial^2}{\partial\varepsilon_1\partial\varepsilon_2}F(\varepsilon_1, \varepsilon_2, x)|_{\varepsilon_1=\varepsilon_2=0} = [V_1, V_2](f).$$

- **Example.** Consider the three-dimensional Lie group **Nil** of 3×3 upper triangular matrices

$$g(m_1, m_2, m_3) = \begin{pmatrix} 1 & m_1 & m_3 \\ 0 & 1 & m_2 \\ 0 & 0 & 1 \end{pmatrix}$$

acting on $\mathbb{R}^3$ by matrix multiplication

$$\tilde{\mathbf{x}} = g(m_1, m_2, m_3)\mathbf{x} = (x + m_1 y + m_3 z,\ y + m_2 z,\ z).$$

The corresponding vector fields[4] V_1, V_2, V_3 are

$$V_\alpha = \left(\frac{\partial \tilde{x}}{\partial m_\alpha}\frac{\partial}{\partial \tilde{x}} + \frac{\partial \tilde{y}}{\partial m_\alpha}\frac{\partial}{\partial \tilde{y}} + \frac{\partial \tilde{z}}{\partial m_\alpha}\frac{\partial}{\partial \tilde{z}} \right) |_{(m_1, m_2, m_3)=(0,0,0)}$$

which gives

$$V_1 = y\frac{\partial}{\partial x}, \quad V_2 = z\frac{\partial}{\partial y}, \quad \text{and} \quad V_3 = z\frac{\partial}{\partial x}.$$

The Lie brackets of these vector fields are

$$[V_1, V_2] = -V_3, \quad [V_1, V_3] = 0, \quad \text{and} \quad [V_2, V_3] = 0.$$

Thus we have obtained the representation of the Lie algebra of **Nil** by vector fields on $\mathbb{R}^3$. Comparing this with the commutators of the matrices (4.1.2) we see that the structure constants only differ by an overall sign. The Lie algebra spanned by the vector fields V_α is isomorphic to the Lie algebra spanned by the matrices M_α.

- **Example.** This is taken from [27]. A driver of a car has two transformations at his disposal. These are generated by vector fields

$$\text{STEER} = \frac{\partial}{\partial \phi}, \quad \text{and} \quad \text{DRIVE} = \cos\theta\frac{\partial}{\partial x} + \sin\theta\frac{\partial}{\partial y} + \frac{1}{L}\tan\phi\frac{\partial}{\partial \theta}, \qquad L = \text{const},$$

where (x, y) are coordinates of the center of the rear axle, θ specifies the direction of the car, and ϕ is the angle between the front wheels and the direction of the car. These two flows do not commute, and

$$[\text{STEER}, \text{DRIVE}] = \text{ROTATE},$$

where the vector field

$$\text{ROTATE} = \frac{1}{L\cos^2\phi}\frac{\partial}{\partial \theta}$$

generates the manoeuvre steer, drive, steer back, and drive back. This manoeuvre alone does not guarantee that the driver parks his car in a tight space. The commutator

$$[\text{DRIVE}, \text{ROTATE}] = \frac{1}{L\cos^2\phi}\left(\sin\theta\frac{\partial}{\partial x} - \cos\theta\frac{\partial}{\partial y}\right) = \text{SLIDE}$$

[4] Note that the lower index labels the vector fields while the upper index labels the components. Thus $V_\alpha = V_\alpha^a \partial/\partial x^a$.

is the key to successful parallel parking. One needs to perform the following sequence steer, drive, steer back, drive, steer, drive back, steer back, and drive back!

In general the Lie bracket is a closed operation in a set of the vector fields generating a group. The vector space of vector fields generating the group action gives a representation of the corresponding Lie algebra. The structure constants $f^{\gamma}_{\alpha\beta}$ do not depend on which of the representations (matrices or vector fields) are used.

4.3 Symmetries of differential equations

Let $u = u(x, t)$ be a solution to the KdV equation (2.1.1). Consider the vector field

$$V = \xi(x, t, u)\frac{\partial}{\partial x} + \tau(x, t, u)\frac{\partial}{\partial t} + \eta(x, t, u)\frac{\partial}{\partial u}$$

on the space of dependent and independent variables $\mathbb{R} \times \mathbb{R}^2$. This vector field generates a one-parameter group of transformations

$$\tilde{x} = \tilde{x}(x, t, u, \varepsilon), \quad \tilde{t} = \tilde{t}(x, t, u, \varepsilon), \quad \text{and} \quad \tilde{u} = \tilde{u}(x, t, u, \varepsilon).$$

This group is called a symmetry of the KdV equation if

$$\frac{\partial \tilde{u}}{\partial \tilde{t}} - 6\tilde{u}\frac{\partial \tilde{u}}{\partial \tilde{x}} + \frac{\partial^3 \tilde{u}}{\partial \tilde{x}^3} = 0.$$

The common abuse of terminology is to refer to the corresponding vector field as a symmetry, although the term infinitesimal symmetry is more appropriate.

- **Example.** An example of a symmetry of the KdV is given by

$$\tilde{x} = x, \quad \tilde{t} = t + \varepsilon, \quad \text{and} \quad \tilde{u} = u.$$

  It is a symmetry as there is no explicit time dependence in the KdV. Its generating vector field is

$$V = \frac{\partial}{\partial t}.$$

Of course there is nothing special about KdV in this definition and the concept of a symmetry applies generally to PDEs and ODEs.

Definition 4.3.1 *Let $X = \mathbb{R}^n \times \mathbb{R}$ be the space of independent and dependent variables in a PDE. A one-parameter group of transformations of this space*

$$\tilde{u} = \tilde{u}(x^a, u, \varepsilon), \quad \tilde{x}^b = \tilde{x}^b(x^a, u, \varepsilon)$$

is called a Lie point symmetry (or symmetry for short) group of the PDE

$$F\left[u, \frac{\partial u}{\partial x^a}, \frac{\partial^2 u}{\partial x^a \partial x^b}, \dots\right] = 0 \tag{4.3.7}$$

if its action transforms solutions to other solutions, that is,

$$F\left[\tilde{u}, \frac{\partial \tilde{u}}{\partial \tilde{x}^a}, \frac{\partial^2 \tilde{u}}{\partial \tilde{x}^a \partial \tilde{x}^b}, \dots\right] = 0.$$

This definition naturally extends to multi-parameter groups of transformation. A Lie group G is a symmetry of a PDE if any of its one-parameter subgroups is a symmetry in the sense of Definition 4.3.1.

A knowledge of Lie point symmetries is useful for the following reasons:

- It allows us to use known solutions to construct new solutions.

  Example. The Lorentz group

  $$(\tilde{x}, \tilde{t}) = \left(\frac{x - \varepsilon t}{\sqrt{1-\varepsilon^2}}, \frac{t - \varepsilon x}{\sqrt{1-\varepsilon^2}}\right), \quad \varepsilon \in (-1, 1)$$

  is the symmetry group of the Sine-Gordon equation (2.1.2). Any t-independent solution $\phi_S(x)$ to (2.1.2) can be used to obtain a time-dependent solution

  $$\phi(x, t) = \phi_S\left(\frac{x - \varepsilon t}{\sqrt{1-\varepsilon^2}}\right), \quad \varepsilon \in (-1, 1).$$

  In physics this procedure is known as 'Lorentz boost'. The parameter ε is usually denoted by v and called velocity. For example, the Lorentz boost of a static kink is a moving kink.
- For ODEs each symmetry reduces the order by 1. So a knowledge of sufficiently many symmetries allows a construction of the most general solution.

  Example. An ODE

  $$\frac{du}{dx} = F\left(\frac{u}{x}\right)$$

  admits a scaling symmetry

  $$(x, u) \longrightarrow (e^{\varepsilon} x, e^{\varepsilon} u), \quad \varepsilon \in \mathbb{R}.$$

  This one-dimensional group is generated by the vector field

  $$V = x\frac{\partial}{\partial x} + u\frac{\partial}{\partial u}.$$

  Introduce the coordinates

  $$r = \frac{u}{x} \quad \text{and} \quad s = \log|x|$$

so that

$$V(r) = 0 \quad \text{and} \quad V(s) = 1.$$

If $F(r) = r$ the general solution is $r = \text{const}$. Otherwise

$$\frac{ds}{dr} = \frac{1}{F(r) - r}$$

and the general implicit solution is

$$\log|x| + c = \int^{\frac{u}{x}} \frac{dr}{F(r) - r}.$$

- For PDEs the knowledge of the symmetry group is not sufficient to construct the most general solution, but it can be used to find special solutions which admit symmetry.

  Example. Consider the one-parameter group of transformations

$$(\tilde{x}, \tilde{t}, \tilde{u}) = (x + c\varepsilon, t + \varepsilon, u)$$

where $c \in \mathbb{R}$ is a constant. It is straightforward to verify that this group is a Lie point symmetry of the KdV equation (2.1.1). It is generated by the vector field

$$V = \frac{\partial}{\partial t} + c\frac{\partial}{\partial x}$$

and the corresponding invariants are u and $\xi = x - ct$. To find the group-invariant solutions assume that a solution of the KdV equation is of the form

$$u(x, t) = f(\xi).$$

Substituting this to the KdV yields a third-order ODE which easily integrates to

$$\frac{1}{2}\left(\frac{df}{d\xi}\right)^2 = f^3 + \frac{1}{2}cf^2 + \alpha f + \beta$$

where (α, β) are arbitrary constants. This ODE is solvable in terms of an elliptic integral, which gives all group-invariant solutions in the implicit form

$$\int \frac{df}{\sqrt{f^3 + \frac{1}{2}cf^2 + \alpha f + \beta}} = \sqrt{2}\xi.$$

Thus we have recovered the cnoidal wave which in Section 3.4 arose from the finite-gap integration. In fact the one-soliton solution (2.1.3) falls into this category: if f and its first two derivatives tend to zero as $|\xi| \to \infty$ then α, β are both zero and the elliptic integral reduces to an elementary one.

Finally we obtain

$$u(x,t) = -\frac{2\chi^2}{\cosh^2 \chi(x - 4\chi^2 t - \phi_0)}$$

which is the one-soliton solution (2.1.3) to the KdV equation.

4.3.1 How to find symmetries

Some of them can be guessed. For example, if there is no explicit dependence on the independent coordinates in the equation then the translations $\tilde{x}^a = x^a + c^a$ are symmetries. All translations form an n-parameter abelian group generated by n vector fields $\partial/\partial x^a$.

In the general case of (4.3.7) we could substitute

$$\tilde{u} = u + \varepsilon\eta(x^a, u) + O(\varepsilon^2), \quad \tilde{x}^b = x^b + \varepsilon\xi^b(x^a, u) + O(\varepsilon^2)$$

into the equation (4.3.7) and keep the terms linear in ε. A more systematic method is given by the *prolongation* of vector field. Assume that the space of independent variables is coordinatized by (x, t) and the equation (4.3.7) is of the form

$$F(u, u_x, u_{xx}, u_{xxx}, u_t) = 0.$$

(e.g. KdV is of that form). The prolongation of the vector field

$$V = \xi(x,t,u)\frac{\partial}{\partial x} + \tau(x,t,u)\frac{\partial}{\partial t} + \eta(x,t,u)\frac{\partial}{\partial u}$$

is

$$\mathbf{pr}(V) = V + \eta^t\frac{\partial}{\partial u_t} + \eta^x\frac{\partial}{\partial u_x} + \eta^{xx}\frac{\partial}{\partial u_{xx}} + \eta^{xxx}\frac{\partial}{\partial u_{xxx}},$$

where $(\eta^t, \eta^x, \eta^{xx}, \eta^{xxx})$ are certain functions of (u, x, t) which can be determined algorithmically in terms of (ξ, τ, η) and their derivatives (we will do this in the next section). The prolongation $\mathbf{pr}(V)$ generates a one-parameter group of transformations on the seven-dimensional space with coordinates

$$(x, t, u, u_t, u_x, u_{xx}, u_{xxx}).$$

(This is an example of a jet space. The symbols $(u_t, u_x, u_{xx}, u_{xxx})$ should be regarded as independent coordinates and not as derivatives of u. See Appendix C for discussion of jets.) The vector field V is a symmetry of the PDE if

$$\mathbf{pr}(V)(F)|_{F=0} = 0. \tag{4.3.8}$$

This condition gives a linear system of PDEs for (ξ, τ, η). Solving this system yields the most general symmetry of a given PDE. The important point is that (4.3.8) is only required to hold when (4.3.7) is satisfied ('on shell' as a physicist would put it).

4.3.2 Prolongation formulae

The first step in implementing the prolongation procedure is to determine the functions

$$\eta^t, \eta^x, \eta^{xx}, \ldots$$

in the prolonged vector field. For simplicity we shall assume that we want to determine a symmetry of an Nth-order ODE:

$$\frac{d^N u}{dx^N} = F\left(x, u, \frac{du}{dx}, \ldots, \frac{d^{N-1}u}{dx^{N-1}}\right).$$

Consider the vector field

$$V = \xi \frac{\partial}{\partial x} + \eta \frac{\partial}{\partial u}.$$

Its prolongation

$$\mathbf{pr}(V) = V + \sum_{k=1}^{N} \eta^{(k)} \frac{\partial}{\partial u^{(k)}}$$

generates a one-parameter transformation group

$$\tilde{x} = x + \varepsilon\xi + O(\varepsilon^2), \quad \tilde{u} = u + \varepsilon\eta + O(\varepsilon^2), \quad \text{and} \quad \tilde{u}^{(k)} = u^{(k)} + \varepsilon\eta^{(k)} + O(\varepsilon^2)$$

of the $(N+2)$-dimensional jet space with coordinates $(x, u, u', \ldots, u^N)$.

The prolongation is an algorithm for the calculation of the functions $\eta^{(k)}$. Set

$$D_x = \frac{\partial}{\partial x} + u' \frac{\partial}{\partial u} + u'' \frac{\partial}{\partial u'} + \cdots + u^{(N)} \frac{\partial}{\partial u^{(N-1)}}.$$

The chain rule gives

$$\tilde{u}^{(k)} = \frac{d\tilde{u}^{(k-1)}}{d\tilde{x}} = \frac{D_x \tilde{u}^{(k-1)}}{D_x \tilde{x}},$$

so

$$\tilde{u}^{(1)} = \frac{D_x \tilde{u}}{D_x \tilde{x}} = \frac{\frac{du}{dx} + \varepsilon D_x(\eta) + \cdots}{1 + \varepsilon D_x(\xi) + \cdots} = \frac{du}{dx} + \varepsilon\left(D_x\eta - \frac{du}{dx} D_x\xi\right) + O(\varepsilon^2).$$

Thus

$$\eta^{(1)} = D_x\eta - \frac{du}{dx} D_x\xi.$$

The remaining prolongation coefficients can now be constructed recursively. The relation

$$\tilde{u}^{(k)} = \frac{u^{(k)} + \varepsilon D_x\left[\eta^{(k-1)}\right]}{1 + \varepsilon D_x(\xi)}$$

yields the general prolongation formula

$$\eta^{(k)} = D_x\left[\eta^{(k-1)}\right] - \frac{d^k u}{dx^k} D_x \xi. \tag{4.3.9}$$

The procedure is entirely analogous for PDEs, where $u = u(x^a)$ but one has to keep track of the index a labelling the independent variables. Set

$$D_a = \frac{\partial}{\partial x^a} + (\partial_a u)\frac{\partial}{\partial u} + (\partial_a^2 u)\frac{\partial}{\partial(\partial_a u)} + \cdots + (\partial_a^N u)\frac{\partial}{\partial(\partial_a^{N-1} u)},$$

where

$$\partial_a^k u = \frac{\partial^k u}{\partial(x^a)^k}.$$

The first prolongation is

$$\eta^{(a)} = D_a \eta - \sum_{b=1}^{n} (D_a \xi^b)\frac{\partial u}{\partial x^b}$$

and the higher prolongations are given recursively by the formula

$$\eta^{A,a} = D_a \eta^A - \sum_{b=1}^{n} (D_a \xi^b)\frac{\partial u^A}{\partial x^b}$$

where $A = (a_1, \ldots, a_k)$ is a multi-index and

$$u^A = \frac{\partial^k u}{\partial x^{a_1} \partial x^{a_2} \cdots \partial x^{a_k}}.$$

- **Example.** Let us follow the prolongation procedure to find the most general Lie-point symmetry of the second-order ODE

$$\frac{d^2 u}{dx^2} = 0.$$

We first need to compute the second prolongation

$$\mathbf{pr}(V) = \xi\frac{\partial}{\partial x} + \eta\frac{\partial}{\partial u} + \eta^x\frac{\partial}{\partial u_x} + \eta^{xx}\frac{\partial}{\partial u_{xx}}.$$

This computation does not depend on the details of the equation but only on the prolongation formulae (4.3.9). The result is

$$\begin{aligned}
\eta^x &= \eta_x + (\eta_u - \xi_x)u_x - \xi_u u_x^2, \\
\eta^{xx} &= \eta_{xx} + (2\eta_{xu} - \xi_{xx})u_x + (\eta_{uu} - 2\xi_{xu})u_x^2 - \xi_{uu}u_x^3 \\
&\quad + (\eta_u - 2\xi_x)u_{xx} - 3\xi_u u_x u_{xx}.
\end{aligned}$$

Now we substitute this, and the ODE to the symmetry criterion (4.3.8)

$$\mathbf{pr}(V)(u_{xx}) = \eta^{xx} = 0.$$

Thus

$$\eta_{xx} + (2\eta_{xu} - \xi_{xx})u_x + (\eta_{uu} - 2\xi_{xu})u_x^2 - \xi_{uu}u_x^3 = 0$$

where we have used the ODE to set $u_{xx} = 0$. In the second-order equation the value of u_x can be prescribed in an arbitrary way at each point (initial condition). Therefore the coefficients of u_x, u_x^2, and u_x^3 all vanish

$$\eta_{xx} = 0, \quad 2\eta_{xu} - \xi_{xx} = 0, \quad \eta_{uu} - 2\xi_{xu} = 0, \quad \text{and} \quad \xi_{uu} = 0.$$

The general solution of these linear PDEs is

$$\xi(x, u) = \varepsilon_1 x^2 + \varepsilon_2 xu + \varepsilon_3 x + \varepsilon_4 u + \varepsilon_5,$$

$$\eta(x, u) = \varepsilon_1 xu + \varepsilon_2 u^2 + \varepsilon_6 x + \varepsilon_7 u + \varepsilon_8.$$

Therefore the trivial ODE in our example admits an eight-dimensional group of symmetries.

Let $V_\alpha, \alpha = 1, \ldots, 8$, be the corresponding vector fields obtained by setting $\varepsilon_\alpha = 1$ and $\varepsilon_\beta = 0$ if $\beta \neq \alpha$

$$V_1 = x^2\frac{\partial}{\partial x} + xu\frac{\partial}{\partial u}, \quad V_2 = xu\frac{\partial}{\partial x} + u^2\frac{\partial}{\partial u}, \quad V_3 = x\frac{\partial}{\partial x}, \quad V_4 = u\frac{\partial}{\partial x},$$

$$V_5 = \frac{\partial}{\partial x}, \quad V_6 = x\frac{\partial}{\partial u}, \quad V_7 = u\frac{\partial}{\partial u}, \quad \text{and} \quad V_8 = \frac{\partial}{\partial u}.$$

Each of the eight vector fields generates a one-parameter group of transformations. Calculating the Lie brackets of these vector fields verifies that they form the Lie algebra of PGL(3, $\mathbb{R}$).

It is possible to show that the Lie-point symmetry group of a general second-order ODE has dimension at most 8. If this dimension is 8 then the ODE is equivalent to $u_{xx} = 0$ by a coordinate transformation $u \to U(u, x), x \to X(u, x)$.

This example shows that the process of prolonging the vector fields and writing down the linear PDEs characterizing the symmetries is tedious but algorithmic. It is worth doing a few examples by hand to familiarize oneself with the method but in practice it is best to use computer programmes like MAPLE or MATHEMATICA to do symbolic computations.

- **Example.** Lie-point symmetries of KdV. The vector fields

$$V_1 = \frac{\partial}{\partial x}, \quad V_2 = \frac{\partial}{\partial t}, \quad V_3 = \frac{\partial}{\partial u} - 6t\frac{\partial}{\partial x}, \quad \text{and} \quad V_4 = x\frac{\partial}{\partial x} + 3t\frac{\partial}{\partial t} - 2u\frac{\partial}{\partial u}$$

generate a four-parameter symmetry group of KdV. The group is non–abelian as the structure constants of the Lie algebra spanned by V_α are non-zero:

$$[V_2, V_3] = -6V_1, \quad [V_1, V_4] = V_1, \quad [V_2, V_4] = 3V_2, \quad \text{and} \quad [V_3, V_4] = -2V_3$$

and all other Lie brackets vanish.

One can use the prolongation procedure to show that this is in fact the most general symmetry group of KdV. One needs to find the third prolongation of a general vector field on $\mathbb{R}^3$ – this can be done 'by hand' but it is best to use `MAPLE` package `liesymm` with the command `determine`. Type `help(determine);` and take it from there.

4.4 Painlevé equations

In this section we shall consider ODEs in the complex domain. This means that both the dependent and independent variables are complex. Let us first discuss linear ODEs of the form

$$\frac{d^N w}{dz^N} + p_{N-1}(z)\frac{d^{N-1}w}{dz^{N-1}} + \cdots + p_1(z)\frac{dw}{dz} + p_0(z)w = 0 \qquad (4.4.10)$$

where $w = w(z)$. If the functions $p_0, \ldots, p_{N-1}$ are analytic at $z = z_0$, then z_0 is called a regular point and for a given initial data there exist a unique analytic solution in the form of a power series

$$w(z) = \sum_k a_k(z - z_0)^k.$$

The singular points of the ODE (4.4.10) can be located only at the singularities of p_k. Thus the singularities are fixed – their location does not depend on the initial conditions. Non-linear ODEs lose this property.

- **Example.** Consider a simple non-linear ODE and its general solution

$$\frac{dw}{dz} + w^2 = 0, \quad w(z) = \frac{1}{z - z_0}.$$

The location of the singularity depends on the constant of integration z_0. This is a movable singularity.

A singularity of a non-linear ODE can be a pole (of arbitrary order), a branch point, or an essential singularity.

- **Example.** The ODE with the general solution

$$\frac{dw}{dz} + w^3 = 0, \quad w(z) = \frac{1}{\sqrt{2(z - z_0)}}.$$

has a movable singularity which is a branch point. Another example with a movable logarithmic branch point is

$$\frac{dw}{dz} + e^w = 0, \quad w(z) = \ln(z - z_0).$$

Definition 4.4.1 *The ODE*

$$\frac{d^N w}{dz^N} = F\left(\frac{d^{N-1}w}{dz^{N-1}}, \ldots, \frac{dw}{dz}, w, z\right)$$

where F is rational in w and its derivatives has the Painlevé property (PP) if its movable singularities are at worst poles.

In nineteenth century Painlevé, Gambier, and Kowalewskaya aimed to classify all second-order ODEs with the PP up to the change of variables

$$\tilde{w}(w, z) = \frac{a(z)w + b(z)}{c(z)w + d(z)}, \quad \tilde{z}(z) = \phi(z)$$

where the functions a, b, c, d, and ϕ are analytic in z. There exist 50 canonical types, 44 of which are solvable in terms of 'known' functions (sine, cosine, elliptic functions, or in general solutions to linear ODEs) [85]. The remaining 6 equations define new transcendental functions

$$\frac{d^2w}{dz^2} = 6w^2 + z \quad \text{PI,} \tag{4.4.11}$$

$$\frac{d^2w}{dz^2} = 2w^3 + wz + \alpha \quad \text{PII,}$$

$$\frac{d^2w}{dz^2} = \frac{1}{w}\left(\frac{dw}{dz}\right)^2 - \frac{1}{z}\frac{dw}{dz} + \frac{\alpha w^2 + \beta}{z} + \gamma w^3 + \frac{\delta}{w} \quad \text{PIII,}$$

$$\frac{d^2w}{dz^2} = \frac{1}{2w}\left(\frac{dw}{dz}\right)^2 + \frac{3}{2}w^3 + 4zw^2 + 2(z^2 - \alpha)w + \frac{\beta}{w} \quad \text{PIV,}$$

$$\frac{d^2w}{dz^2} = \left(\frac{1}{2w} + \frac{1}{w-1}\right)\left(\frac{dw}{dz}\right)^2 - \frac{1}{z}\frac{dw}{dz} + \frac{(w-1)^2}{z^2}\left(\alpha w + \frac{\beta}{w}\right)$$
$$+ \frac{\gamma w}{z} + \frac{\delta w(w+1)}{w-1} \quad \text{PV, and}$$

$$\frac{d^2w}{dz^2} = \frac{1}{2}\left(\frac{1}{w} + \frac{1}{w-1} + \frac{1}{w-z}\right)\left(\frac{dw}{dz}\right)^2 - \left(\frac{1}{z} + \frac{1}{z-1} + \frac{1}{w-z}\right)\frac{dw}{dz}$$
$$+ \frac{w(w-1)(w-z)}{z^2(z-1)^2}\left[\alpha + \beta\frac{z}{w^2} + \gamma\frac{z-1}{(w-1)^2} + \delta\frac{z(z-1)}{(w-z)^2}\right] \quad \text{PVI.}$$

Here α, β, γ, and δ are constants. Thus PVI belongs to a four-parameter family of ODEs but PI is rigid up to coordinate transformations.

How do we check the PP for a given ODE? If a second-order equation possesses the PP then it is either linearizable or can be put into one of the six Painlevé types by appropriate coordinate transformation. Exhibiting such a transformation is often the most straightforward way of establishing the PP.

Otherwise, especially if we suspect that the equation does not have PP, the singular-point analysis may be performed. If a general Nth-order ODE possesses the PP then the general solution admits a Laurent expansion with a finite number of terms with negative powers. This expansion must contain N arbitrary constants so that the initial data consisting of w and its first $(N-1)$ derivatives can be specified at any point. Assume that the leading term in the expansion of the solution is of the form

$$w(z) \sim a(z-z_0)^p, \quad a \neq 0, \quad a, p \in \mathbb{C}$$

as $z \to z_0$. Substitute this into the ODE and require the maximal balance condition. This means that two (or more) terms must be of equal maximally small order as $(z-z_0) \to 0$. This should determine a and p and finally the form of a solution around z_0. If z_0 is a singularity we should also be able to determine if it is movable or fixed.

- **Example.** Consider the ODE

$$\frac{dw}{dz} = w^3 + z.$$

The maximal balance condition gives

$$ap(z-z_0)^{p-1} \sim a^3(z-z_0)^{3p}.$$

Thus $p = -1/2, a = \pm i\sqrt{2}^{-1}$ and

$$w(z) \sim \pm i\frac{\sqrt{2}}{2}(z-z_0)^{-1/2}$$

possesses a movable branch point as z_0 depends on the initial conditions. The ODE does not have PP.

- **Example.** Consider the first Painlevé equation

$$\frac{d^2w}{dz^2} = 6w^2 + z.$$

The orders of the three terms in this equations are

$$p-2, \quad 2p, \quad \text{and} \quad 0.$$

Balancing the last two terms gives $p = 0$ but this is not the maximal balance as the first term is then of order -2. Balancing the first and last terms gives $p = 2$. This is a maximal balance and the corresponding solution is analytic around z_0. Finally balancing the first two terms gives $p = -2$ which again is the maximal balance: the 'balanced' terms behave like $(z - z_0)^{-4}$ and the remaining term is of order 0. Now we find that $a = a^2$ and so $a = 1$ and

$$w(z) \sim \frac{1}{(z - z_0)^2}.$$

Thus the movable singularity is a second-order pole.

This singular-point analysis is useful to rule out PP, but does not give sufficient conditions (at least not in the heuristic form in which we presented it), as some singularities may have been missed or the Laurent series may be divergent. The analysis of sufficient conditions is tedious and complicated – we shall leave it out.

The PP guarantees that the solutions of the six Painlevé equations are single valued thus giving rise to proper functions. The importance of the Painlevé equations is that they define new transcendental functions in the following way. Any sufficiently smooth function can be defined as a solution to a certain DE. For example, we can define the exponential function as the general solution to

$$\frac{dw}{dz} = w$$

such that $w(0) = 1$. Similarly we define the function PI from the general solution of the first Painlevé equation. From this point of view the exponential and PI functions are on equal footing. Of course we know more about the exponential as it possesses simple properties and arises in a wide range of problems in natural sciences.

The irreducibility of the Painlevé equations is a more subtle issue. It roughly means the following. One can define a field of classical functions by starting off with the rational functions $\mathbb{Q}[z]$ and adjoining those functions which arise as solutions of algebraic or linear DEs with coefficients in $\mathbb{Q}[z]$. For example, the exponential, Bessel function, and hyper-geometric function are all solutions of linear DEs, and thus are classical. A function is called irreducible (or transcendental) if it is not classical. Painlevé himself anticipated that the Painlevé equations define irreducible functions but the rigorous proofs for PI and PII appeared only recently. They use a far-reaching extension of Galois theory from number fields to differential fields of functions. The irreducibility problem is analogous to the existence of non-algebraic numbers (numbers which are not roots of any polynomial equations with rational coefficients). Thus the the appearance of Galois theory is not that surprising.

4.4.1 Painlevé test

Can we determine whether a given PDE is integrable? This is to a large extent an open problem as the satisfactory definition of integrability of PDEs is still missing. The following algorithm is based on the observation of Ablowitz, Ramani, and Segur [1] (see also [2]) that PDEs integrable by the IST reduce (when the solutions are required to be invariant under some Lie symmetries) to ODEs with PP.

- **Example.** Consider the Lie-point symmetry

$$(\tilde{\rho}, \tilde{\tau}) = (c\rho, \frac{1}{c}\tau), \quad c \neq 0$$

of the Sine-Gordon equation

$$\frac{\partial^2 \phi}{\partial \rho \partial \tau} = \sin \phi.$$

The group-invariant solutions are of the form $\phi(\rho, \tau) = F(z)$ where $z = \rho\tau$ is an invariant of the symmetry. Substituting $w(z) = \exp[iF(z)]$ into the Sine-Gordon yields

$$\frac{d^2 w}{dz^2} = \frac{1}{w}\left(\frac{dw}{dz}\right)^2 - \frac{1}{z}\frac{dw}{dz} + \frac{1}{2}\frac{w^2}{z} - \frac{1}{2z},$$

which is the third Painlevé equation PIII with the special values of parameters

$$\alpha = \frac{1}{2}, \quad \beta = -\frac{1}{2}, \quad \gamma = 0, \text{ and } \delta = 0.$$

- **Example.** Consider the modified KdV equation

$$v_t - 6v^2 v_x + v_{xxx} = 0,$$

and look for a Lie-point symmetry of the form

$$(\tilde{v}, \tilde{x}, \tilde{t}) = (c^\alpha v, c^\beta x, c^\gamma t), \quad c \neq 0.$$

The symmetry condition will hold if all three terms in the equation have equal weight

$$\alpha - \gamma = 3\alpha - \beta = \alpha - 3\beta.$$

This gives $\beta = -\alpha, \gamma = -3\alpha$ where α can be chosen arbitrarily. The corresponding symmetry group depends on one parameter c^α and is generated by

$$V = v\frac{\partial}{\partial v} - x\frac{\partial}{\partial x} - 3t\frac{\partial}{\partial t}.$$

This has two independent invariants which we may take to be

$$z = (3)^{-1/3}\, x\, t^{-1/3} \quad \text{and} \quad w = (3)^{1/3}\, v\, t^{1/3}$$

where the constant factor $(3)^{-1/3}$ has been added for convenience. The group-invariant solutions are of the form $w = w(z)$ which gives

$$v(x, t) = (3t)^{-1/3} w(z).$$

Substituting this into the modified KdV equation leads to a third-order ODE for $w(z)$

$$w_{zzz} - 6w^2 w_z - w - z w_z = 0.$$

Integrating this ODE once shows that $w(z)$ satisfies the second Painlevé equation PII with a general value of the parameter α.

The general Painlevé test comes down to the following algorithm: Given a PDE

1. Find all its Lie-point symmetries
2. Construct ODEs characterizing the group-invariant solutions
3. Check for the PP

This procedure only gives necessary conditions for integrability. If all reductions possess PP the PDE does not have to be integrable in general.

Exercises

1. Consider three one-parameter groups of transformations of $\mathbb{R}$

$$x \to x + \varepsilon_1, \quad x \to e^{\varepsilon_2} x, \quad \text{and} \quad x \to \frac{x}{1 - \varepsilon_3 x},$$

and find the vector fields V_1, V_2, V_3 generating these groups. Deduce that these vector fields generate a three-parameter group of transformations

$$x \to \frac{ax + b}{cx + d}, \quad ad - bc = 1.$$

Show that

$$[V_\alpha, V_\beta] = \sum_{\gamma=1}^{3} f_{\alpha\beta}^{\gamma} V_\gamma, \quad \alpha, \beta = 1, 2, 3$$

for some constants $f_{\alpha\beta}^{\gamma}$ which should be determined.

2. Consider the vector field

$$V = x\frac{\partial}{\partial x} - u\frac{\partial}{\partial u}$$

and find the corresponding one-parameter group of transformations of $\mathbb{R}^2$. Sketch the integral curves of this vector field.
Find the invariant coordinates, that is, functions $s(x, u)$, $g(x, u)$ such that

$$V(s) = 1, \quad \text{and} \quad V(g) = 0$$

[These are not unique. Make sure that that s, g are functionally independent in a domain of $\mathbb{R}^2$ which you should specify.]

Use your results to integrate the ODE

$$x^2 \frac{du}{dx} = F(xu)$$

where F is an arbitrary function of one variable.

3. Consider the vector fields

$$V_1 = \frac{\partial}{\partial x}, \quad V_2 = \frac{\partial}{\partial t}, \quad V_3 = \frac{\partial}{\partial u} + \alpha t \frac{\partial}{\partial x}, \quad \text{and} \quad V_4 = \beta x \frac{\partial}{\partial x} + \gamma t \frac{\partial}{\partial t} + \delta u \frac{\partial}{\partial u}$$

where $(\alpha, \beta, \gamma, \delta)$ are constants and find the corresponding one-parameter groups of transformations of $\mathbb{R}^3$ with coordinates (x, t, u).

Find $(\alpha, \beta, \gamma, \delta)$ such that these are symmetries of KdV and deduce the existence of a four-parameter symmetry group.

Determine the structure constants of the corresponding Lie algebra of vector fields.

4. Consider a one-parameter group of transformations of $\mathbb{R}^n \times \mathbb{R}$

$$(\tilde{x}^1, \ldots, \tilde{x}^n, \tilde{u}) = (c^{\alpha_1} x^1, \ldots, c^{\alpha_n} x^n, c^{\alpha} u), \tag{4.4.12}$$

where $c \neq 0$ is the parameter of the transformation and $(\alpha, \alpha_1, \ldots, \alpha_n)$ are fixed constants, and find a vector field generating this group.

Find all Lie-point symmetries of the PDE

$$u_t = u u_x$$

of the form (4.4.12) where $(x^1, x^2) = (x, t)$. Guess two more Lie-point symmetries of this PDE not of the form (4.4.12) and calculate the Lie brackets of the corresponding vector fields.

5. Show that the solutions to the Tzitzeica equations

$$u_{xy} = e^u - e^{-2u}$$

where $u = u(x, y)$ which admit a scaling symmetry $(x, y) \to (cx, c^{-1} y)$ are characterized by the Painlevé III ODE with special values of parameters.

5 Lagrangian formalism and field theory

Our treatment of integrable systems in the first three chapters made essential use of the Hamiltonian formalism both in finite and infinite dimensional settings. In the next two chapters we shall concentrate on classical field theory, where the covariant formulation requires the Lagrangian formalism. It is assumed that the reader has covered the Lagrangian treatment of classical mechanics and classical field theory at the basic level [102, 187]. The aim of this chapter is not to provide a crash course in these subjects, but rather to introduce less standard aspects.

5.1 A variational principle

In the Lagrangian approach to classical mechanics states of dynamical systems are represented by points in n-dimensional configuration space X with local coordinates $q^i, i = 1, \dots, n$. Physically n is the number of degrees of freedom of the system.

A trajectory $q^i(t)$ is determined from the principle of least action. For given initial and final conditions $(\mathbf{q}_1, t_1, \mathbf{q}_2, t_2)$ the action is defined by

$$S[\mathbf{q}] = \int_{t_1}^{t_2} L(\mathbf{q}(t), \dot{\mathbf{q}}(t))dt, \tag{5.1.1}$$

where the Lagrangian L is a smooth function of $\mathbf{q}$ and $\dot{\mathbf{q}}$, that is a function on the tangent bundle TX. A natural example leading to Newton's equations for a particle with a unit mass moving in a potential $U = U(\mathbf{q})$ is given by

$$L = \frac{1}{2}|\dot{\mathbf{q}}|^2 - U(\mathbf{q}), \tag{5.1.2}$$

where $|\mathbf{q}|$ is the Euclidean norm.

Let $\mathbf{q}_s(t)$ be a family of curves depending smoothly on a parameter s such that

$$\mathbf{q}_s(t_1) = \mathbf{q}(t_1), \qquad \mathbf{q}_s(t_2) = \mathbf{q}(t_2), \quad \text{and} \quad \mathbf{q}_0(t) = \mathbf{q}(t) \qquad t \in [t_1, t_2].$$

The variation δq is defined, for each fixed t, by

$$\delta q = \frac{d}{ds}\mathbf{q}_s|_{s=0}.$$

The particle will follow a trajectory for which the action is stationary, that is,

$$\delta S = 0.$$

Integration by parts shows that this condition leads to the Euler–Lagrange equations: The variation of the action

$$\delta S = \frac{d}{ds}S|_{s=0} = \int_{t_1}^{t_2}\left(\frac{\partial L}{\partial q^i}\delta q^i + \frac{\partial L}{\partial \dot{q}^i}\delta \dot{q}^i\right)dt$$

vanishes if the Euler–Lagrange equations

$$\frac{d}{dt}\frac{\partial L}{\partial \dot{q}^i} - \frac{\partial L}{\partial q^i} = 0 \tag{5.1.3}$$

hold, as δq^i vanishes on the boundary of $[t_1, t_2]$.

The Euler–Lagrange equations are usually non-linear and exact solutions are difficult (or impossible) to obtain. In some cases linearization leads to satisfactory approximate solutions. To find these one chooses an equilibrium position, that is, a point $\mathbf{q}_0 \in X$ such that

$$\frac{\partial U}{\partial q}|_{\mathbf{q}=\mathbf{q}_0} = 0,$$

and expands U around this equilibrium neglecting terms of order higher than 2

$$U = \text{const} + \frac{1}{2}b_{ij}q^iq^j, \qquad \text{where} \qquad b_{ij} = \frac{\partial^2 U}{\partial q^i \partial q^j}|_{\mathbf{q}=\mathbf{q}_0}.$$

The symmetric quadratic form b can be diagonalized, and the system undergoes small oscillations with frequencies given by the eigenvalues of b.

Most problems treated in this book owe their interesting physical and mathematical properties to the non-linearity of the underlying equations, and resorting to the method of small oscillation is not appropriate. There is a less well-known alternative: Consider a particle in $\mathbb{R}^{n+1}$ with the Lagrangian (5.1.2) where $U : \mathbb{R}^{n+1} \to \mathbb{R}$ is a potential whose minimum value is 0. The equilibrium positions are on a subspace $X \subset \mathbb{R}^{n+1}$ given by $U = 0$. If the kinetic energy of the particle is small, and the initial velocity is tangent to X, the exact motion will be approximated by a motion on X with the Lagrangian L' given by a restriction of L to X

$$L' = \frac{1}{2}h_{rs}\dot{\gamma}^r\dot{\gamma}^s, \qquad r, s = 1, \dots, \dim X. \tag{5.1.4}$$

Here, the γ's are local coordinates on X, and the metric $h = h_{rs}(\gamma)d\gamma^r d\gamma^s$ is induced on X from the Euclidean inner product on $\mathbb{R}^{n+1}$. The Euler–Lagrange equations for (5.1.4) are

$$\ddot{\gamma}^r + \Gamma^r_{su}\dot{\gamma}^s\dot{\gamma}^u = 0, \tag{5.1.5}$$

where the functions $\Gamma^r_{su} = \Gamma^r_{su}(\gamma)$ are the Chrisoffel symbols of the Levi-Civita connection of h (given by (9.2.5)). If, for example, $U = (1 - r^2)^2$, where $r = |\mathbf{q}|$, then the motion with small energy is approximated by the motion on the unit sphere in $\mathbb{R}^{n+1}$ where trajectories are great circles, that is, a circular motion at $r = 1$ with constant speed. The true motion will have small oscillations in the direction transverse to X, with the approximation becoming exact at the limit of zero initial velocity [142].

In Arnold's treatment [5] of constrained mechanical systems the constraints are replaced by a potential which becomes large away from the surface of constraints. The method leading to (5.1.4) does the converse: A slow motion in a potential becomes a free motion on the manifold of constraints.

5.1.1 Legendre transform

Given a configuration space X, and a Lagrangian $L : TX \longrightarrow \mathbb{R}$, define n functions, called conjugate momenta, by

$$p_i = \frac{\partial L}{\partial \dot{q}^i}.$$

We will assume that (q, p) can be used as coordinates in place of $(q, \dot{q})$. The Hamiltonian $H = H(p, q, t)$ is then defined by the Legendre transform

$$H(q, p, t) = p_i\dot{q}^i - L(q, \dot{q}, t) \tag{5.1.6}$$

where in the above formula $\dot{q}$ must be expressed in terms of (p, q), and we assume that the Lagrangian can explicitly depend on t. Comparing the two differentials

$$\begin{aligned} dH &= \frac{\partial H}{\partial p_i}dp_i + \frac{\partial H}{\partial q^i}dq^i + \frac{\partial H}{\partial t}dt = d\left[p_i\dot{q}^i - L(q, \dot{q}, t)\right] \\ &= \dot{q}^i dp_i - \frac{\partial L}{\partial q^i}dq^i - \frac{\partial L}{\partial t}dt \end{aligned}$$

leads to

$$\frac{\partial H}{\partial t} = -\frac{\partial L}{\partial t}, \tag{5.1.7}$$

and the Hamilton canonical equations (1.1.2) where in this chapter we use upper indices and lower indices for position and momenta, respectively.

The reader will have noticed that we have abused the notation. In general $\partial(H+L)/\partial t \neq 0$ despite that (5.1.7) suggests otherwise. The apparent paradox (which Nick Woodhouse calls the second fundamental confusion of calculus [187]) serves as a warning. The meaning of $\partial/\partial t$ in (5.1.7) depends on what variables we hold fixed. These variables are different in the Lagrangian and Hamiltonian formulations, although the time coordinate t is unchanged by the Legendre transform.

5.1.2 Symplectic structures

In Section 1.3 we considered Poisson structures as a general arena for the Hamiltonian formalism. Here we shall concentrate on symplectic structures which arise as special cases of Poisson structures.

A symplectic manifold is a smooth manifold M of dimension $2n$ with a closed two-form $\omega \in \Lambda^2(M)$ which is non-degenerate at each point, that is, $\omega^n \neq 0$. The symplectic two-form restricted to a point in M gives an isomorphism between the tangent and cotangent spaces given by

$$V \longrightarrow V \lrcorner\, \omega,$$

where V is a vector field, and $\lrcorner$ denotes a contraction of a differential form with a vector field. Using index notation $(V \lrcorner\, \omega)_b = V^a \omega_{ab}$. In particular a function f on M gives rise to a Hamiltonian vector field X_f given by

$$X_f \lrcorner\, \omega = -df. \tag{5.1.8}$$

The Poisson bracket (1.3.8) of two functions f, g can be defined as

$$\{f, g\} = X_g(f) = \omega(X_g, X_f) = -\{g, f\},$$

where $\omega(X_g, X_f) = X_f \lrcorner\, (X_g \lrcorner\, \omega)$. It automatically satisfies the Jacobi identity as

$$0 = d\omega(X_f, X_g, X_k) = \{f, \{g, k\}\} + \{g, \{k, f\}\} + \{k, \{f, g\}\}.$$

It also satisfies

$$[X_f, X_g] = -X_{\{f,g\}},$$

which follows from calculating the RHS on an arbitrary function and using the Jacobi identity.

Hamiltonian vector fields preserve the symplectic form as

$$\mathrm{Lie}_{X_f}(\omega) = d(X_f \lrcorner\, \omega) + X_f \lrcorner\, d\omega = -ddf = 0,$$

where Lie is the Lie derivative (A3) defined in the Appendix A. Conversely if a vector field Lie derives ω then it is always Hamiltonian provided that M is simply connected. The one-parameter group of transformations generated by a Hamiltonian vector field is called a symplectomorphism.

The Darboux theorem stated in Section 1.3 implies that symplectic manifolds are locally isomorphic to $\mathbb{R}^{2n}$ with its canonical symplectic structure

$$\omega = \sum_{i=1}^{n} dp_i \wedge dq^i. \tag{5.1.9}$$

The formula (5.1.9) is also valid if $M = T^*X$ and p_j are local coordinates on the fibres of the cotangent bundle.

If ω is given in the Darboux atlas, then the Poisson bracket is given by (1.1.1), and the Hamiltonian vector field corresponding to the function H is

$$X_H = \sum_{i=1}^{n} \frac{\partial H}{\partial p_i}\frac{\partial}{\partial q^i} - \frac{\partial H}{\partial q^i}\frac{\partial}{\partial p_i}.$$

In general

$$X_H = \sum_{a,b=1}^{2n} \omega^{ab}\frac{\partial H}{\partial \xi^b}\frac{\partial}{\partial \xi^a}$$

where ξ^a, $a = 1, \dots, 2n$, are local coordinates on M.

5.1.3 Solution space

Let M be a solution space of a second-order Euler–Lagrange equations (5.1.3). We shall assume that no boundary conditions are imposed on the variation δq, and derive the symplectic structure on M from the boundary term in (5.1.3). Let

$$S_{12} = \int_{t_1}^{t_2} L(\mathbf{q}(t), \dot{\mathbf{q}}(t))dt$$

be a function on M. Consider a one-parameter family of paths $\mathbf{q}_s(t)$. Then

$$\frac{dS_{12}}{ds}|_{s=0} = \frac{\partial L}{\partial \dot{q}}\delta q|_{t_1}^{t_2}$$

because equations (5.1.3) are satisfied. Rewrite the last formula as

$$dS_{12} = P_{t_2} - P_{t_1}$$

where

$$P_t = p_i dq^i|_t$$

is the canonical one-form on T^*X. The identity $ddS_{12} = 0$ implies that

$$\omega = dP_{t_1} = dP_{t_2} \tag{5.1.10}$$

is a two-form on $M = T^*X$ which does not depend on the choice of points t_1, t_2.

5.2 Field theory

Let $\mathbb{R}^{1,D}$ denote a $(D+1)$-dimensional Minkowski space-time with coordinates

$$x^\mu = (x^0, x^1, \ldots, x^D) = (t, \mathbf{x}),$$

and the flat metric of signature $(+ - - \cdots -)$

$$ds^2 = \eta_{\mu\nu} dx^\mu dx^\nu = dt^2 - |d\mathbf{x}|^2.$$

This metric will be used to raise and lower indices. We shall discuss a relativistic field theory of N scalar fields. Let $\phi : \mathbb{R}^{D+1} \to Y \subset \mathbb{R}^N$ be a scalar field with components $\phi^a, a = 1, \ldots, N$, on $\mathbb{R}^{1,D}$.

We assume that the Lagrangian density $\mathcal{L} = \mathcal{L}(\phi^a, \partial_\mu \phi^a)$, where $\partial_\mu = \partial/\partial_\mu$ depends only on fields and their first derivatives. The action is given by

$$S = \int_{\mathbb{R}^D \times \mathbb{R}} \mathcal{L} \, d^D x dt. \tag{5.2.11}$$

The field equations are derived from the least-action principle

$$\frac{\partial \mathcal{L}}{\partial \phi^a} - \frac{\partial}{\partial x^\mu} \frac{\partial \mathcal{L}}{\partial(\partial_\mu \phi^a)} = 0.$$

The natural Lorentz-invariant Lagrangian density

$$\mathcal{L} = \frac{1}{2} \partial_\mu \phi^a \partial^\mu \phi^a - U(\phi). \tag{5.2.12}$$

leads to the second-order field equations

$$\partial^\mu \partial_\mu \phi^a = -\frac{\partial U}{\partial \phi^a}. \tag{5.2.13}$$

Suppose that a Lie group G acts on the spaces of dependent and independent variables in the way described in Chapter 4. Suppose that the infinitesimal group action changes the Lagrangian by a total divergence

$$\mathcal{L} \longrightarrow \mathcal{L} + \varepsilon \partial_\mu B^\mu$$

for some $B^\mu(x)$ (this condition must hold before the field equations are imposed). If G acts only on the target space Y we talk about internal symmetries. In a neighbourhood of the identity transformation we have

$$\phi^a(x) \longrightarrow \phi^a(x) + \varepsilon W^a(x).$$

We calculate the corresponding infinitesimal change in the Lagrangian density

$$\mathcal{L} \longrightarrow \mathcal{L} + \varepsilon \left[\frac{\partial \mathcal{L}}{\partial(\partial_\mu \phi^a)} \partial_\mu (W^a) + \frac{\partial \mathcal{L}}{\partial \phi^a} W^a \right] = \mathcal{L} + \varepsilon \partial_\mu \left[\frac{\partial \mathcal{L}}{\partial(\partial_\mu \phi^a)} W^a \right]$$
$$+ \varepsilon \left[\frac{\partial \mathcal{L}}{\partial \phi^a} - \partial_\mu \frac{\partial \mathcal{L}}{\partial(\partial_\mu \phi^a)} \right] W^a .$$

We now use the Euler–Lagrange equations to set the last term to zero. Therefore

$$J^\mu = \frac{\partial \mathcal{L}}{\partial(\partial_\mu \phi^a)} W^a - B^\mu = (J^0, \mathbf{J})$$

is the conserved current. The divergence free condition

$$\frac{\partial J^0}{\partial t} + \nabla \cdot \mathbf{J} = 0$$

implies the conservation of the Noether charge

$$Q = \int_{\mathbb{R}^D} J^0 d^D x.$$

An application of the divergence theorem shows that this charge is independent of time if $\mathbf{J}$ vanishes at spatial infinity

$$\frac{dQ}{dt} = \int_{\mathbb{R}^D} \frac{\partial J^0}{\partial t} d^D x = - \int_{\mathbb{R}^D} \nabla \cdot \mathbf{J} d^D x = 0.$$

If the action of G on Y is trivial, and G acts on $\mathbb{R}^{D+1}$ isometrically we talk about space-time symmetries. The Lagrangian (5.2.12) is invariant under the transformation

$$x^\mu \longrightarrow x^\mu + \varepsilon V^\mu (x^\nu).$$

Infinitesimally the field and the Lagrangian density transform by the Lie derivative along the vector $V = V^\mu \partial_\mu$

$$\phi^a \longrightarrow \phi^a (x^\mu + \varepsilon V^\mu) = \phi^a + \varepsilon \mathrm{Lie}_V \phi^a \quad \text{and} \quad \mathcal{L} \longrightarrow \mathcal{L} + \varepsilon \mathrm{Lie}_V \mathcal{L}.$$

Assume that V^μ is a constant vector, so that G is the group of space-time translations. The conserved current is in this case given by the energy–momentum tensor

$$T^\mu_\nu = \frac{\partial \mathcal{L}}{\partial(\partial_\mu \phi^a)} \partial_\nu \phi^a - \eta^\mu_\nu \mathcal{L}.$$

This tensor is divergence free, $\partial_\mu T^\mu_\nu = 0$, and the associated conserved charges are the energy E and momentum P_i:

$$E = \int_{\mathbb{R}^D} T^0_0 d^D x \quad \text{and} \quad P_i = -\int_{\mathbb{R}^D} T^0_i d^D x, \qquad i = 1, \dots, D.$$

The conservation of E and P_i is therefore related, by Noether's theorem, to the invariance of $\mathcal{L}$ under the time and spatial translations, respectively. In the special case, when $\mathcal{L}$ is given by (5.2.12) we can define the kinetic and the potential energy by

$$T = \frac{1}{2}\int_{\mathbb{R}^D} \phi^a_t \phi^a_t d^D x \quad \text{and} \quad V = \int_{\mathbb{R}^D}\left[\frac{1}{2}\nabla\phi^a \cdot \nabla\phi^a + U(\phi)\right] d^D x \tag{5.2.14}$$

so that $E = T + V$.

5.2.1 Solution space and the geodesic approximation

A solution space $\mathcal{S}$ of the Euler–Lagrange equations (5.2.13) is an infinite-dimensional manifold, and formally we can equip it with a symplectic structure, which arises from the boundary term in the variational principle. Let ϕ be a solution to (5.2.13). A tangent vector $\delta\phi$ to $\mathcal{S}$ at a given solution ϕ is the linearization of (5.2.13) around ϕ

$$\partial_\mu \partial^\mu \delta\phi^a = -\frac{\partial^2 U}{\partial\phi^a \partial\phi^b}|_{\phi=\phi_0} \delta\phi^b. \tag{5.2.15}$$

Analysing the variation of the action along the lines leading to (5.1.10) we find a closed two-form Ω on $\mathcal{S}$

$$\Omega(\delta_1\phi, \delta_2\phi) = \int_{\mathbb{R}^D}\left[\delta_1\phi^a \frac{\partial}{\partial t}(\delta_2\phi^a) - \delta_2\phi^a \frac{\partial}{\partial t}(\delta_1\phi^a)\right] d^D x,$$

where $\delta_1\phi$ and $\delta_2\phi$ are two solutions to (5.2.15) which implies that integrand does not depend on t.

The dynamics of finite-energy solutions to (5.2.13) with small initial velocity can be reduced to a finite-dimensional dynamical system. The idea goes back to Manton [113], and the method is analogous to the argument leading to (5.1.4) with $\mathbb{R}^{n+1}$ replaced by an infinite-dimensional configuration space of the fields ϕ, and X replaced by the the moduli space $\mathcal{M}$ of static finite-energy solutions to (5.2.13).

Assume that all finite energy static solutions $\phi_S = \phi_S(x, \gamma)$ to (5.2.13) are parameterized by points in some finite-dimensional manifold $\mathcal{M}$ with local coordinates γ. These solution give the absolute minimum of the potential energy. The time-dependent solutions to (5.2.13) with small total energy (hence small potential energy) above the absolute minimum will be approximated by a sequence of static states, that is, a free motion in $\mathcal{M}$. This free motion is geodesic with respect to a natural Riemannian metric on $\mathcal{M}$

$$h_{rs}(\gamma) = \int_{\mathbb{R}^D} \frac{\partial \phi^a_S}{\partial \gamma^r} \frac{\partial \phi^a_S}{\partial \gamma^s} d^D x \tag{5.2.16}$$

which arises from the kinetic energy (5.2.14) by

$$\begin{aligned} T &= \frac{1}{2} \int_{\mathbb{R}^D} \frac{\partial \phi^a_S}{\partial t} \frac{\partial \phi^a_S}{\partial t} d^D x = \frac{1}{2} \int_{\mathbb{R}^D} \frac{\partial \phi^a_S}{\partial \gamma^r} \frac{\partial \phi^a_S}{\partial \gamma^s} \dot{\gamma}^r \dot{\gamma}^s d^D x \\ &= \frac{1}{2} h_{rs}(\gamma) \dot{\gamma}^r \dot{\gamma}^s. \end{aligned}$$

5.3 Scalar kinks

In this section we shall study solitons in the context of Lagrangian field theory. The term 'soliton' here has a different meaning to that which we used in the first three chapters. The solitons are necessarily static and the inverse scattering theory is not used in general.

Definition 5.3.1 *Solitons are non-singular, static, finite energy solutions of the classical field equations.*

At the quantum level solitons correspond to localized extended objects (particles): Kinks in one-dimension, Vortices or Lumps in two dimensions, and Monopoles in three dimensions. One finds solitons by solving classical non-linear equations exactly. Sometimes time-dependent solitons are considered and it is required that they are non-dispersive and preserve their shape after collisions. These are 'rare' in the sense that they only appear in integrable field theories which we studied in Chapter 2.

Consider a single scalar field on two-dimensional space time. The Lagrangian density (5.2.12) with $D = 1$ gives

$$L = \int_{\mathbb{R}} \left[\frac{1}{2}\phi_t^2 - \frac{1}{2}\phi_x^2 - U(\phi) \right] dx = T - V,$$

where $\phi_t = \partial_t \phi$, $\phi_x = \partial_x \phi$ and

$$T = \int_{\mathbb{R}} \frac{1}{2}\phi_t^2 dx, \qquad V = \int_{\mathbb{R}} \left[\frac{1}{2}\phi_x^2 + U(\phi) \right] dx$$

are the kinetic and the potential energies, respectively. The field equations (5.2.13) are

$$\phi_{tt} - \phi_{xx} = -\frac{dU}{d\phi}. \tag{5.3.17}$$

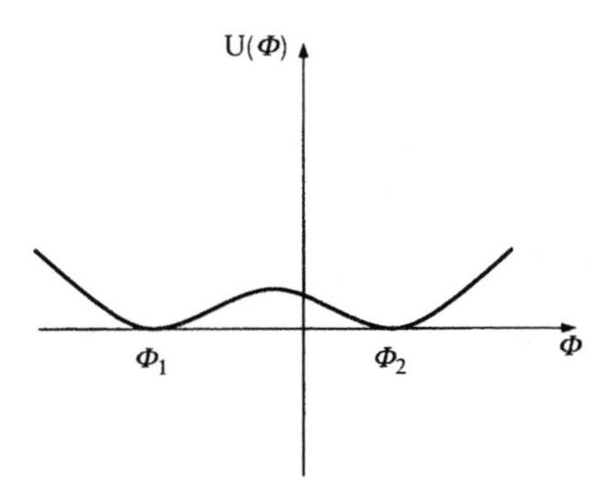

Figure 5.1 *Multiple vacuum*

We need $U(\phi) \geq U_0$ for a stable vacuum, and we choose the normalization $U_0 = 0$. Assume that the set $U^{-1}(0) = \{\phi_1, \phi_2, ...\}$ is non-empty and discrete (Figure 5.1). In perturbation theory ϕ undergoes small oscillations around one of the minima, $\phi = \phi_1 + \delta\phi$. The basic perturbative excitation is a scalar boson with a squared mass equal to the quadratic part of U when expanded about a minimum. This is because (5.2.12) is the Lagrangian density for the Klein–Gordon equation $(\Box + m^2)\delta\phi = 0$.

The finite-energy solutions must asymptotically approach an element of $U^{-1}(0)$. This element can however be different at different ends of a real line. The simplest topological solitons are characterized by the boundary conditions

$$\phi \cong \phi_1 \quad \text{as} \quad x \to -\infty \quad \text{and} \quad \phi \cong \phi_2 \quad \text{as} \quad x \to \infty,$$

and cannot be treated within the perturbation theory. These are the *Kink solutions* connecting neighbouring vacua. The static field equation

$$\phi_{xx} = \frac{dU}{d\phi}$$

formally resembles the Newton equations in classical mechanics. It integrates to

$$\frac{1}{2}\phi_x^2 = U(\phi) + c, \qquad c = \text{const.}$$

The boundary conditions yield $U(\phi_1) = U(\phi_2) = 0$, and therefore $c = 0$. The kink solution is implicitly given by

$$x - x_0 = \pm \int^{\phi} \frac{1}{\sqrt{2U(\tilde{\phi})}} d\tilde{\phi}. \tag{5.3.18}$$

The RHS diverges near a minimum of U. Here the constant x_0 is the location of the kink and the sign on the RHS corresponds to the direction of the kink.

The potential energy of the kink is

$$E = \int \left(\frac{1}{2}\phi_x^2 + U\right) dx = \int 2U dx.$$

- **Example.** Let $U = \lambda^2(\phi^2 - a^2)^2/2$. In perturbation theory this describes a scalar boson with mass $2\lambda a$, because

$$U(\phi) = \frac{1}{2}\lambda^2(\phi + a)^2(\phi - a)^2 = \frac{1}{2}(2\lambda a)^2(\phi - a)^2 + O[(\phi - a)^3]$$

and we regard a and ϕ as dimensionless. The integration (5.3.18) gives

$$x - x_0 = \pm\frac{1}{\lambda}\int_0^{\phi} \frac{d\tilde{\phi}}{a^2 - \tilde{\phi}^2} = \pm\frac{1}{\lambda a}\tanh^{-1}(\phi/a).$$

Therefore

$$\phi_K(x) = \pm a \tanh[\lambda a(x - x_0)] \tag{5.3.19}$$

which approaches $\pm a$ as $x \to \pm\infty$. This is a truly non-perturbative solution as for the fixed boson mass $m = 2\lambda a$ the RHS of (5.3.19) is not analytic in λ, and therefore it can not be obtained by starting from a solution to the $(1 + 1)$-dimensional wave equation for ϕ and expanding in λ.

The energy of the kink is given by

$$E = \lambda^2 a^4 \int_{\mathbb{R}} \text{sech}^4[\lambda a(x - x_0)]dx = \frac{4}{3}\lambda a^3.$$

We identify E with the mass of the kink. This is because the solution has non-zero energy density in a small region of order a^{-1}. The energy density has its maximum at $x = x_0$ which justifies the interpretation of x_0 as the position of the kink.

The mass of the kink is therefore much larger than the boson mass if a^2 is large with λa fixed, which is the perturbative regime of quantized theory. In quantum theory the field fluctuations around the kink can contribute to the kink mass. The higher order corrections are calculated in [141].

A *moving kink* can be obtained by a Lorentz boost of a static kink:

$$\phi(x, t) = \phi_K[\gamma(x - vt)], \qquad \gamma = (1 - v^2)^{-1/2}.$$

It solves the field equation (5.3.17). The moving kink has conserved energy $E = T + V$ related to its mass by $E = \gamma M$. It also has a conserved momentum

$$P = -\int_{-\infty}^{\infty} \phi_t\phi_x dx = \gamma M v.$$

5.3.1 Topology and Bogomolny equations

All *finite energy field configurations* approach vacuum at $\pm\infty$. Asymptotic values cannot change (Figures 5.2–5.4). *Topological conserved currents* are properties of a finite energy continuous field. For (5.2.13) with $D = 1$ topological conserved quantities are

$$\phi_- = \lim_{x\to-\infty} \phi \quad \text{and} \quad \phi_+ = \lim_{x\to\infty} \phi.$$

If $\phi_+ = \phi_-$ then the field $\phi(x)$ can be continuously deformed into the zero-energy vacuum $\phi = \phi_+$. If on the other hand $\phi_+ \neq \phi_-$ then ϕ cannot be continuously deformed into a vacuum, which is the reason for topological stability of kinks.

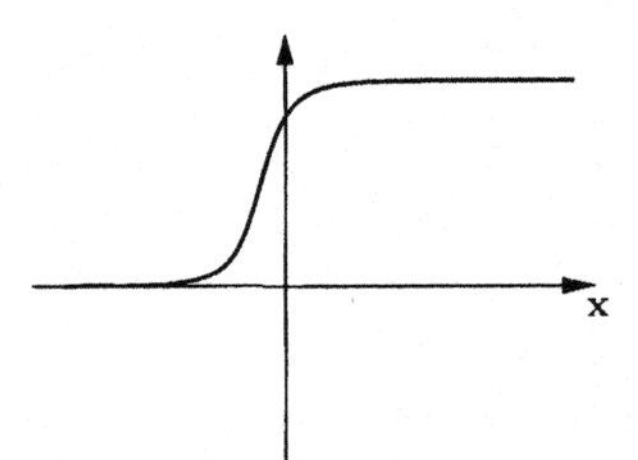

Figure 5.2 *Kink,* $N = 1$

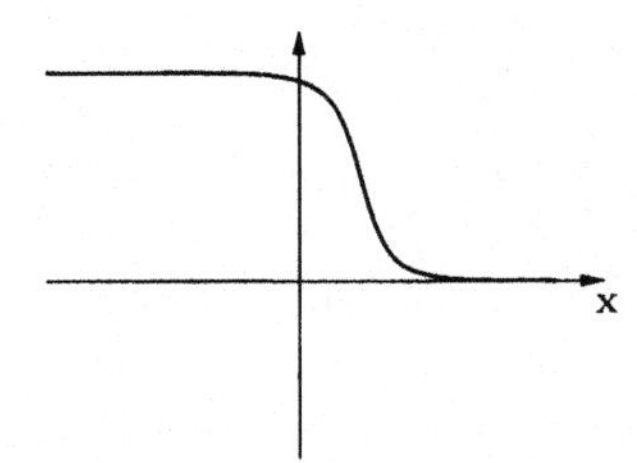

Figure 5.3 *Anti-kink,* $N = -1$

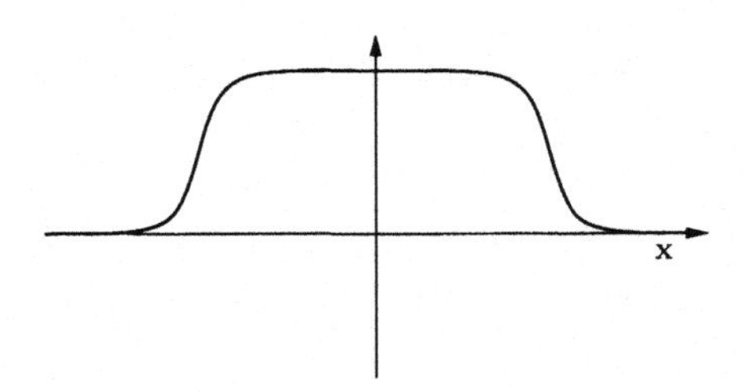

Figure 5.4 *Kink–anti-kink pair,* $N = 0$

The associated conserved current

$$N = \phi_+ - \phi_- = \int_{\mathbb{R}} \phi_x dx$$

is an integral of a total derivative which depends only on boundary conditions. It is conserved because we insisted on the finiteness of the energy. Note that the field equations have not entered the discussion at this stage. Topological conserved currents are in this sense different from the Noether currents which result from continuous symmetries of the Lagrangian.

We shall look for minimal energy configurations. If $U \geq 0$ we can always find $W(\phi)$ such that

$$U(\phi) = \frac{1}{2}\left[\frac{dW(\phi)}{d\phi}\right]^2.$$

Now

$$\begin{aligned} E &= \frac{1}{2}\int_{\mathbb{R}} dx(\phi_t^2 + \phi_x^2 + W_\phi^2) = \frac{1}{2}\int_{\mathbb{R}} dx\left[\phi_t^2 + (\phi_x \pm W_\phi)^2 \mp 2\phi_x W_\phi\right] \\ &= \frac{1}{2}\int_{\mathbb{R}} dx\left[\phi_t^2 + (\phi_x \pm W_\phi)^2\right] \mp [W(\phi(\infty)] - W[\phi(-\infty)]. \end{aligned}$$

Therefore

$$E \geq |W[\phi(\infty)] - W[\phi(-\infty)]|. \tag{5.3.20}$$

This is the *Bogomolny bound*. It depends only on the topological data at $\pm\infty$. Say

$$\phi(\infty) = \phi_2 > \phi(-\infty) = \phi_1.$$

The minimum energy configurations satisfy $\phi_t = 0$ (the static condition), $E = W(\phi_2) - W(\phi_1)$ and

$$\frac{d\phi}{dx} = \frac{dW}{d\phi} \tag{5.3.21}$$

which is the *Bogomolny equation*. Its solutions are kinks (5.3.18).

The field equation (5.3.17) is a second-order PDE, and its special solutions arise form the first-order ODE (the Bogomolny equation). A time-dependent solution to (5.3.17) can be found by the Lorentz boost. In general (any dimension and Lagrangian) the full field equations are usually not integrable, but the Bogomolny equations are often integrable (and have lower order).

- **Example.** $U = \lambda^2(\phi^2 - a^2)^2/2$ gives $W = \lambda(a^2\phi - \phi^3/3)$. The Bogomolny equations (5.3.21) yield

$$\phi_x = \lambda(a^2 - \phi^2),$$

and ϕ is the static kink solutions (5.3.19) with energy

$$E = W(a) - W(-a) = \frac{4}{3}\lambda a^3.$$

This kink is stable as it would take infinite energy to change this solution into a constant vacuum solution $\phi = 0$. In general kinks minimize the energy within their topological class. The absolute minimum is of course 0 which corresponds to $\phi_1 = \phi_2$.

- **Example.** $U = 1 - \cos\beta\phi$. This theory has infinitely many vacua,

$$\phi_n = \frac{2\pi n}{\beta},$$

parameterized by $n \in \mathbb{Z}$, and kinks interpolate between adjacent vacua. The field equation (5.3.17) is the Sine-Gordon equation.

$$\phi_{tt} - \phi_{xx} + \beta \sin\beta\phi = 0 \qquad (5.3.22)$$

which is essentially equivalent to (2.1.2) as the parameter β can be set to 1 by scalings of (x, t, ϕ).
The kink solution with $\phi(-\infty) = 0$, $\phi(\infty) = 2\pi/\beta$ is given by

$$\phi = \frac{4}{\beta}\arctan^{-1} e^{\beta(x-x_0)}.$$

The solution is multivalued and we get all possible kinks depending on which branch we choose. In this case time-dependent solutions to the full equations (multi-kinks and solitons) can be constructed using the integrability of the Sine-Gordon equation. The simplest solution generating technique is the Bäcklund transformation described in Section 2.1.2.
The Sine-Gordon equation admits time-dependent solutions such that ϕ tends to the same limit at $\pm\infty$. These so-called breathers have trivial topological charge and owe their stability to the complete integrability of (5.3.22), and the existence of an infinite number of conservation laws preventing annihilation into radiation. By contrast $U = \lambda^2(\phi^2 - a^2)^2/2$ does not posses such solutions, as the corresponding field equations are not integrable.

5.3.2 Higher dimensions and a scaling argument

Can there exist finite-energy static critical points of (5.2.12) in more than one spatial dimension? In this section we shall examine a scaling argument, originally due to Derrick [37] and rule out all dimensions higher than 2. In Section 6.1.1 we shall return to Derrick's argument in the context of gauge theory where the spatial dimensions three and four are also allowed.

If $\phi(x)$ is a static solution to (5.2.13) in D spatial dimensions, then

$$\nabla^2 \phi = \frac{dU}{d\phi},$$

and ϕ is a critical point of the energy functional

$$E(\phi) = \int_{\mathbb{R}^D} d^D x \left[\frac{1}{2} |\nabla \phi|^2 + U(\phi) \right] = E_{\text{grad}} + E_U.$$

Consider a one-parameter family of configurations $\phi_{(c)}(x) = \phi_{(1)}(cx)$, where $\phi_1(x)$ is a static finite-energy solution. Then

$$E[\phi_{(c)}] = \frac{1}{c^{D-2}} E_{\text{grad}} + \frac{1}{c^D} E_U.$$

Since $E(\phi_1)$ is a minimum of E we have

$$dE[\phi_{(c)}]/dc|_{c=1} = 0$$

which implies

$$(D-2)E_{\text{grad}} + DE_U = 0. \tag{5.3.23}$$

$D = 1$. Static solutions are possible with $E_{\text{grad}} = E_U$. These are the kinks (5.3.21).

$D = 2$. Static solutions are possible with $E_U = 0$. This can still lead to non-linear field theories if the target space is a manifold without a linear structure, $\phi : \mathbb{R}^{2+1} \to \Sigma$. These so-called sigma models will be studied in Section 5.4.

$D = 3$. Finite-energy static solutions do not exist. Adding a Skyrme term $|\nabla \phi|^4$ to the Lagrangian density allows static solutions via the scaling argument. See [114] for a complete discussion of the Skyrme model.

Although the $D = 1$ kinks do not generalize to solitons in $D > 1$, they can be trivially lifted to translationally invariant solutions of scalar field theory (5.2.13) in any dimension. These lifted solutions have infinite energy as a result of the integration along the $(D-1)$ spatial directions on which the kinks do not depend. The energy is however finite per unit volume. This type of solution is called a domain wall. Finally we note that the Derrick argument breaks down for time-dependent configurations.

5.3.3 Homotopy in field theory

Any static, smooth field configuration $\phi : \mathbb{R}^{D+1} \longrightarrow \mathbb{R}^N$ is topologically trivial, as it can be transformed to zero by a homotopy $(1 - \tau)\phi$ (see Appendix A for discussion of homotopy). Non-trivial field configurations appear if we assume that the energy density

$$\mathcal{E} = \frac{1}{2}\nabla\phi^a \cdot \nabla\phi^a + U(\phi^1, \dots, \phi^N)$$

of a static field decays as $r \longrightarrow \infty$. This condition alone does not necessarily lead to finite energy, but it gives rise to a topological classification. Let $M \subset \mathbb{R}^N$ be a submanifold of the target space implicitly defined by

$$U(\phi^1, \dots, \phi^N) = U_{min},$$

where U_{min} is the minimal value of U. At spatial infinity $\phi = \phi_\infty$ must take its values in M, or the density $\mathcal{E}$ will not vanish. Therefore

$$\phi_\infty : S^{D-1} \longrightarrow M,$$

and smooth field configurations are classified by elements of the homotopy group $\pi_{D-1}(M)$.

Later we will meet other ways of classifying smooth field configurations: In sigma models $U = 0$, and $\phi : \mathbb{R}^{D+1} \longrightarrow \Sigma$. Fields are classified by elements of $\pi_D(\Sigma)$ as any static field with finite $\mathcal{E}$ must smoothly extend to the one-point compactification S^D of $\mathbb{R}^D$. In pure gauge theories gauge fields are classified by Chern numbers which arise from integrating various powers of the field tensor. This will be discussed in Section 6.4. Finally in gauge theories with Higgs fields, the Higgs fields at infinity carry all topological information (see Section 6.3.1).

5.4 Sigma model lumps

Sigma models are non-linear in a fundamental way: the target space is not a linear space. Consider a field $\phi : \mathbb{R} \times \mathbb{R}^2 \longrightarrow S^{N-1}$, with components $\phi^a(x^\mu) \in \mathbb{R}^N$ which satisfy the non-linear relation

$$\sum_{a=1}^{N} \phi^a\phi^a = 1. \tag{5.4.24}$$

The kinetic Lagrangian density

$$\mathcal{L} = \frac{1}{2}\partial_\mu\phi^a\partial^\mu\phi^a$$

gives rise to a non-linear equation. To see this introduce a Lagrange multiplier $\lambda(x^\mu)$, and consider the Euler–Lagrange equations of

$$\mathcal{L}' = \mathcal{L} - (1/2)\lambda(x^\mu)(1 - \sum \phi^a\phi^a).$$

This yields

$$\Box\phi^a - \lambda\phi^a = 0, \qquad \text{where} \qquad \Box = \eta^{\mu\nu}\partial_\mu\partial_\nu,$$

and the constraint (5.4.24). Multiplying the relation above by ϕ^a, summing over a, and eliminating λ yield the non-linear field equations

$$\Box\phi^a - (\phi^b\Box\phi^b)\phi^a = 0. \tag{5.4.25}$$

The Lagrangian and the equation (5.4.25) are invariant with respect to global $O(N)$ rotations of the field ϕ. This is an example of a model with an internal symmetry.

Solving the constraint (5.4.24) yields $\phi^N = \pm\sqrt{1 - \phi^p\phi^p}$, where $p, q, r = 1, \ldots, N-1$, and allows us to write the Lagrangian density as

$$\mathcal{L} = \frac{1}{2} g_{pq}(\phi)\eta_{\mu\nu}\partial^\mu\phi^p\partial^\nu\phi^q, \tag{5.4.26}$$

where

$$g_{pq} = \delta_{pq} + \frac{\phi^p\phi^q}{1 - \sum\phi^r\phi^r}$$

is the metric on S^{N-1} induced by the Euclidean inner product in $\mathbb{R}^N$. Lagrange densities of the form (5.4.26) where η and g are arbitrary metrics on a space time and a target space, respectively, define more general sigma models. For example, superstring theory can be viewed as a sigma model where η is a metric on a 'string world-sheet' which is a Riemann surface, and g is a metric on a 10-dimensional target which plays the role of 'space time'.

From now on we restrict our attention to (5.4.25) with $N = 3$. The corresponding model describes the Heisenberg ferromagnet in low temperatures, when the local magnets line up [66]. The system is characterized by the direction of the spin vector, that is, a unit vector ϕ^a.

We will be interested in time-independent solutions with a finite-energy functional $\int \mathcal{L} d^2x$. This condition implies that $r|\nabla\phi^a| \to 0$ as $r \to \infty$, therefore $\phi(x^i)$ tends to a constant field ϕ^∞ at spatial infinity, which we choose to be the north pole $(0, 0, 1)$. This means that the finite-energy static solutions extend[1] to S^2. This sphere has infinite radius, and is a one-point compactification of $\mathbb{R}^2$. The Laplacian Δ in two dimensions is conformally invariant is the sense that

$$c\Delta_{cg} = \Delta_g$$

[1] This is not strictly true, as the finiteness of the L^2 norm of $\nabla\phi$ does not imply that $\nabla\phi \longrightarrow 0$. There could exist finite-energy maps which do not extend to S^2. Sacks and Uhlenbeck [143] show that this does not happen. Their proof uses the equations of motion and their conformal invariance.

for two conformally related metrics g and $c(x^i)g$. This can be verified from the definition

$$\Delta_g = |g|^{-1/2}\partial_i(|g|^{1/2}g^{ij}\partial_j).$$

Therefore the static equations

$$\Delta\phi^a - (\phi^b\Delta\phi^b)\phi^a = 0 \tag{5.4.27}$$

are satisfied on S^2.

Another consequence of the conformal invariance is that the spatial rescaling does not change the energy, and can be used to shrink any static solution down to zero. Therefore the solitons we are just about to describe do not fully deserve their name, and are called lumps by some authors.

Continuous maps $\phi : S^2 \longrightarrow S^2$ are classified by their topological degree (in this context also called topological charge) given by (A7)

$$Q = \deg\phi = \frac{1}{8\pi}\int_{S^2} \varepsilon^{ij}\varepsilon^{abc}\phi^a\partial_i\phi^b\partial_j\phi^c d^2x$$

which partially characterizes static solutions. A field with a given Q cannot be continuously deformed into a field with different Q. It is now clear why we have focused on $N = 3$. The spheres with $N > 3$ as target spaces would not lead to non-trivial topological configurations, as the relevant homotopy group $\pi_2(S^{N-1})$ vanishes.

The degree Q carries only global information and fields can have different energies within one topological sector.

Proposition 5.4.1 *The energy*

$$E = \frac{1}{2}\int_{S^2} \partial_i\phi^a\partial_i\phi^a d^2x \geq 4\pi|Q| \tag{5.4.28}$$

is bounded from below with equality when the first-order Bogomolny equations

$$\partial_i\phi^a = \pm\varepsilon_{ij}\varepsilon^{abc}\phi^b\partial_j\phi^c \tag{5.4.29}$$

are satisfied.

Proof 1 Consider the identity

$$\int (\partial_i\phi^a \pm \varepsilon_{ij}\varepsilon^{abc}\phi^b\partial_j\phi^c)(\partial_i\phi^a \pm \varepsilon_{ij}\varepsilon^{abc}\phi^b\partial_j\phi^c)d^2x \geq 0$$

and use the relations

$$\varepsilon_{ij}\varepsilon_{ik} = \delta_{jk}, \qquad \varepsilon^{abc}\varepsilon^{ade} = \delta^{bd}\delta^{ce} - \delta^{be}\delta^{cd}, \quad \text{and} \quad \phi^a\partial_j\phi^a = 0$$

to deduce (5.4.28) and (5.4.29). □

The solutions to the Bogomolny equations (5.4.29) are critical points of the energy functionals, and so are also solutions to the second-order static-field equations (5.4.27). It can be shown [166] that all finite-energy solutions to (5.4.27) are solutions to the Bogomolny equations. In the context of the Heisenberg ferromagnet the solutions of (5.4.29) are called spin waves. The degree of ϕ describes how often the spins aligned along some axis twist around this axis. More generally, the degree can be interpreted as the number of lumps, because generically the energy density is concentrated in Q localized regions.

The Bogomolny equations (5.4.29) can be easily solved with the help of complex numbers. Identify S^2 with a complex projective line $\mathbb{CP}^1$ (the corresponding model is sometimes called the $\mathbb{CP}^1$ model, or O(3) model). Let $f : \mathbb{R}^{2,1} \longrightarrow \mathbb{CP}^1$ be given by

$$\phi^1 + i\phi^2 = \frac{2f}{1 + |f|^2} \quad \text{and} \quad \phi^3 = \frac{|f|^2 - 1}{|f|^2 + 1}, \quad \text{so that} \quad f = \frac{\phi_1 + i\phi_2}{1 - \phi_3}. \tag{5.4.30}$$

The Bogomolny equations now imply that f is holomorphic, or anti-holomorphic in $z = x^1 + ix^2$. The total energy (5.4.28) of static fields in terms of f is given by

$$E = \frac{1}{4}\int_{S^2} \frac{df \wedge \overline{df}}{(1 + |f|^2)^2}$$

and the rational function

$$f = c\frac{(z - p_1)\cdots(z - p_Q)}{(z - r_1)\cdots(z - r_Q)},$$

gives finite-energy solutions which saturate the Bogomolny bound with $\deg(\phi) = Q$. This is because only the rational functions give Q-fold coverings of an extended complex plane with finite Q. The overall factor c can be set to 1 by a global rotation of the field. The space of static solutions is isomorphic to the space-based rational maps because $\lim_{z\to\infty}\phi = \phi^\infty$. It is the complement in $\mathbb{C}^{2Q}$ of a hypersurface where the enumerator and denominator have common poles.

Exercises

1. Starting with the Lagrangian of the Sine-Gordon theory

$$\mathcal{L} = \frac{1}{2}(\phi_t^2 - \phi_x^2) - (1 - \cos\beta\phi)$$

derive the Sine-Gordon equation. Find a kink solution of the Sine-Gordon theory, and use the Bogomolny bound to find its energy. How many types of kinks are there?

2. The Sine-Gordon equation is

$$\phi_{xx} - \phi_{tt} = \sin\phi, \quad \text{where} \quad \phi = \phi(x, t).$$

Set $\tau = (x + t)/2$, $\rho = (x - t)/2$ and consider the Bäcklund transformations

$$\partial_\rho(\phi_1 - \phi_0) = 2b \sin\left(\frac{\phi_1 + \phi_0}{2}\right) \quad \text{and} \quad \partial_\tau(\phi_1 + \phi_0) = 2b^{-1} \sin\left(\frac{\phi_1 - \phi_0}{2}\right),$$

where $b = \text{const}$ and ϕ_0, ϕ_1 are functions of (τ, ρ). Take $\phi_0 = 0$ and construct the one-soliton (kink) solution ϕ_1.

3. Let $\phi = \phi(x, t)$ be a scalar field and let $U = U(\phi) \geq 0$. Define the energy of solutions to the Euler–Lagrange equations with the Lagrangian density

$$\mathcal{L} = \frac{1}{2}\phi_t^2 - \frac{1}{2}\phi_x^2 - U(\phi).$$

Assume that $U = (1/2)\phi^2(\phi^2 - \beta^2)^2$, where $\beta \in \mathbb{R}$. How many static kink solutions are there? Find a moving kink solution for the model if $\beta \neq 0$.

4. The Lagrangian density for a complex scalar field ϕ on the two-dimensional Minkowski space $\mathbb{R}^{1,1}$ is

$$\mathcal{L} = \frac{1}{2}|\phi_t|^2 - \frac{1}{2}|\phi_x|^2 - \frac{1}{2}\lambda^2(a^2 - |\phi|^2)^2, \qquad a \in \mathbb{R}.$$

Find the field equations, and verify that the real kink $\phi_0(x) = a \tanh(\lambda a x)$ is a solution. Now consider a small pure imaginary perturbation $\phi(x, t) = \phi_0(x) + i\eta(x, t)$ with η real and find the linear equation satisfied by η. By considering $\eta = \text{sech}\,(\alpha x)e^{\omega t}$ show that the kink is unstable.

5. Let $\phi : \mathbb{R}^{2,1} \to S^2$. Set

$$\phi^1 + i\phi^2 = \frac{2f}{1 + |f|^2} \quad \text{and} \quad \phi^3 = \frac{|f|^2 - 1}{|f|^2 + 1},$$

and deduce that the Bogomolny equations

$$\partial_i \phi^a = \pm\varepsilon_{ij}\varepsilon^{abc}\phi^b\partial_j\phi^c \quad \text{and} \quad \phi_t = 0$$

imply that f is holomorphic or anti-holomorphic in $z = x^1 + ix^2$. Find an expression for the total energy

$$E = \frac{1}{2}\int \partial_j\phi^a\partial_j\phi^a \, d^2x$$

in terms of f.

6 Gauge field theory

In Chapter 5 we have given examples of Lagrangians invariant under the action of symmetry groups. These symmetry transformations were identical at every point of space time. This is referred to as *global symmetry* in the physics literature. In this chapter we shall introduce a concept of gauge symmetry, where the symmetry transformation is allowed to depend on a space-time point. This is what physicists call *local symmetry*. This type of symmetry is already present in Maxwell's electrodynamics. The kinetic term $i\overline{\psi}\gamma^{\mu}\partial_{\mu}\psi$ in the Lagrangian involving the matter field (electron) ψ is unchanged if we replace ψ by $e^{i\mathrm{e}\theta}\psi$, where e is the electric charge of an electron and γ^{μ} are the Dirac matrices. If θ is a constant, we talk about global symmetry. Gauging, or localizing, this symmetry comes down to allowing $\theta = \theta(x^{\mu})$. The Lagrangian is no longer invariant, unless we replace the ordinary derivative $\partial/\partial x^{\mu}$ by a covariant derivative $D_{\mu} = \partial_{\mu} - i\,\mathrm{e}A_{\mu}$, where the gauge potential A_{μ} transforms as $A_{\mu} \to A_{\mu} + \partial_{\mu}\theta$.

The matter Lagrangian is then complemented by adding a gauge term $-(1/4)F_{\mu\nu}F^{\mu\nu}$ where $F_{\mu\nu} = \partial_{\mu}A_{\nu} - \partial_{\nu}A_{\mu}$ is the gauge field. The whole Lagrangian is invariant under the local gauge transformations and the gauge potential is promoted to a dynamical variable. It corresponds to a gauge boson which in the case of electrodynamics is identified with a photon. The symmetry group which has been gauged in this example is $U(1)$. Thus electrodynamics is a $U(1)$ gauge theory. The abelian nature of $U(1)$ implies that there are no interactions between the photons.

The breakthrough made by Yang and Mills [188] was to replace $U(1)$ by a non-abelian Lie group G. In the gauging process one needs to introduce one gauge boson for each generator of G. The bosons are particles which 'carry interactions' between the matter fields, and the form of the interactions is dictated by the gauge symmetry. The bosons take values in the Lie algebra of G.

If $G = SU(3)$ there are eight gauge bosons generalizing one photon. They are called gluons. The matter fields generalizing the electron are called quarks. The quarks are charged with colour, which generalizes the electric charge. The quantum $SU(3)$ gauge theory is called quantum chromodynamics (QCD). It is a theory of strong nuclear interactions.

The standard model unifying electromagnetic and nuclear interactions is an example of gauge theory with $G = SU(3) \times SU(2) \times U(1)$. The 12 gauge bosons in this model consist of a massless photon, 8 gluons, and 3 massive W and Z bosons which carry the weak nuclear force. The electromagnetic interaction is long ranged as the photon is massles, but both nuclear forces are short ranged. In the case of the weak interaction this is due to the corresponding gauge bosons being massive. In the case of strong interaction there is a still poorly understood mechanism, called confinement, which implies that the forces between quarks increase when the quarks become separated. This effectively limits the range of strong interactions to 10^{-15} meters. See [168] for a very good presentation of QCD and other gauge theories in the context of particle physics.

It is fair to say that the concept of gauge symmetry gave rise to the greatest revolution in physics in the second half of the twentieth century. It has lead to several Nobel Prizes awarded for theoretical work: In 1979 to Glashow, Salam, and Weinberg for their work on gauge theory of electroweak interactions done in the 1960s. In 1999 to t'Hooft and Veltman for their work on renormalizability of quantum gauge theories done in the early 1970s. In 2004 to Gross, Politzer, and Wilczek for their work on asymptotic freedom done in the early 1970s. In 2008 to Nambu for his discovery in the 1960s of the mechanism of spontaneously broken symmetry in particle physics and to Kobayashi and Maskawa for their work on CP violation done in the 1970s. More prizes are likely to follow if the LHC discovers the Higgs particle and other forms of matter (see the footnote on page 26).

The pure mathematical studies of gauge theory initiated by the Oxford school of Atiyah led to advances in differential geometry and eventually to solutions of several long-standing problems in topology of lower dimensional manifolds [39, 40]. The twistor techniques, proposed by Penrose [129] and used by Ward [169] to solve the anti-self-dual sector of the gauge field equations, proved to be a universal language for most lower dimensional integrable systems describing solitons. The gauge theory lead to Fields medals which carry the weight of Nobel Prize in mathematics: In 1986 to Donaldson for his gauge-inspired work on topology of four manifolds. In 1990 to Witten (the first physicist to be awarded the medal) for his work on mathematical aspects of quantum gauge theories. In 1998 to Kontsevich for a rigorous formulation of the Feynman integral in topological field theories.

6.1 Gauge potential and Higgs field

In this section we consider gauge theory in $(D+1)$-dimensional Minkowski space M, ... with a preferred volume from. Let the gauge potential $A = A_\mu dx^\mu$

be a one form with values in the Lie algebra $\mathfrak{g}$ of some Lie group G and let $D = d + A$ be the covariant derivative. Let

$$F = \frac{1}{2}F_{\mu\nu}dx^{\mu} \wedge dx^{\nu} = dA + A \wedge A$$

be the gauge field of A. Its components are given by

$$F_{\mu\nu} = \partial_{\mu} A_{\nu} - \partial_{\nu} A_{\mu} + [A_{\mu}, A_{\nu}] = [D_{\mu}, D_{\nu}].$$

Gauge transformations identify (A, A') and (F, F'), where[1]

$$A' = gAg^{-1} - dgg^{-1}, \quad F' = gFg^{-1}, \quad \text{and} \quad g = g(x^{\mu}) \in G. \tag{6.1.1}$$

The field satisfies the Bianchi identity

$$\begin{aligned} DF = dF + [A, F] &= d^2 A + dA \wedge A - A \wedge dA \\ &+ A \wedge dA + A^3 - dA \wedge A - A^3 = 0. \end{aligned} \tag{6.1.2}$$

In addition to the gauge potential we also introduce a Higgs field $\Phi : \mathbb{R}^{D+1} \to \mathfrak{g}$ in the adjoint representation, that is, $D\Phi = d\Phi + [A, \Phi]$. Its gauge transformation is

$$\Phi' = g\Phi g^{-1}. \tag{6.1.3}$$

We should note that the standard model of elementary particles uses a different set-up for the Higgs field. There one regards Φ as a multiplet with complex scalar fields transforming in the fundamental representation of G. In particular in the electroweak theory one takes $G = U(2)$, and the Higgs field is a complex doublet. This theory does not admit solitons. We shall see that choosing the adjoint representation (6.1.3) will allow solitons in the form of non-abelian monopoles. Solitons also play a role in supersymmetric gauge theories which we do not discuss. The reader should consult [41], [162].

Notation

Let $* : \Lambda^p \to \Lambda^{D+1-p}$ be the linear map defined by

$$*(dx^{\mu_1} \wedge \cdots \wedge dx^{\mu_p}) = \frac{\sqrt{|\det(\eta)|}}{(D+1-p)!}\varepsilon^{\mu_1 \cdots \mu_p}{}_{\mu_{p+1} \cdots \mu_{D+1}} dx^{\mu_{p+1}} \wedge \cdots \wedge dx^{\mu_{D+1}}.$$

In this chapter we shall use the letters $a, b, c, \ldots$ to denote the Lie algebra indices. If $G = SU(2)$ we choose a basis $T_a, a = 1, 2, 3$ for the Lie algebra of $SU(2)$ (anti-Hermitian, traceless 2×2 matrices) such that

$$[T_a, T_b] = -\varepsilon_{abc}T_c, \quad T_a = \frac{1}{2}i\sigma_a, \quad \text{and} \quad \mathrm{Tr}(T_a T_b) = -\frac{1}{2}\delta_{ab}.$$

[1] Note the sign difference between the inhomogeneous terms in (6.1.1) and (3.3.13). This is consistent with $A = -Ud\rho - Vd\tau$ on $\mathbb{R}^2$.

Here σ_a are the Pauli matrices, and a general group element is $g = \exp(\alpha^a T_a)$ with α^a real. The components of $D\Phi$ and F with respect to this basis are given by

$$(D_\mu \Phi)^a = \partial_\mu \Phi^a - \varepsilon^{abc} A^b_\mu \Phi^c \quad \text{and} \quad F^a_{\mu\nu} = \partial_\mu A^a_\nu - \partial_\nu A^a_\mu - \varepsilon^{abc} A^b_\mu A^c_\nu. \tag{6.1.4}$$

If $D + 1 = 4$, the dual of the field tensor is given by

$$(*F)_{\mu\nu} = (1/2)\varepsilon_{\mu\nu\alpha\beta} F^{\alpha\beta}. \tag{6.1.5}$$

A two form $F = (1/2)F_{\mu\nu} dx^\mu \wedge dx^\nu$ is called self-dual (SD) or anti-self-dual (ASD) if $*F = F$ or $*F = -F$, respectively.

In terms of differential forms

$$-\mathrm{Tr}(F \wedge *F) = -\frac{1}{2}\mathrm{Tr}(F_{\mu\nu} F^{\mu\nu}) d^4x = \frac{1}{4} F^a_{\mu\nu} F^{\mu\nu\, a} d^4x,$$

where $d^4x = \frac{1}{24}\varepsilon_{\mu\nu\alpha\beta} dx^\mu \wedge dx^\nu \wedge dx^\alpha \wedge dx^\beta$, and we have used the identities

$$\varepsilon^{\mu\nu\alpha\beta} d^4x = -dx^\mu \wedge dx^\nu \wedge dx^\alpha \wedge dx^\beta \quad \text{and} \quad \varepsilon_{\alpha\beta\rho\sigma}\varepsilon^{\mu\nu\rho\sigma} = -4\delta^{[\mu}_{[\alpha}\delta^{\nu]}_{\beta]}.$$

6.1.1 Scaling argument

Let us now examine how Derrick's scaling argument used in Section 5.3.2 applies to gauge fields. Consider the fields (A, Φ) given by a potential one-form and a scalar Higgs field, respectively, with energy functional

$$E = \int_{\mathbb{R}} d^D x [|F|^2 + |D\Phi|^2 + U(\Phi)] = E_F + E_{D\Phi} + E_U.$$

The terms like $|F|^2$ here are positive definite and correspond to energy in the $D + 1$ splitting, that is,

$$F = \frac{1}{2} B_{ij} dx^i \wedge dx^j + E_i dx^i \wedge dt \quad \text{and} \quad |F|^2 \sim -\mathrm{Tr}(B^2 + E^2) \neq -\mathrm{Tr}(F_{\mu\nu} F^{\mu\nu}).$$

(Note that the trace is negative definite for $\mathfrak{su}(n)$.) We are interested in static, finite-energy critical points of this functional. Let $A(x)$, $\Phi(x)$ be such a critical point, and let

$$\Phi_{(c)}(x) = \Phi(cx), \quad A_{(c)}(x) = cA(cx), \quad F_{(c)}(x) = c^2 F(cx), \quad \text{and}$$
$$D_{(c)}\Phi_{(c)} = cD\Phi(cx).$$

This leads to

$$E_{(c)} = \frac{1}{c^{D-4}} E_F + \frac{1}{c^{D-2}} E_{D\Phi} + \frac{1}{c^D} E_U,$$

and $dE_{(c)}/dc|_{c=1} = 0$ yields

$$(D - 4)E_F + (D - 2)E_{D\Phi} + DE_U = 0. \tag{6.1.6}$$

Therefore $E_{(c)}$ can be stationary provided that $0 \leq D \leq 4$. Derrick's scaling argument can only rule out some dimensions and configurations. It is not necessarily true that topological solitons exists if (6.1.6) holds. The following solutions can nevertheless be shown to exist:

$D = 1$. Gauged kinks generalizing solution (5.3.18).

$D = 2$. Vortices with $E_F = E_U$ in the Ginsburg–Landau model. Consult the monograph [114] for detailed discussion of these solitons.

$D = 3$. Non-abelian monopoles with $E_F = E_{D\Phi}$.

$D = 4$. Solutions are possible with $E_{D\Phi} = E_U = 0$. These are instantons in pure gauge theory.

6.1.2 Principal bundles

The mathematical formalism behind gauge theory is that of a connection on a principal bundle $\pi : P \rightarrow M$ with a structure group G. This section is not meant to be an introduction to this formalism – the definitions and proofs can be found in [43, 62, 94] – but is included for more mathematically inclined readers who want to place gauge theory in a geometric context. Other readers can skip it.

Let ω be a connection one-form with values in $\mathfrak{g}$ whose vertical component is the Maurer–Cartan one-form, and let Ω be its curvature. In local coordinates

$$\omega = \gamma^{-1} A \gamma + \gamma^{-1} d\gamma \quad \text{and} \quad \Omega = d\omega + \omega \wedge \omega = \gamma^{-1} F \gamma, \tag{6.1.7}$$

where (A, F) are the gauge potential and gauge field on M and $\gamma : M \rightarrow G$ takes values in the gauge group. The G-valued transition functions act on the fibres by left multiplication. If U and U' are two overlapping open sets in M, and $g_{UU'} = g$ is the transition function, then the local fibre coordinates γ and γ' are related by $\gamma' = g\gamma$. The connection and the curvature will be well defined in the overlap region if

$$\gamma^{-1} A \gamma + \gamma^{-1} d\gamma = \gamma'^{-1} A' \gamma' + \gamma'^{-1} d\gamma' \quad \text{and} \quad \gamma^{-1} F \gamma = \gamma'^{-1} F' \gamma'.$$

These relations hold if A, A', F, F' are related by the gauge transformations (6.1.1).

Any section $\gamma = \gamma(x)$ can be used to pull back ω and Ω to the base space, so that the pulled-back connection $A = \gamma^*(\omega)$ is the gauge potential and the pulled-back curvature $F = \gamma^*(\Omega)$ is the gauge field. The gauge transformations (6.1.1) correspond to changes of the section. If the bundle is non-trivial (e.g. the Dirac monopole), the global section does not exist and the gauge potentials can be only defined locally.

Given ω we can define the splitting of TP into the horizontal and vertical components:

$$TP = H(P) \oplus V(P),$$

where the horizontal vectors belong to the kernel of ω, that is, $H(P)$ consists of vector fields X on P such that $X \lrcorner\, \omega = 0$. The basis of the distribution $V(P)$ is given by left-invariant vector fields, and the horizontal distribution is spanned by

$$D_\mu = \frac{\partial}{\partial x^\mu} - A^a_\mu(x) R_a, \quad a = 1, \ldots, \dim \mathfrak{g}$$

where R_a are the right invariant vector fields on G such that

$$[R_a, R_b] = -f^c_{ab} R_c.$$

The curvature is the obstruction to the integrability of the horizontal distribution as

$$[D_\mu, D_\nu] = -F^a_{\mu\nu} R_a.$$

The negative sign on the RHS is consistent with

$$F = \frac{1}{2} F^a_{\mu\nu} T_a dx^\mu \wedge dx^\nu \quad \text{and} \quad A = A^a_\mu T_a dx^\mu,$$

where $[T_a, T_b] = f^c_{ab} T_c$.

6.2 Dirac monopole and flux quantization

Consider Maxwell electrodynamics as a $U(1)$-gauge theory on $\mathbb{R}^{3,1}$ with a field given by

$$F = E_i dx^i \wedge dt + \frac{1}{2} \varepsilon_{ijk} B_i dx^j \wedge dx^k.$$

The Maxwell equations with a source one-form $J = J_\mu dx^\mu$ are

$$d * F = *J \quad \text{and} \quad dF = 0.$$

If there is no source the duality $F \to *F$ corresponds to the symmetry between the electric and magnetic fields E and B.

The lack of magnetic charges is a consequence of the contractibility of the space and follows from the Bianchi identity. Allowing the $U(1)$ bundle to be defined on the complement of a point will allow magnetic charges. Introduce two coordinate patches U_+ and U_- in $\mathbb{R}^3 - \{0\}$ covering the regions $z > -\varepsilon$

and $z < \varepsilon$, respectively. Topologically $\mathbb{R}^3 - \{0\}$ is $S^2 \times \mathbb{R}$, and we can think of $U_\pm$ as a two-patch covering of the sphere. Consider the gauge potentials which are regular in the overlap region:

$$A_\pm = \frac{g}{4\pi r}\frac{1}{z \pm r}(xdy - ydx) = \frac{g}{4\pi}(\pm 1 - \cos\theta)d\phi, \quad g = \text{const.}$$

They are related by a gauge transformation

$$A_+ = A_- + \frac{g}{2\pi}d\tan^{-1}(y/x) = A_- + \frac{g}{2\pi}d\phi,$$

where

$$x = r\sin\theta\cos\phi, \quad y = r\sin\theta\sin\phi, \quad \text{and} \quad z = r\cos\theta, \quad 0 \leq \theta < \pi, 0 \leq \phi < 2\pi.$$

The field is given by $F = dA_\pm$ in $U_\pm$, so

$$F = \frac{g}{4\pi r^3}(xdy \wedge dz + ydz \wedge dx + zdx \wedge dy) = \frac{g}{4\pi}\sin\theta d\theta \wedge d\phi,$$

or

$$\mathbf{E} = 0 \quad \text{and} \quad \mathbf{B} = \frac{g\mathbf{r}}{4\pi r^3}. \tag{6.2.8}$$

The two-form F is closed but not exact.

We interpret this solution as a *magnetic monopole*, because it is analogous to the electric field of a point charge q given by

$$\mathbf{E} = \frac{q\mathbf{r}}{4\pi r^3} \quad \text{and} \quad \mathbf{B} = 0$$

(with $A = -qdt/r$). In Dirac's approach [38] the coordinate patches were not used which led to the appearance of non-physical 'string singularities' along the z-axis (at $\theta = \pi$ or $\theta = 0$). A gauge transformation moves the singular string to any other half-axis which starts at $r = 0$ and ends at ∞.

The magnetic flux through a sphere around $\mathbf{r} = 0$ is given by

$$Q = \int_{S^2} F = \int_{U_+} dA_+ + \int_{U_-} dA_- = \int_C (A_+ - A_-) = \frac{g}{2\pi}\int_C d\phi = \frac{g}{2\pi}(\text{increase of } \phi)$$

because the equator C has opposite orientations as the boundary in the two regions $U_\pm$.

Dirac has argued from the QM insight: The potential A couples to particles/fields of electric charge q via covariant derivative:

$$D\Psi = d\Psi - iqA\Psi,$$

where Ψ is a wave function, and $\Psi_+ = \exp(iqg\phi/2\pi)\Psi_-$ so that

$$D\Psi_+ = (d - iqA_+)\Psi_+ = \left(d - iqA_- - \frac{iqg}{2\pi}d\phi\right)\exp\left(\frac{iqg\phi}{2\pi}\right)\Psi_-$$

$$= \exp\left(\frac{iqg\phi}{2\pi}\right)D\Psi_- .$$

We require $\exp(igq\phi/2\pi)$ to be well defined on C so that the gauge transformations make sense, and we obtain the Dirac quantization condition

$$gq = 2\pi N, \qquad N \in \mathbb{Z}, \tag{6.2.9}$$

where g and q are any magnetic and electric charges of particles (there would be a factor $\hbar$ on the RHS if we did not set it to 1). If we accept that all electric charges are integer multiples of the electron charge e then magnetic charges g must satisfy $ge = 2\pi N$, and the minimal magnetic charge is $g = 2\pi/e$. If there existed just one magnetic monopole in the universe, we would understand electric charge quantization. Each electric charge would be an integer multiple of $2\pi/g$.

The Dirac monopole is not a soliton because of its singularity at $r = 0$. Its energy density decays like $1/r^4$, and therefore the monopole mass diverges linearly. This divergence can be regularized, and leads to a large finite mass. So far no magnetic monopoles have been detected experimentally – all magnets seem to have two poles.

6.2.1 Hopf fibration

From the the differential geometric perspective the monopole number classifies the principal $U(1)$-bundles over a two-sphere S^2. The two-sphere is homotopy equivalent to $\mathbb{R}^3 - \{0\}$ in the following sense: Let $j : S^2 \longrightarrow \mathbb{R}^3$ be the inclusion, and let $p : \mathbb{R}^3 - \{0\} \longrightarrow S^2$ be the projection $\mathbf{x} \longrightarrow \mathbf{x}/|\mathbf{x}|$. Then $p \circ j = Id$ and $j \circ p$ is homotopic to the identity map by

$$f(\mathbf{x}, t) = t\mathbf{x} - \frac{(1-t)\mathbf{x}}{|\mathbf{x}|}.$$

Transition functions for $U(1)$-bundles over S^2 are continuous maps from $S^1 \times \mathbb{R}$ to $S^1 \times \mathbb{R}$ and the bundles are classified by the degree of these maps restricted to S^1. The connection one-form (6.1.7) ω is given by

$$\omega = \begin{cases} A_+ + d\psi_+ & \text{on } U_+ \\ A_- + d\psi_- & \text{on } U_-. \end{cases}$$

This is globally defined on P which gives the transition relations $e^{i\psi_-} = ge^{i\psi_+}$. The bundle will be a manifold if the transition function is of the form $e^{(iN\phi)}$ for

$N \in \mathbb{Z}$. This gives a bundle with the first Chern number $-N$ given by

$$c_1 = -\frac{1}{2\pi}\int_{S^2} F,$$

as in general $C_1 = i\Omega/2\pi$, and the curvature is purely imaginary for the $U(1)$ principal bundle, that is, $\Omega = iF$ (see Section 6.4.1 for the definition of Chern numbers). This number (which is the negative of the monopole number) only depends on the transition function, and not on the fact that A satisfies the Maxwell equations.

The case $N = 0$ corresponds to a trivial bundle, and $N = -1$ gives the Hopf bundle $S^3 \to S^2$. We shall present a description of the Hopf bundle in terms of complex numbers. The total space of the bundle is $P = S^3 \subset \mathbb{C}^2$ which is explicitly given by

$$Z^0\overline{Z^0} + Z^1\overline{Z^1} = 1, \quad \text{where} \quad (Z^0, Z^1) \in \mathbb{C}^2.$$

Any complex line $A_0 Z^0 + A_1 Z^1 = 0$ through the origin in $\mathbb{C}^2$ intersects the three-sphere in a circle S^1. Each circle is a one-dimensional fibre over a point (A_0, A_1) in the space of complex lines in $\mathbb{C}^2$. To specify such a point one only needs a ratio of complex numbers A_0 and A_1 which is allowed to be infinite. The space of these ratios is a Riemann sphere $\mathbb{CP}^1$ (see Appendix B). As a real manifold it is diffeomorphic to S^2, with the explicit map given by (5.4.30). The projection $\pi : S^3 \to S^2$ is

$$(Z^0, Z^1) \longrightarrow \frac{Z^1}{Z^0}$$

in a patch of S^2 where $Z^0 \neq 0$. The bundle has $N = -1$ and therefore is not diffeomorphic to $S^2 \times S^1$. To see that $N = -1$ set $\lambda = Z^1/Z^0 = \tan(\theta/2)\exp(i\phi)$ and $\tilde{\lambda} = Z^0/Z^1$, so that the Riemannian metric on the sphere is

$$4\frac{|d\lambda|^2}{(1+|\lambda|^2)^2} = d\theta^2 + \sin^2\theta d\phi^2.$$

Parameterize the three-sphere by

$$Z^0 = \cos(\theta/2)e^{i\xi_0} \quad \text{and} \quad Z^1 = \sin(\theta/2)e^{i\xi_1},$$

where $(\xi_0, \xi_1, \theta) \in \mathbb{R}^2 \times S^1$. The projection is

$$\pi(\xi_0, \xi_1, \theta) = \left(\frac{\lambda + \overline{\lambda}}{1+|\lambda|^2}, \frac{\lambda - \overline{\lambda}}{i(1+|\lambda|^2)}, \frac{-1+|\lambda|^2}{1+|\lambda|^2}\right) = (\sin\theta\cos\phi, \sin\theta\sin\phi, \cos\theta),$$

where $\phi = \xi_1 - \xi_0$. We shall read off the patching function from two local sections

$$s(\lambda) = \frac{(1, \lambda)}{\sqrt{1+|\lambda|^2}} \quad \text{and} \quad \tilde{s}(\tilde{\lambda}) = \frac{(\tilde{\lambda}, 1)}{\sqrt{1+|\tilde{\lambda}|^2}},$$

where $(Z^0, Z^1)/\sqrt{|Z^0|^2 + |Z^1|^2}$ is a unit vector in $\mathbb{C}^2$ corresponding to a point in S^3. Now

$$\tilde{s}(\tilde{\lambda}) = \frac{\tilde{\lambda}(1, \lambda)}{\sqrt{1 + |\tilde{\lambda}|^2}} = \left(\frac{\lambda}{|\lambda|}\right)^{-1} s(\lambda)$$

so the transition function is $e^{-i\phi}$, and $N = -1$. In Appendix B the Chern number N has a holomorphic interpretation as the Hopf bundle is the restriction of the tautological line bundle $\mathcal{O}(-1)$ over $\mathbb{CP}^1$ to real fibres.

The family of S^1's in the Hopf bundle consists of circles (sometimes called the Clifford parallels) which twists around each other with linking number equal to one. This can be visualized by a stereographic projection of S^3 to $\mathbb{R}^3$, where the fibres of the Hopf bundle map to circles in the Euclidean three-space. This picture has motivated the early development of twistor theory [132].

An alternative description of the Hopf bundle is as the principal fibre bundle with a total space $SU(2)$, and the fibration

$$SU(2) \to SU(2)/U(1) = S^2,$$

where the $U(1)$ is identified with a subgroup of $SU(2)$ consisting of diagonal matrices. The right action of S^1 on $SU(2)$ is $(Z^0, Z^1)e^{i\alpha} = (Z^0 e^{i\alpha}, Z^1 e^{i\alpha})$, where $|Z^0|^2 + |Z^1|^2 = 1$. This action fixes the ratio $Z^0/Z^1 \in \mathbb{CP}^1 = S^2$.

6.3 Non-abelian monopoles

Choose the gauge group $G = SU(2)$ and consider the Lagrangian density

$$\mathcal{L} = -\frac{1}{4} F^a_{\mu\nu} F^{\mu\nu a} + \frac{1}{2} D_\mu \Phi^a D^\mu \Phi^a - U(\Phi) \tag{6.3.10}$$

which is gauge invariant if $U(\Phi)$ is gauge invariant. Choose

$$U(\Phi) = \frac{1}{4} c(|\Phi|^2 - v^2)^2, \quad \text{where} \quad |\Phi|^2 = \Phi^a \Phi^a, \quad v \in \mathbb{R}^+,$$

and c is a constant.

The finiteness of the energy is assured by

$$|\Phi| \longrightarrow v, \quad D_\mu \Phi \longrightarrow 0, \quad \text{and} \quad F_{\mu\nu} \longrightarrow 0, \quad \text{as} \quad r \to \infty. \tag{6.3.11}$$

The field equations in components relative to the standard basis T^a can be derived using (6.1.4):

$$(D_\nu F^{\mu\nu})^a = -\varepsilon^{abc} \Phi^b (D^\mu \Phi)^c \quad \text{and} \quad (D_\mu D^\mu \Phi)^a = -c(|\Phi|^2 - v^2)\Phi^a. \tag{6.3.12}$$

One solution is $A = 0$, $\Phi = \Phi_0 = \text{const}$, with $|\Phi_0| = v$. This ground state is not unique if $v \neq 0$, and as a consequence the gauge group is spontaneously broken to $U(1)$. This is because we need $\Phi_0 = g\Phi_0 g^{-1}$ to preserve the ground state.

The reasons for introducing the Higgs field have to do with a description of short-range interactions in gauge theory carried by massive bosons. Pure Yang–Mills (YM) fields describe massless particles, with a number of gluons given by the dimension of the gauge group. Adding a mass term of the form $m^2 Q(A)$ which is quadratic in A to the Lagrangian would necessarily break the gauge invariance. A way around this is to include an additional field Φ such that $\Phi \longrightarrow \Phi_0$ with $\Phi_0 = \text{const}$ at spatial infinity. Strictly speaking the minimum of the potential is not unique, but can be transformed to Φ_0 by a gauge transformation. Then $D\Phi_0 = A\Phi_0$, and a gauge-invariant term $|D\Phi|^2$ induces a term quadratic in A with the mass determined by the constant Φ_0.

6.3.1 Topology of monopoles

In this section we shall consider a soliton – static finite-energy solution to (6.3.12). We shall chose a gauge $A_0 = 0$, so that $D_0(f) = 0$ where f is any Lie algebra–valued field which does not depend on time. For notational convenience we shall set $v = 1$, which is always possible if $v \neq 0$ by a rescaling of Φ. Let

$$\hat{\Phi} = \frac{\Phi}{|\Phi|}.$$

The Higgs field at infinity carries all the topological information. The boundary conditions imply that

$$\hat{\Phi}_\infty = \lim_{r \to \infty} \Phi$$

defines a map from S^2_∞ (the two-sphere at the spatial infinity) to the unit two-sphere in the Lie algebra (as $|\Phi| = v$ asymptotically to ensure finite energy). This field is topologically classified by its degree N (see formula (A7)) related to the outward magnetic flux at infinity.

The non-abelian monopoles are different from the Dirac monopole in that the fields are smooth everywhere in $\mathbb{R}^3$, but there is a connection: the gauge group $SU(2)$ is broken down to $U(1)$ at infinity in $\mathbb{R}^3$ by the field $\hat{\Phi}$. A non-abelian monopole looks like a Dirac N-pole when viewed from a distance. The only difference is that now we have two long-range fields Φ and F unlike the one field F (6.2.8) for the Dirac monopole. For short distances the theory is however non-abelian which guarantees finite energy and smoothness.

The non-vanishing part of the gauge field at infinity is the $U(1)$ magnetic field:

$$B_k{}^{(m)} = \frac{1}{2}\varepsilon_{ijk} F^a_{ij} \hat{\Phi}^a.$$

To find the magnetic charge of $B^{(m)}$ note that asymptotically $D_i \hat{\Phi} = 0$ with $|\hat{\Phi}| = 1$ which yields

$$A^a_i = -\varepsilon^{abc} \partial_i \hat{\Phi}^b \hat{\Phi}^c + k_i \hat{\Phi}^a \tag{6.3.13}$$

for some vector k_i. The last term is globally well defined and gives no net flux. We shall calculate the corresponding field

$$\begin{aligned} F^a_{ij} &= \partial_i A^a_j - \partial_j A^a_i - \varepsilon^{abc} A^b_i A^c_j \\ &= 2\varepsilon^{abc} \partial_i \hat{\Phi}^b \partial_j \hat{\Phi}^c - (\varepsilon^{pqr} \partial_i \hat{\Phi}^p \partial_j \hat{\Phi}^q \hat{\Phi}^r) \hat{\Phi}^a. \end{aligned}$$

The corresponding magnetic charge is

$$\begin{aligned} Q &= \int_{S^2_\infty} \frac{1}{2} \varepsilon^{ijk} (F^a_{jk} \hat{\Phi}^a) n^i \, d^2S \\ &= \int_{S^2_\infty} \frac{1}{2} \varepsilon^{ijk} (\varepsilon^{pqr} \partial_i \hat{\Phi}^p \partial_j \hat{\Phi}^q) \hat{\Phi}^r n^k d^2S \\ &= 4\pi N, \end{aligned} \tag{6.3.14}$$

where $\mathbf{n}$ is a unit normal to S^2_∞ since the integrand is the area form on S^2_{vac} pulled back to S^2_∞, 4π is the area of the unit sphere and

$$N = \deg(\hat{\Phi}_\infty)$$

is the topological degree (A7). The integer N is called the monopole number, and 4π is the unit of magnetic charge.

6.3.2 Bogomolny–Prasad–Sommerfeld (BPS) limit

Consider the limit $c = 0$, $|\Phi_\infty| = 1$ of the field equations (6.3.12), and define the non-abelian magnetic field by

$$B_i = \frac{1}{2}\varepsilon_{ijk} F_{jk}.$$

Theorem 6.3.1 *The energy of a non-abelian magnetic monopole is bounded from below:*

$$E \geq 4\pi |N|, \tag{6.3.15}$$

where N is the topological degree of the asymptotic electromagnetic field at infinity. For positive N the bound is saturated if

$$*_3 F = D\Phi, \tag{6.3.16}$$

where $*_3 : \Lambda^k \longrightarrow \Lambda^{3-k}$ *is the three-dimensional Hodge operator given by*

$$*_3 dx^k = \frac{1}{2}\varepsilon^{ijk} dx^i \wedge dx^k.$$

Proof The energy functional is given by

$$\begin{aligned} E &= \int_{\mathbb{R}^3} \left[\frac{1}{4} F^a_{ij} F^a_{ij} + \frac{1}{2}(D_k\Phi)^a (D_k\Phi)^a\right] d^3x \\ &= \frac{1}{2}\int_{\mathbb{R}^3} \left[B^a_k B^a_k + (D_k\Phi)^a (D_k\Phi)^a\right] d^3x \\ &= \frac{1}{2}\int_{\mathbb{R}^3} (B_k - D_k\Phi)^a (B_k - D_k\Phi)^a d^3x + \int_{\mathbb{R}^3} B^a_k (D_k\Phi)^a d^3x \\ &= E_1 + E_2, \end{aligned} \tag{6.3.17}$$

where E_1 is non-negative (recall that E = 0 as $A_0 = 0$). The spatial Bianchi identity $D_{[i} F_{jk]} = 0$ is equivalent to $D_k B_k = 0$. We use this identity, Stoke's theorem, and the cyclic property of trace to rewrite the second term on the RHS as

$$\begin{aligned} E_2 &= -2\int_{\mathbb{R}^3} \mathrm{Tr}(B_k D_k \Phi) d^3x = -2\int_{\mathbb{R}^3} \mathrm{Tr}[D_k(B_k\Phi)] d^3x \\ &= -2\int_{\mathbb{R}^3} \partial_k \,\mathrm{Tr}(B_k\Phi) d^3x = \int_{S^2_\infty} B^a_k \Phi^a n^k d^2S = 4\pi N, \end{aligned}$$

where n^k is the unit outward normal to the sphere at infinity and we have identified the flux of the asymptotic electromagnetic field (6.3.14). This yields the Bogomolny bound (6.3.15) which (for a positive N) is saturated if $B_k = D_k\Phi$ or, in terms of differential forms, if (6.3.16) holds. □

Equations (6.3.16) are nine coupled non-linear PDEs known as the Bogomolny equations. They imply the static field equations as (A, Φ) is a critical point of the energy functional. For a given monopole number N the space of finite-energy solutions (6.3.16) modulo the gauge transformations (6.1.1) is $4N - 4$ dimensional up to an overall rotation [180]. In Section 8.1 we shall see that the system (6.3.16) is integrable but explicit formulae can only be found for $N = 1$ and (to some extend) for $N = 2$.

In practice the monopole number can be read off from the asymptotic behaviour of the Higgs field as

$$|\Phi| = 1 - \frac{N}{r} + O(r^{-2}), \quad \text{where} \quad r \longrightarrow \infty. \tag{6.3.18}$$

To verify this calculate the topological term E_2 in (6.3.17) on solutions to the Bogomolny equations:

$$\begin{aligned} E_2 &= \int_{\mathbb{R}^3} B_k^a (D_k\Phi)^a d^3x = \int_{\mathbb{R}^3} (D_k\Phi)^a (D_k\Phi)^a d^3x \\ &= \frac{1}{2}\int_{\mathbb{R}^3} \Delta|\Phi|^2 d^3x = \frac{1}{2}\int_{S^2_\infty} \nabla\left[1 - \frac{2N}{r} + O(r^{-2})\right] \cdot \mathbf{dS} = 4\pi N, \end{aligned}$$

where we have used the static field equations $D_k D_k \Phi = 0$ and the divergence theorem. This is in agreement with the calculation leading to (6.3.15).

- **Example.** To find explicit solutions we make the spherically symmetric ansatz:

$$\Phi^a = h(r)\frac{x^a}{r} \quad \text{and} \quad A_i{}^a = -\varepsilon^{aij}\frac{x^j}{r^2}[1 - k(r)].$$

This makes use of the isomorphism $\mathfrak{su}(2) \cong \mathbb{R}^3$ as the ansatz replaces the Lie algebra indices by the space-time indices. The Bogomolny equations reduce to a pair of ODEs:

$$\frac{dh}{dr} = r^{-2}(1 - k^2) \quad \text{and} \quad \frac{dk}{dr} = -kh.$$

Using the change of variables $H = h + r^{-1}$ and $K = k/r$ one finds the Prasad–Sommerfield solution [138]:

$$\Phi^a = \frac{x^a}{r}\left[\coth(r) - \frac{1}{r}\right] \quad \text{and} \quad A_i^a = -\varepsilon^{aij}\frac{x^j}{r^2}\left[1 - \frac{r}{\sinh(r)}\right]. \tag{6.3.19}$$

This solution has $N = 1$ which follows from (6.3.18) as

$$1 - |\Phi|^2 = \frac{1}{r} - 2e^{-r} + O(e^{-3r}).$$

Asymptotically the solution approaches the Dirac one-monopole (6.2.8) with $g = 4\pi$ and mass equal to 4π. It is the lowest energy one–monopole configuration, therefore the solution is stable.

The corresponding energy density is given by a spherically symmetric function concentrated around the origin $r = 0$, thus supporting the interpretation of the monopole as a particle located at the origin.

6.4 Yang–Mills equations and instantons

Let us start with the following:

Definition 6.4.1 *Instantons are non-singular solutions of classical equations of motion in Euclidean space whose action is finite.*

In this section we shall study instantons in pure YM theory. Our interest in such configurations is motivated by QM, where in the WKB approximation the tunnelling amplitudes are controlled by the exponentially small factor $e^{-S/\hbar}$, where S is the minimal *Euclidean* action to pass form initial to final state. In this section we use the Euclidean metric $\eta_{\mu\nu} = \text{diag}(1, 1, 1, 1)$, and the coordinates x^μ, where $\mu = 1, \ldots, 4$. The term 'instanton' is used because a solution localized in $\mathbb{R}^4$ with a Euclidean metric $d\mathbf{x}^2 + d\tau^2$ is simultaneously localized in space and in an instant of Euclidean time.

The Euclidean YM action

$$S = -\int_{\mathbb{R}^4} \text{Tr}(F \wedge *F)$$

yields the YM equations

$$D * F = 0. \tag{6.4.20}$$

Finiteness of the action is ensured by

$$F_{\mu\nu}(x) \sim O(1/r^3) \quad \text{and} \quad A_\mu(x) \sim -\partial_\mu g g^{-1} + O(1/r^2), \quad \text{as} \quad r \to \infty, \tag{6.4.21}$$

the important point being that the gauge transformation $g(x)$ needs only to be defined asymptotically, so that $g : S^3_\infty \to SU(2)$. This function can be continuously extended to $\mathbb{R}^4$ if its degree (A8) vanishes. Making another gauge transformation of A_μ at infinity will change g, but not its homotopy class.

The boundary conditions can be understood in terms of the one-point compactification $S^4 = \mathbb{R}^4 \cup \{\infty\}$, which has a metric conformally equivalent to the flat metric on $\mathbb{R}^4$. The YM equations are conformally invariant and solutions extend from $\mathbb{R}^4$ to S^4. Any smooth solution of YM equations on S^4 project stereographically to a connection on $\mathbb{R}^4$ with a curvature which vanishes at infinity with the rate (6.4.21). Uhlenbeck [165] established a converse of this result: For any finite action smooth solution A to the YM equations on $\mathbb{R}^4$ there exists a bundle over S^4 which stereographic projects to A. The proof uses the conformal invariance of the YM equations and of the Hodge operator in four dimensions. In this approach the base space S^4 is not contractible, so the principal YM bundles need not be topologically trivial (in fact they are classified by the same integer which classified the gauge equivalence classes of A at ∞ in $\mathbb{R}^4$). Let ω be a connection one-form on a principal bundle $P \longrightarrow S^4$, and let

$$f_N : S^4 - (N = \{0, 0, 0, 1\}) \quad \text{and} \quad f_S : S^4 - (S = \{0, 0, 0, -1\})$$

be stereographic projections from the north pole and south pole, respectively. The connection projects down to $(d + A_N)$ and $(d + A_S)$ on $\mathbb{R}^4$, but the pull back $(f_N)^*(A_N)$ to S^4 does not extend smoothly to N unless P is trivial, and the similar statement can be made about the pull–back of A_S by f_S. The two one-forms $(f_N)^*(A_N)$ and $(f_S)^*(A_S)$ are defined on $S^4 - \{N \cup S\}$, and related by the gauge transformation $g : S^3 \times \mathbb{R} \to SU(2) = S^3$:

$$(f_N)^*(A_N) = g(f_S)^*(A_S)g^{-1} - (dg)g^{-1}.$$

Let F_N and F_S be the gauge fields of A_N and A_S on $\mathbb{R}^4$. The gauge invariance implies that the four forms

$$(f_N)^*\mathrm{Tr}(F_N \wedge *F_N) = (f_S)^*\mathrm{Tr}(F_S \wedge *F_S)$$

agree on $S^4 - \{N \cup S\}$, and so they are equal and well defined everywhere on S^4.

6.4.1 Chern and Chern–Simons forms

Consider a pure gauge theory on $\mathbb{R}^D$. If F takes values in $\mathfrak{su}(n)$, then the Chern class [62, 175] is given by

$$C(F) = \det\left(1 + \frac{i}{2\pi}F\right) = 1 + C_1(F) + C_2(F) + \cdots,$$

where $C_p(F)$ is a $2p$-form (a polynomial in $\mathrm{Tr}(F^k)$). The pth Chern form $C_p(F)$ is gauge invariant (which is why we can work with F and not Ω defined in Section 6.1.2) and closed because the Bianchi identity implies that $\mathrm{Tr}(F^k)$ is closed for all k. From now on take $G = SU(2)$. We have

$$C_1(F) = \frac{i}{2\pi}\mathrm{Tr}(F) = 0 \quad \text{and}$$

$$C_2(F) = \frac{1}{8\pi^2}\left[\mathrm{Tr}(F \wedge F) - \mathrm{Tr}(F) \wedge \mathrm{Tr}(F)\right] = \frac{1}{8\pi^2}\mathrm{Tr}(F \wedge F).$$

Explicitly and in any dimension

$$dC_2 = \frac{1}{4\pi^2}\mathrm{Tr}(dF \wedge F) = \frac{1}{4\pi^2}\mathrm{Tr}(DF \wedge F - A \wedge F \wedge F + F \wedge A \wedge F) = 0,$$

where we used the Bianchi identity (6.1.2) and the cyclic property of the trace. Therefore, since $\mathbb{R}^D$ is contractible, $C_2 = dY_3$, where Y_3 is the so-called Chern–Simons three-form given by

$$Y_3 = \frac{1}{8\pi^2}\mathrm{Tr}(dA \wedge A + \frac{2}{3}A^3).$$

This can be verified using $\mathrm{Tr}(A^4) = 0$.

Integrating the Chern classes over manifolds of appropriate dimensions gives integer Chern numbers. If $D = 4$

$$c_2 = \int_{\mathbb{R}^4} C_2 = \frac{1}{8\pi^2}\int_{\mathbb{R}^4} dY_3 = \frac{1}{8\pi^2}\int_{\mathbb{R}^4} d\mathrm{Tr}\left(F \wedge A - \frac{1}{3}A^3\right)$$
$$= -\frac{1}{24\pi^2}\int_{S^3_\infty} \mathrm{Tr}(A^3) \in \mathbb{Z}, \tag{6.4.22}$$

since

$$A_\infty = -(dg)\, g^{-1}, \quad F_\infty = 0, \quad \text{and} \quad \frac{1}{24\pi^2}\int_{S^3_\infty} \mathrm{Tr}\left\{\left[(dg)\, g^{-1}\right]^3\right\} = \deg(g) \in \mathbb{Z},$$

where we used (A8).

We shall now explain how to understand this result from the point of view of bundles over S^4 with no boundary conditions. Cover S^4 by two hemispheres U_+ and U_- with the overlap being a cylinder $U_+ \cap U_- = S^3 \times [-\epsilon, \epsilon]$, where S^3 is the equatorial three-sphere and $[-\epsilon, \epsilon]$ is a line segment. The connection and the curvature (6.1.7) are given by

$$\omega = \begin{cases} \gamma_+{}^{-1} A_+ \gamma_+ + \gamma_+{}^{-1} d\gamma_+ & \text{on } U_+ \\ \gamma_-{}^{-1} A_- \gamma_- + \gamma_-{}^{-1} d\gamma_- & \text{on } U_- \end{cases}$$

and

$$\Omega = \begin{cases} \gamma_+{}^{-1} F_+ \gamma_+ & \text{on } U_+ \\ \gamma_-{}^{-1} F_- \gamma_- & \text{on } U_-, \end{cases}$$

where $\gamma_- = g\gamma_+$ and $(A_+, A_-), (F_+, F_-)$ are related by the gauge transformations:

$$A_- = gA_+g^{-1} - (dg)g^{-1} \quad \text{and} \quad F_- = gF_+g^{-1}, \quad \text{where } g = g(x^\mu) \in G.$$

We claim that the second Chern number characterizing the bundle is given by (6.4.22), albeit with a different interpretation of g. We shall use the fact that both hemispheres $U_\pm$ are separately contractible, so Chern–Simons three-form can be introduced on each of them:

$$k = -\frac{1}{8\pi^2}\int_{S^4} \mathrm{Tr}(\Omega \wedge \Omega) = -\frac{1}{8\pi^2}\left[\int_{U_+} \mathrm{Tr}(F_+ \wedge F_+) + \int_{U_-} \mathrm{Tr}(F_- \wedge F_-)\right]$$
$$= -\frac{1}{8\pi^2}\int_{S^3}\left\{\mathrm{Tr}\left[F_+ \wedge A_+ - \frac{1}{3}(A_+)^3\right] - \mathrm{Tr}\left[F_- \wedge A_- - \frac{1}{3}(A_-)^3\right]\right\}$$

$$= -\frac{1}{8\pi^2}\int_{S^3} \mathrm{Tr}\left\{\frac{1}{3}\left[(dg)g^{-1}\right]^3 - d\left[A_+ \wedge (dg)g^{-1}\right]\right\}$$
$$= -\frac{1}{24\pi^2}\int_{S^3} \mathrm{Tr}\left\{\left[(dg)g^{-1}\right]^3\right\}$$

in agreement with (6.4.22). In this calculation we interpret g as the transition function of the bundle $P \to S^4$. The bundle is trivial when restricted to U_+ or U_-, and the non-triviality arises when the hemispheres are patched together using $g : U_+ \cap U_- \to SU(2)$. In the case of the one instanton solution $k = 1$ the total space of P is S^7 [163]. The argument we have just given readily generalizes to show that principal G bundles over S^n are classified by elements of the homotopy group $\pi_{n-1}(G)$.

6.4.2 Minimal action solutions and the anti-self-duality condition

Theorem 6.4.2 *The YM action S within a given topological sector*

$$c_2 = \frac{1}{8\pi^2}\int_{\mathbb{R}^4} Tr(F \wedge F) > 0$$

is bounded from below by $8\pi^2 c_2$. The bound is saturated if the anti-self-dual Yang–Mills (ASDYM) equations

$$F = - * F \quad or \quad F_{12} = -F_{34}, \quad F_{13} = -F_{42}, \quad F_{14} = -F_{23} \tag{6.4.23}$$

hold.

Proof Note that $F \wedge F = *F \wedge *F$ and calculate

$$S = -\frac{1}{2}\int_{\mathbb{R}^4} \mathrm{Tr}[(F + *F) \wedge (F + *F)] + \int_{\mathbb{R}^4} \mathrm{Tr}(F \wedge F)$$
$$= -\frac{1}{2}\int_{\mathbb{R}^4} \mathrm{Tr}[(F + *F) \wedge *(F + *F)] + 8\pi^2 c_2$$
$$\geq 8\pi^2 c_2, \tag{6.4.24}$$

as the first integral is non-negative. This gives the Bogomolny bound for S, with the equality iff the field tensor is ASD and the ASDYM equations (6.4.23) hold. □

A similar calculation with $c_2 \leq 0$ would lead to the SD equations $F = *F$. If the YM connection satisfies the SD or ASD conditions, then the YM equations (6.4.20) are satisfied by virtue of the Bianchi identity (6.1.2). Changing the sign of the volume form (reversing the orientation) interchanges ASD and SD fields.

The ASDYM fields in $\mathbb{R}^4$ which satisfy the boundary conditions (6.4.21) are called instantons, and $k = -c_2$ is called the instanton number. There is an $8|k| - 3$-dimensional manifold of instantons of charge k [7].

6.4.3 Ansatz for ASD fields

In this section we shall present an ansatz, originally due to Corrigan and Fairlie [34], which reduces the non-linear ASDYM equations to the Laplace equation in $\mathbb{R}^4$. The ansatz will lead to explicit instantons with arbitrarily large instanton number.

Introduce the antisymmetric objects $\sigma_{\mu\nu}$ on $\mathbb{R}^4$ by

$$\sigma_{ab} = \varepsilon_{abc} T_c, \quad \sigma_{a4} = -\sigma_{4a} = T_a,$$

where $T_a, a = 1, 2, 3$, form a basis of $\mathfrak{su}(2)$, that is, $[T_a, T_b] = -\varepsilon_{abc} T_c$. Note $\sigma_{12} = \sigma_{34}$ etc. One can check the relation

$$[\sigma_{\mu\kappa}, \sigma_{\nu\lambda}] = -\delta_{\mu\nu}\sigma_{\kappa\lambda} + \delta_{\mu\lambda}\sigma_{\kappa\nu} + \delta_{\kappa\nu}\sigma_{\mu\lambda} - \delta_{\kappa\lambda}\sigma_{\mu\nu}$$

which implies the identities

$$\sigma_{\mu\nu}\sigma_{\mu\kappa} = -\frac{3}{4}1\delta_{\nu\kappa} - \sigma_{\nu\kappa}, \quad \sigma_{\mu\nu}\sigma_{\mu\nu} = -31. \tag{6.4.25}$$

Note that

$$\frac{1}{2}\varepsilon_{\mu\nu\kappa\lambda}\sigma_{\kappa\lambda} = \sigma_{\mu\nu}$$

so the objects $\sigma_{\mu\nu}$ are SD. This has the following interpretation. The Lie algebra $\mathfrak{so}(4) = \mathfrak{su}(2) \oplus \mathfrak{su}(2)$ regarded as a six-dimensional vector space is isomorphic to the space Λ^2 of two-forms on $\mathbb{R}^4$. The forms $\sigma_{\mu\nu}$ select the three-dimensional space of SD two-forms Λ^2_+ from the six-dimensional space $\Lambda^2 = \Lambda^2_+ \oplus \Lambda^2_-$ and project it onto $\mathfrak{su}(2)$. The three-dimensional vector spaces $\mathfrak{su}(2)$ and Λ^2_+ are isomorphic.

Proposition 6.4.3 *Let $\rho : \mathbb{R}^4 \to \mathbb{R}$ be a function. The potential*

$$A = \sigma_{\mu\nu}\frac{\partial_\nu \rho}{\rho} dx^\mu \tag{6.4.26}$$

satisfies the ASDYM equations (6.4.23) *iff the Laplace equation*

$$\Box\rho = 0 \tag{6.4.27}$$

holds, where $\Box = \eta^{\mu\nu}\partial_\mu\partial_\nu$.

Proof The YM field corresponding to (6.4.26) will be ASD iff

$$\frac{1}{2}\sigma_{\mu\nu}F_{\mu\nu} = \sigma_{\mu\nu}(\partial_\mu A_\nu + A_\mu A_\nu) = 0,$$

which follows from

$$\sigma_{12}F_{12} + \cdots + \sigma_{34}F_{34} + \cdots = \sigma_{12}(F_{12} + F_{34}) + \cdots .$$

Now compute

$$\begin{aligned}
&\sigma_{\mu\nu}(\partial_\mu A_\nu + A_\mu A_\nu) \\
&= \sigma_{\mu\nu}\left(\sigma_{\nu\lambda}\partial_\mu \frac{\partial_\lambda \rho}{\rho} + \sigma_{\mu\lambda}\sigma_{\nu\kappa}\frac{\partial_\lambda \rho \partial_\kappa \rho}{\rho^2}\right) \\
&= \left(\frac{3}{4}\delta_{\mu\lambda} + \sigma_{\mu\lambda}\right)\left(\frac{\partial_\mu \partial_\lambda \rho}{\rho} - \frac{\partial_\lambda \rho \partial_\mu \rho}{\rho^2}\right) + \left(-\frac{3}{4}\delta_{\nu\lambda} - \sigma_{\nu\lambda}\right)\sigma_{\nu\kappa}\frac{\partial_\lambda \rho \partial_\kappa \rho}{\rho^2} \\
&= \frac{3}{4}\frac{\partial_\mu \partial_\mu \rho}{\rho}.
\end{aligned}$$

So F is ASD iff $\partial_\mu \partial_\mu \rho = 0$ which is (6.4.27). □

The basic solution $\rho = r^{-2}$ is a pure gauge and gives $F = 0$. The Jackiw–Nohl–Rebbi (JNR) N-instanton solutions [91] are obtained by superposing $(N+1)$ fundamental solutions to the Laplace equation:

$$\rho = \sum_{p=0}^{N} \frac{\lambda_p}{|x - x_p|^2}. \tag{6.4.28}$$

This family of instantons depends on $5N+4$ parameters consisting of the choice of $N+1$ points in $\mathbb{R}^4$ or S^4 as the overall scaling of ρ has no effect. The instanton number is N, because each of the $(N+1)$ singularities can be removed by a gauge transformation $g_p : (S^3)_p \to SU(2)$ of degree one defined on a sphere S^3_p surrounding the point x_p and no other singularities, and one unit is taken away because of the asymptotic behaviour of the gauge field.

The general N-instanton solution is known to depend on $8N - 3$ parameters, and the JNR ansatz gives all solutions for $N = 1$ and 2. In the limiting case, where one of the fixed points is at ∞ we recover the t'Hooft ansatz [156]:

$$\rho = 1 + \sum_{p=1}^{N} \frac{\lambda_p}{|x - x_p|^2}.$$

This depends on $5N$ parameters and, unlike (6.4.28), is not conformally invariant, as we have selected a point in $\mathbb{R}^4$. Although the gauge potential has singularities in both JNR and t'Hooft solutions, the field itself is regular.

6.4.4 Gradient flow and classical mechanics

This section is intended as a physical motivation for studying instanton solutions of the YM equations in the Euclidean signature. We shall follow the treatment of [40] and present a reformulation of Euclidean YM theory as classical mechanics on an infinite-dimensional manifold of connections on $\mathbb{R}^3$.

Consider a trajectory $q^k(t)$ of a particle with unit mass moving in a potential $V(q)$ in $\mathbb{R}^D$, with coordinates q^k, $k = 1, \ldots, D$. The equations of motion are

$$\ddot{q}^k = -\frac{\partial V(q)}{\partial q^k}. \tag{6.4.29}$$

Assume that $V = -|\nabla W|^2/2$, where $W = W(q)$. Then any solution to the first-order gradient-flow equations

$$\dot{q}^k = \frac{\partial W}{\partial q^k}$$

is automatically a solution to (6.4.29) as

$$\ddot{q}^k = \frac{d}{dt}\left(\frac{\partial W}{\partial q^k}\right) = \frac{\partial^2 W}{\partial q^k \partial q^j}\frac{\partial W}{\partial q^j} = \frac{\partial}{\partial q^k}\left(\frac{1}{2}|\nabla W|^2\right).$$

The total energy of these special solutions is

$$\frac{1}{2}|\nabla W|^2 + V(q) = 0,$$

and adding a constant E to the potential generalizes this to

$$\frac{1}{2}|\nabla W|^2 + V(q) = E. \tag{6.4.30}$$

Now consider a corresponding QM problem governed by the Schrödinger equation

$$-\frac{\hbar^2}{2}\nabla^2\Psi + V\Psi = E\Psi,$$

where $\Psi(q)$ is the complex wave function. Represent the wave function as

$$\Psi = a(q)e^{iW(q)/\hbar},$$

where $a(q)$ and $W(q)$ are real valued. In the WKB approximation one analyses the leading terms in the $\hbar$ expansion of the Schrödinger equation. The lowest order $\hbar$ coincides with the gradient flow (6.4.30). In the classically forbidden region $V > E$ the appropriate asymptotic form of the wave function is

$$\Psi = a(q)e^{W(q)/\hbar},$$

and one finds that the quantum system is approximated by a classical motion along the gradient lines with a reversed potential, that is,

$$\frac{1}{2}|\nabla W|^2 = V(q) - E. \tag{6.4.31}$$

Thus the QM tunnelling between two minima q_0 and q_1 of V separated by an energy barrier is governed by a classical gradient-flow motion with a reversed potential.

A far-reaching extrapolation of this example is that solutions of classical equations of motion with a reversed potential (or equivalently in imaginary time) are relevant in quantum field theory. In particular the reversed gradient-flow trajectories approximate the QM tunnelling between topologically inequivalent classical vacua.

Consider the YM equation over a 'space-time' $\mathbb{R}^3 \times \mathbb{R}$, with its Euclidean metric. Let $Y = \mathbb{R}^3$, and let t be a local coordinate on $\mathbb{R}$. The ASDYM equations (6.4.23) can be considered as an evolution equation for a one-parameter family of connections $\mathbf{A}(t)$ on Y. Let $A = A_4 dt + \mathbf{A}$. In a gauge where $A_4 = 0$ we have

$$F_{4i} = \frac{\partial A_i}{\partial t}$$

and the ASD equations (6.4.23) become

$$\frac{\partial A_i}{\partial t} = \frac{1}{2}\varepsilon_{ijk} F_{jk}$$

or

$$\frac{d\mathbf{A}(t)}{dt} = *_3 \mathbf{F}[\mathbf{A}(t)], \tag{6.4.32}$$

where $\mathbf{F}[\mathbf{A}(t)]$ is the curvature of a one-parameter family of connections $\mathbf{A}(t)$ on Y, and $*_3$ is the Hodge operator on Y related to the Hodge operator on $\mathbb{R}^4$ by $*_4(dt \wedge \phi) = *_3\phi$ for any one-form ϕ on Y. These equations can be further rewritten in a gradient-flow form

$$\frac{d\mathbf{A}_i}{dt} = \frac{\delta W[\mathbf{A}]}{\delta \mathbf{A}_i},$$

where

$$W[\mathbf{A}] = \int_Y \mathrm{Tr}(\mathbf{A} \wedge d\mathbf{A} + \frac{2}{3}\mathbf{A} \wedge \mathbf{A} \wedge \mathbf{A})$$

is the Chern–Simons functional in three dimensions.

Now consider the Lorentzian YM equations in the temporal gauge. These equations can be formally regarded as the motion of a particle in an infinite-dimensional space of connections on Y because the Lorentzian YM Lagrangian is

$$\int_{\mathbb{R}} \left(\frac{1}{2}||\dot{\mathbf{A}}||^2 - V[\mathbf{A}]\right) dt,$$

where the potential $V[\mathbf{A}]$ is given by the magnetic part of the YM curvature, that is,

$$||\dot{\mathbf{A}}||^2 = \int_{\mathbb{R}^3} \mathrm{Tr}\left(\frac{d\mathbf{A}_i}{dt}\frac{d\mathbf{A}_i}{dt}\right) d^3x \quad \text{and}$$

$$V[\mathbf{A}] = \frac{1}{4}\int_{\mathbb{R}^3} \mathrm{Tr}(\mathbf{F}_{ij}\mathbf{F}_{ij})d^3x = \frac{1}{2}\int_{\mathbb{R}^3} \mathrm{Tr}\left(\frac{\delta W}{\delta \mathbf{A}_i}\frac{\delta W}{\delta \mathbf{A}_i}\right) d^3x = \frac{1}{2}\left|\frac{\delta W}{\delta \mathbf{A}}\right|^2.$$

The analogy with classical mechanics is achieved by making the following formal replacements:

$$\mathbb{R}^D \longrightarrow \text{space of connections on } Y$$

$$|q|^2 = q^k q^k \longrightarrow ||\mathbf{A}||^2 = \int_Y \mathrm{Tr}(\mathbf{A}_i\mathbf{A}_i)d^3x$$

$$\nabla \longrightarrow \frac{\delta}{\delta \mathbf{A}}.$$

The Euclidean YM equations correspond to a motion with a reversed potential and the gradient lines (6.4.32) of $W[\mathbf{A}]$ are the YM instantons – finite action solutions to the Euclidean YM equations. The YM quantum field theory can be regarded as a QM on the space of connections on Y, and the QM tunnelling takes place between different flat connections on Y.

Exercises

1. Derive the Yang–Mills–Higgs equations of motion (6.3.12) from the Lagrangian (6.3.10).
2. Show that in $SU(2)$ Yang–Mills–Higgs theory the general solution to the equation $D_i\hat{\Phi} = 0$ with $|\hat{\Phi}| = 1$ is

$$A_i^a = -\varepsilon^{abc}\partial_i\hat{\Phi}^b\hat{\Phi}^c + k_i\hat{\Phi}^a$$

for some k_i, and calculate the gauge field corresponding to this potential. What can you deduce about the solution of the equation $D_i\Phi = 0$?

[Hint: Write $\Phi = |\Phi|\hat{\Phi}$ and use the covariant Leibniz rule].

3. The Higgs field $\hat{\Phi}$ at infinity defines a map from S^2 to S^2. In polar coordinates the asymptotic magnetic field has non-zero components:

$$F_{\theta\phi} = \varepsilon_{abc}\partial_\theta\hat{\Phi}^a\partial_\phi\hat{\Phi}^b\hat{\Phi}^c.$$

By writing

$$\hat{\Phi} = (\sin\nu\cos\mu, \sin\nu\sin\mu, \cos\nu),$$

where $\nu = \nu(\theta,\phi)$ and $\mu = \mu(\theta,\phi)$ show that the magnetic charge satisfies

$$g = \int_{S^2} F_{\theta\phi}d\theta d\phi = 4\pi\deg(\hat{\Phi}).$$

4. Make the ansatz

$$\Phi^a = h(r)\frac{x^a}{r} \quad \text{and} \quad A_i{}^a = -\varepsilon^{aij}\frac{x^j}{r^2}[1 - k(r)]$$

and show that the Bogomolny equations for the non-abelian magnetic monopole reduce to

$$h' = r^{-2}(1 - k^2) \quad \text{and} \quad k' = -kh.$$

Use the change of variables $H = h + r^{-1}$ and $K = k/r$ to find the one-monopole solution (6.3.19).

5. Derive the pure $SU(2)$ YM theory on $\mathbb{R}^4$ from the action. Let $A_\mu(x)$ be a solution to these equations. Show that $\widetilde{A}_\mu(x) = cA_\mu(cx)$ is also a solution and that it has the same action.
6. Show that any two-form F in four dimensions satisfies $F \wedge F = *F \wedge *F$.
7. Let A be a one-form gauge potential with values in $\mathfrak{su}(2)$, and let F be its curvature. Verify that $\mathrm{Tr}(A)$, $\mathrm{Tr}(A \wedge A)$, $\mathrm{Tr}(A \wedge A \wedge A \wedge A)$, and $\mathrm{Tr}(F)$ all vanish.
8. Show that the harmonic function $\rho = r^{-2}$ in the ansatz (6.4.26) gives a pure gauge potential and implies $F = 0$.

   The ansatz (6.4.26) with the harmonic function $\rho = 1 + r^{-2}$ determines a one-instanton solution. Use the explicit integration to find the second Chern number of the corresponding bundle.
9. Consider the map $g : S^3 \to SU(2)$ defined by

$$g(x_1, x_2, x_3, x_4) = x_4 1 + i(x_1\sigma_1 + x_2\sigma_2 + x_3\sigma_3),$$

where σ_i are Pauli matrices and $x_1^2 + x_2^2 + x_3^3 + x_4^2 = 1$ and find its degree. By calculating $\mathrm{Tr}\{[(dg)\,g^{-1}]^3\}$ at the point on S^3 where $x_4 = 1$, or otherwise deduce that the formula

$$\deg(g) = \frac{1}{24\pi^2}\int_{S^3} \mathrm{Tr}\left\{[(dg)\,g^{-1}]^3\right\}$$

is correctly normalized.

7 Integrability of ASDYM and twistor theory

The ASDYM equations played an important role in the last chapter because of their connection with the YM instantons. In this chapter we shall explore the integrability of these equations using the twistor methods. The twistor transform described in Section 7.2 is a far reaching generalization of the inverse scattering transform studied in Chapter 2. All local solutions to the ASDYM equations will be parameterized by certain holomorphic vector bundles over a three-dimensional complex manifold called the twistor space. Some solutions to ASDYM can be written down explicitly as the equations can be reduced to a linear problem. The class (6.4.26) is one example. While one cannot hope to write the most general solution in terms of 'known' functions, the twistor methods will allow to reduce the problem to a number of algebraic operations like the Riemann–Hilbert factorization.

We shall start by introducing the Lax pair for the ASDYM equations, as it plays a pivotal role in the twistor correspondence.

7.1 Lax pair

In this chapter we shall consider complex solutions to the ASDYM equations on the complexified Minkowski space. Once the integrability of ASDYM is understood in this setting, the reality conditions can be imposed.

Consider the complexified Minkowski space $M_{\mathbb{C}} = \mathbb{C}^4$ with coordinates $w, z, \tilde{w}, \tilde{z}$, and the metric

$$ds^2 = 2(dzd\tilde{z} - dwd\tilde{w}). \tag{7.1.1}$$

The signature of the metric in $\mathbb{C}^4$ is not well defined, as it can be changed by a complex coordinate transformation.

Let $x^1, x^2, x^3, x^4 \in \mathbb{R}$. There are three different reality conditions one can impose on $M_{\mathbb{C}}$ which lead to $\mathbb{R}^{p,q}$, that is real flat metrics of signature (p, q) with $p + q = 4$ on $\mathbb{R}^4$.

- Euclidean slice

$$z = \frac{x^1 + ix^4}{\sqrt{2}}, \quad \tilde{z} = \bar{z}, \quad w = -\frac{x^2 + ix^3}{\sqrt{2}}, \quad \text{and} \quad \tilde{w} = -\bar{w}.$$

- Lorentzian slice

$$z = \frac{x^1 + x^4}{\sqrt{2}}, \quad \tilde{z} = \frac{x^1 - x^4}{\sqrt{2}}, \quad w = -\frac{x^2 + ix^3}{\sqrt{2}}, \quad \text{and} \quad \tilde{w} = \bar{w}.$$

In the context of integrable systems it is also interesting to consider the neutral reality conditions resulting in a real metric in $(2, 2)$ signature. There are two inequivalent ways to do it:

- Neutral slice (a). Take $z, \tilde{z}, w,$ and $\tilde{w} \in \mathbb{R}$.
- Neutral slice (b)

$$z = \frac{x^1 + ix^4}{\sqrt{2}}, \quad \tilde{z} = \bar{z}, \quad w = -\frac{x^2 + ix^3}{\sqrt{2}}, \quad \text{and} \quad \tilde{w} = \bar{w}.$$

Choose the orientation given by the volume form

$$\text{vol} = dw \wedge d\tilde{w} \wedge dz \wedge d\tilde{z}.$$

The two-forms

$$\omega_1 = dw \wedge dz, \quad \omega_2 = dw \wedge d\tilde{w} - dz \wedge d\tilde{z}, \quad \text{and} \quad \omega_3 = d\tilde{z} \wedge d\tilde{w} \tag{7.1.2}$$

span the space of SD two-forms. We write $D_w = \partial_w + A_w$, etc. The ASD condition (6.4.23) becomes $F \wedge \omega_i = 0$, or

$$F_{wz} = 0, \quad F_{w\tilde{w}} - F_{z\tilde{z}}, \quad \text{and} \quad F_{\tilde{w}\tilde{z}} = 0. \tag{7.1.3}$$

The ASDYM equations arise as the compatibility condition for an overdetermined linear system. This is an important concept which underlies the integrability of the ASDYM (and other equations). Let us motivate it with an example which is essentially the zero-curvature representation (3.3.10) presented in a gauge-theoretic context.

- **Example.** Let A_x, A_y be $\mathfrak{gl}(2, \mathbb{R})$-valued functions on $\mathbb{R}^2$ which depend on (x, y). Assume that we want to find a two-component vector v depending on (x, y) which satisfies

$$D_x v := \partial_x v + A_x v = 0 \quad \text{and} \quad D_y v := \partial_y v + A_y v = 0. \tag{7.1.4}$$

This is an overdetermined system as there are twice as many equations as unknowns (the general discussion of overdetermined systems and their solutions is given in Appendix C). The compatibility conditions coming from the Frobenius theorem (Theorem C.2.5) can be obtained by cross-differentiating

$$\partial_y \partial_x v - \partial_x \partial_y v = -\partial_y (A_x v) + \partial_x (A_y v) = (\partial_x A_y - \partial_y A_x + [A_x, A_y])v = 0$$

as the partial derivatives commute. Therefore the linear system (7.1.4) is consistent iff the nonlinear equation

$$\partial_x A_y - \partial_y A_x + [A_x, A_y] = 0 \tag{7.1.5}$$

holds. This is just the flatness $F = 0$ of the connection $A = A_x dx + A_y dy$, and we could have obtained this result directly by commuting the covariant derivatives

$$F_{xy} := [D_x, D_y] = 0$$

in (7.1.4). Let us assume that $F_{xy} = 0$, and let g be a fundamental matrix solution to (7.1.4), that is, a matrix whose columns are two linearly independent vectors satisfying (7.1.4). Then multiplying (7.1.4) by g^{-1} yields the general solution to (7.1.5)

$$A_x = -(\partial_x g)g^{-1} \quad \text{and} \quad A_y = -(\partial_y g)g^{-1},$$

and $A = -(dg)g^{-1}$ is a pure gauge. Using the Jacobi identity we could show that the calculation yields the same result (7.1.5) if solutions of (7.1.4) are in the adjoint representation.

In Section 3.3 we have seen that many non-linear integrable equations admit a zero-curvature representation analogous to (7.1.4). To make the whole picture non-trivial one needs to introduce a parameter into the picture. In the case of ASDYM equations one proceeds as follows: Consider the pair of operators

$$L = D_{\tilde{z}} - \lambda D_w \quad \text{and} \quad M = D_{\tilde{w}} - \lambda D_z \tag{7.1.6}$$

which commute for every value of the complex spectral parameter $\lambda \in \mathbb{CP}^1$ as a consequence of ASDYM

$$[L, M] = F_{\tilde{z}\tilde{w}} - \lambda(F_{w\tilde{w}} - F_{z\tilde{z}}) + \lambda^2 F_{wz} = 0.$$

Therefore the ASDYM equations arise as the compatibility condition for an overdetermined linear system

$$L\Psi = 0 \quad \text{and} \quad M\Psi = 0,$$

where $\Psi = \Psi(w, z, \tilde{w}, \tilde{z}, \lambda)$ is the fundamental (matrix) solution. A pair of differential operators like (7.1.6) is called a Lax pair. Another terminology used in Section 3.3 and due to Zaharov and Shabat is the 'zero-curvature representation'. This encapsulates the geometric content of (7.1.6) but is not appropriate if L, M are operators of higher order. The existence of a Lax pair with a spectral parameter seems to be the key, if not the defining property of integrable non-linear PDEs.

The representation (7.1.6) can be an effective method of finding solutions if we know $\Psi(x^\mu, \lambda)$ in the first place. This is because the linear system

$$-(\partial_{\tilde{z}}\Psi - \lambda\partial_w\Psi)\Psi^{-1} = A_{\tilde{z}} - \lambda A_w \quad \text{and} \quad -(\partial_{\tilde{w}}\Psi - \lambda\partial_z\Psi)\Psi^{-1} = A_{\tilde{w}} - \lambda A_z, \tag{7.1.7}$$

allows to read off the components of A from the LHS. An example of this procedure will be given in Section 8.2.1.

7.1.1 Geometric interpretation

The vectors

$$l = \partial_{\tilde{z}} - \lambda\partial_w \quad \text{and} \quad m = \partial_{\tilde{w}} - \lambda\partial_z \tag{7.1.8}$$

span a totally null plane in $M_\mathbb{C}$ for each value of λ in the sense that

$$\eta(l, m) = \eta(l, l) = \eta(m, m) = 0$$

where η is the metric. There are two types of totally null planes. The classification is based on a two-form

$$\omega = \text{vol}(l, m, \ldots, \ldots),$$

where l and m are the spanning vectors. This form must be SD or ASD in the sense of (6.1.5) which follows from contracting $\varepsilon_{\mu\nu\alpha\beta}$ with $\omega_{\mu\nu} = l_{[\mu}m_{\nu]}$. More precisely, $*\omega$ is also annihilated by l, m and so is proportional to ω. The proportionality constant must be an eigenvalue of the Hodge operator regarded as a linear map on $\Lambda^2(\mathbb{C}^4)$, that is, ± 1. Therefore each null plane is SD or ASD. If l and m are given by (7.1.8) then

$$\omega(\lambda) = \omega_1 + \lambda\omega_2 + \lambda^2\omega_3 \tag{7.1.9}$$

is a linear combination of the three SD two-forms (7.1.2). The Lax pair (7.1.6) can be expressed as

$$L = l + l \lrcorner A \quad \text{and} \quad M = m + m \lrcorner A.$$

The Lax characterization of the ASDYM condition can now be summarized in the following result.

Proposition 7.1.1 *The ASDYM condition* $[L, M] = 0$ *on a one-form* $A : \mathbb{C}^4 \to \mathfrak{g} \otimes \Lambda^1$ *is equivalent to the vanishing of the YM curvature* $F = dA + A \wedge A$ *on each null SD two-plane.*

Proof This follows from the fact that $[L, M] = 0$ is equivalent to $F(l, m) = 0$, or

$$F \wedge \omega(\lambda) = 0.$$

□

This observation suggests the closer study of the space of all SD null two-planes in the complexified Minkowski space, and underlies the twistor approach to the ASDYM equations. We shall study this subject in the next section.

7.2 Twistor correspondence

7.2.1 History and motivation

Twistor methods appear in various part of this book purely as a tool in solving non-linear DEs. The original motivation behind twistor theory was rather different and this section serves as a historical introduction to the subject. It does not contain detailed proofs and readers interested in the applications of twistor theory to ASDYM and other equations may skip this section at the first reading and go directly to Section 7.2.2.

Twistor theory was created by Roger Penrose [129] in 1967. The original motivation was to unify general relativity and quantum mechanics in a non-local theory based on complex numbers. Twistor theory is based on projective geometry and as such has its roots in the nineteenth century Klein correspondence. It can also be traced back to other areas of mathematics. One such area is a subject now known as integral geometry and can be exemplified by following construction.

7.2.1.1 *John transform*

Let $f : \mathbb{R}^3 \longrightarrow \mathbb{R}$ be a smooth function with suitable decay conditions at ∞ and let $L \subset \mathbb{R}^3$ be an oriented line. Define a function on the space of oriented lines in $\mathbb{R}^3$ by $v(L) := \int_L f$ or

$$v(w, z, \tilde{w}, \tilde{z}) = \int_{-\infty}^{\infty} f(w + s\tilde{z}, z + s\tilde{w}, s)ds \tag{7.2.10}$$

where the real numbers $(w, z, \tilde{w}, \tilde{z})$ parameterize the four-dimensional space M of oriented lines in $\mathbb{R}^3$. (Note that this parameterization misses out the lines parallel to the plane x_3 = const. The whole construction can be done invariantly without choosing any parameterization, but here we choose the explicit approach for clarity.) The space of oriented lines is four-dimensional, and 4 > 3 so expect one condition on v. Differentiating under the integral sign yields the wave equation in the neutral signature:

$$\frac{\partial^2 v}{\partial w \partial \tilde{w}} - \frac{\partial^2 v}{\partial z \partial \tilde{z}} = 0,$$

and John has shown [92] that all smooth solutions to this equation arise from some function on $\mathbb{R}^3$. This is a feature of twistor theory: an unconstrained function on twistor space (which in this case is identified with $\mathbb{R}^3$) yields a solution to a differential equation on space-time (in this case locally $\mathbb{R}^4$ with a metric of (2, 2) signature):

7.2.1.2 *Penrose transform*

In 1969 Penrose gave a formula for solutions to wave equation in Minkowski space [130]:

$$v(x, y, \zeta, t) = \frac{1}{2\pi i} \oint_{\Gamma \subset \mathbb{CP}^1} f(-(x + iy) + \lambda(t - \zeta), (t + \zeta) + \lambda(-x + iy), \lambda) d\lambda. \tag{7.2.11}$$

Here $\Gamma \subset \mathbb{CP}^1$ is a closed contour and the function f is holomorphic on $\mathbb{CP}^1$ except some number of poles. Differentiating the RHS verifies that

$$\frac{\partial^2 v}{\partial t^2} - \frac{\partial^2 v}{\partial x^2} - \frac{\partial^2 v}{\partial y^2} - \frac{\partial^2 v}{\partial \zeta^2} = 0.$$

Despite the superficial similarities the Penrose formula is mathematically much more sophisticated than John's formula (7.2.10). One could modify a contour and add a holomorphic function inside the contour to f without changing the solution v. The proper description uses sheaf cohomology which considers equivalence classes of functions and contours.

7.2.1.3 *Twistor programme*

Penrose's formula (7.2.11) gives real solutions to the wave equation in Minkowski space from holomorphic functions of three arguments. According to the twistor philosophy this appearance of complex numbers should be understood at a fundamental, rather than technical, level. In quantum physics the complex numbers are regarded as fundamental: the complex wave function is an element of a complex Hilbert space. In twistor theory Penrose aimed to bring the classical physics at the equal footing, where the complex numbers play a role from the start. This already takes place in special relativity, where the complex numbers appear on the celestial sphere visible to an observer on a night sky. This is a t = const section of observer's past null cone.

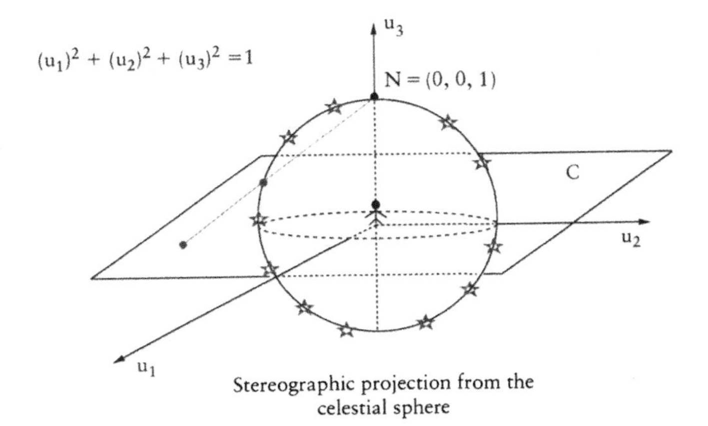

Stereographic projection from the celestial sphere

The two-dimensional sphere is the simplest example of a non-trivial complex manifold (see Appendix B for more details). Stereographic projection from the north pole $(0, 0, 1)$ gives a complex coordinate

$$\lambda = \frac{u_1 + iu_2}{1 - u_3}.$$

Projecting from the south pole $(0, 0, -1)$ gives another coordinate

$$\tilde{\lambda} = \frac{u_1 - iu_2}{1 + u_3}.$$

On the overlap $\tilde{\lambda} = 1/\lambda$. Thus the transition function is holomorphic and this makes S^2 into a complex manifold $\mathbb{CP}^1$ (Riemann sphere). The double-covering $SL(2, \mathbb{C}) \xrightarrow{2:1} SO(3, 1)$ can be understood in this context. If world-lines of two observers travelling with relative constant velocity intersect at a point in space-time, the celestial spheres these observers see are related by a Möbius transformation

$$\lambda \to \frac{\alpha\lambda + \beta}{\gamma\lambda + \delta},$$

where the unit-determinant matrix

$$\begin{pmatrix} \alpha & \beta \\ \gamma & \delta \end{pmatrix} \in SL(2, \mathbb{C})$$

corresponds to the Lorentz transformation relating the two observers.

The celestial sphere is a past light cone of an observer O which consist of light rays through an event O at a given moment. In the twistor approach the light rays are regarded as more fundamental than events in space-time. The five-dimensional space of light rays $\mathcal{PN}$ in the Minkowski space is a hypersurface in a three-dimensional complex manifold $\mathcal{PT} = \mathbb{CP}^3 - \mathbb{CP}^1$ called the projective twistor space.

Let $(Z^0, Z^1, Z^2, Z^3) \sim (cZ^0, cZ^1, cZ^2, cZ^3), c \in \mathbb{C}^*$ with $(Z^2, Z^3) \neq (0, 0)$, be homogeneous coordinates of a twistor (a point in $\mathcal{PT}$). The twistor space and the Minkowski space are linked by the incidence relation

$$\begin{pmatrix} Z^0 \\ Z^1 \end{pmatrix} = \frac{i}{\sqrt{2}} \begin{pmatrix} t - \zeta & -x - iy \\ -x + iy & t + \zeta \end{pmatrix} \begin{pmatrix} Z^2 \\ Z^3 \end{pmatrix}, \tag{7.2.12}$$

where $x^\mu = (t, x, y, \zeta)$ are coordinates of a point in Minkowski space. If two points in Minkowski space are incident with the same twistor, then they are null separated.

Define the Hermitian inner product

$$\Sigma(Z, \overline{Z}) = Z^0\overline{Z^2} + Z^1\overline{Z^3} + Z^2\overline{Z^0} + Z^3\overline{Z^1}$$

on the non-projective twistor space $\mathcal{T} = \mathbb{C}^4 - \mathbb{C}^2$. The signature of Σ is $(++--)$ so that the orientation-preserving endomorphisms of $\mathcal{T}$ preserving Σ form a group $SU(2,2)$. This group has 15 parameters and is locally isomorphic to the conformal group $SO(4,2)$ of the Minkowski space. We divide the twistor space into three parts depending on whether Σ is positive, negative, or zero. This partition descends to the projective twistor space. In particular the hypersurface

$$\mathcal{PN} = \{[Z] \in \mathcal{PT}, \Sigma(Z, \overline{Z}) = 0\} \subset \mathcal{PT}$$

is preserved by the conformal transformations of the Minkowski space which can be verified directly using (7.2.12).

Fixing the coordinates x^μ of a space-time point in (7.2.12) gives a plane in the non-projective twistor space $\mathbb{C}^4 - \mathbb{C}^2$ or a projective line $\mathbb{CP}^1$ in $\mathcal{PT}$. If the coordinates x^μ are real this line lies in the hypersurface $\mathcal{PN}$. Conversely, fixing a twistor in $\mathcal{PN}$ gives a light ray in the Minkowski space.

So far only the null twistors (points in $\mathcal{PN}$) have been relevant in this discussion. General points in $\mathcal{PT}$ can be interpreted in terms of the complexified Minkowski space $\mathbb{C}^4$ where they correspond to null two-dimensional planes with SD tangent bi-vectors (see Section 7.2.3). This is a direct consequence of (7.2.12) where now the coordinates x^μ are complex. There is also an interpretation of non-null twistors in the real Minkowski space, but this is less obvious [129]: The Hermitian inner product Σ defines a vector space $\mathcal{T}^*$ dual to the non-projective twistor space. The elements of the corresponding projective space $\mathcal{PT}^*$ are called dual twistors. Now take a non-null twistor $Z \in \mathcal{PT}$. Its dual $\overline{Z} \in \mathcal{PT}^*$ corresponds to a projective two-plane $\mathbb{CP}^2$ in $\mathcal{PT}$. A holomorphic two-plane intersects the hypersurface $\mathcal{PN}$ in a real three-dimensional locus. This locus corresponds to a three-parameter family of light rays in the real Minkowski space. This family representing a single twistor is called the Robinson congruence. A picture of this configuration which appears in vol 2 on page 62 [132] shows a system of twisted oriented circles in the Euclidean space $\mathbb{R}^3$, the point being that any light ray is represented by a point in $\mathbb{R}^3$ together with an arrow indicating the direction of the ray's motion. This configuration originally gave rise to a name 'twistor'.

Finally we can give a twistor interpretation of the contour integral formula (7.2.11). Consider a function $f = f(Z^0/Z^3, Z^1/Z^3, Z^2/Z^3)$ which is holomorphic on an intersection of two open sets covering $\mathcal{PT}$ (one of this sets is defined by $Z^3 \neq 0$ and the other by $Z^2 \neq 0$) and restrict this function to a rational curve (7.2.12) in $\mathcal{PN}$. Now integrate f along a contour in this curve. This gives (7.2.11) with $\lambda = Z^2/Z^3$.

To sum up, the space-time points are derived objects in twistor theory. They become 'fuzzy' after quantization. This may provide an attractive framework for quantum gravity, but it must be said that despite 40 years of research the twistor theory is still waiting to have its major impact on physics. It has however had surprisingly major impact on pure mathematics: ranging from representation theory and differential geometry to solitons, instantons, and integrable systems.

7.2.2 Spinor notation

The ASD condition in four dimensions can be conveniently expressed in terms of two-component spinor notation [132].

The displacement vector from the origin is identified with a matrix (a two-index spinor)

$$x^{AA'} = \begin{pmatrix} x^{00'} & x^{01'} \\ x^{10'} & x^{11'} \end{pmatrix} = \begin{pmatrix} \tilde{z} & w \\ \tilde{w} & z \end{pmatrix}, \quad \text{where} \quad A = 0, 1, \ A' = 0', 1'.$$

This exhibits a canonical isomorphism

$$T = \mathbb{S} \otimes \mathbb{S}',$$

where T is the space of complex vectors in $\mathbb{C}^4$ and $\mathbb{S}, \mathbb{S}'$ are complex two-dimensional vector spaces whose elements are called two-component spinors.

The closely related isomorphism

$$SO(4, \mathbb{C}) = SL(2, \mathbb{C}) \times SL(2, \mathbb{C})/\mathbb{Z}_2$$

is realized by expressing any rotation as

$$x^{AA'} \longrightarrow \Lambda^A_B \Lambda^{A'}_{B'} x^{BB'}$$

where $\Lambda^A_B \in SL(2, \mathbb{C})$ and $\Lambda^{A'}_{B'} \in SL(2, \mathbb{C})$ act on $\mathbb{S}$ and $\mathbb{S}'$, respectively. We shall regard $\mathbb{S} = \mathbb{C}^2$ as a symplectic vector space (so called spin space), with anti-symmetric product

$$\kappa \cdot \rho = \kappa^0 \rho^1 - \kappa^1 \rho^0 = \varepsilon(\kappa, \rho).$$

The elements of $\mathbb{S}$ are of the form $\kappa^A = (\kappa^0, \kappa^1)$. The constant symplectic form ε is represented by the matrix

$$\varepsilon_{AB} = \begin{pmatrix} 0 & 1 \\ -1 & 0 \end{pmatrix},$$

and can be used to 'raise and lower the indices' according to $\kappa_A = \kappa^B \varepsilon_{BA}$, $\kappa^A = \varepsilon^{AB} \kappa_B$, where $\varepsilon_{AB} \varepsilon^{CB}$ is the identity endomorphism. The analogous symplectic structure $\varepsilon' = \varepsilon_{A'B'}$ is put on $\mathbb{S}' = \mathbb{C}^2$.

- **Important convention.** The projective primed spin space $P(\mathbb{S}')$ is the complex projective line $\mathbb{CP}^1$. The homogeneous coordinates are denoted by $\pi_{A'} = (\pi_{0'}, \pi_{1'})$, and the two set covering of $\mathbb{CP}^1$ is

  $$U = \{\pi_{A'}, \pi_{1'} \neq 0\} \quad \text{and} \quad \tilde{U} = \{\pi_{A'}, \pi_{0'} \neq 0\}.$$

  The functions $\lambda = \pi_{0'}/\pi_{1'}$ and $\tilde{\lambda} = 1/\lambda$ are inhomogeneous coordinates in U and $\tilde{U}$, respectively. It then follows that $\lambda = -\pi^{1'}/\pi^{0'}$.

The holomorphic metric (7.1.1) in $\mathbb{C}^4$ is

$$ds^2 = 2 \det (dx^{AA'}) = \varepsilon_{AB} \varepsilon_{A'B'} dx^{AA'} dx^{BB'}.$$

The decomposition of a two-form into SD and ASD parts is straightforward in spinor notation. Let

$$F = \frac{1}{2} F_{AA'BB'} dx^{AA'} \wedge dx^{BB'}$$

be a two-form. Now

$$\begin{aligned} F_{AA'BB'} &= F_{(AB)(A'B')} + F_{[AB][A'B']} + F_{(AB)[A'B']} + F_{[AB](A'B')} \\ &= F_{(AB)(A'B')} + c\varepsilon_{AB}\varepsilon_{A'B'} + \phi_{AB}\varepsilon_{A'B'} + \tilde{\phi}_{A'B'}\varepsilon_{AB}. \end{aligned}$$

Here we have used the fact that in two dimensions there is a unique anti-symmetric matrix up to scale, so whenever an anti-symmetrized pair of spinor indices occurs we can substitute a multiple of ε_{AB} or $\varepsilon_{A'B'}$ in their place. Now observe that the first two terms are incompatible with F being a two-form, that is, $F_{AA'BB'} = -F_{BB'AA'}$. So we obtain

$$F_{AA'BB'} = \phi_{AB}\varepsilon_{A'B'} + \tilde{\phi}_{A'B'}\varepsilon_{AB}, \tag{7.2.13}$$

where ϕ_{AB} and $\tilde{\phi}_{A'B'}$ are symmetric.

If F is taken to be the YM field, the spinor form of the ASDYM equations is

$$\tilde{\phi}_{A'B'} = 0.$$

These equations can be written as

$$[\pi^{A'} D_{AA'}, \pi^{B'} D_{BB'}] = 0,$$

where $D_{AA'} = \partial_{AA'} + A_{AA'}$, and $L_A = \pi^{A'} D_{AA'}$ is the spinor form of the Lax pair.

7.2.3 Twistor space

For any vector we have $|V|^2 = 2 \det(V^{AA'})$, and so null vectors correspond to matrices $V^{AA'}$ of rank 1. Therefore any null vector is of the form $V^{AA'} = \kappa^A \pi^{A'}$. Fixing $\pi^{A'}$ and varying κ^A gives a null two-dimensional plane in $\mathbb{C}^4$ called an α-plane.

Definition 7.2.1 *The twistor space $\mathcal{PT}$ of complexified Minkowski space $M_\mathbb{C}$ is set of all α-planes in $M_\mathbb{C}$.*

The twistor space is a three-dimensional complex manifold (according to Penrose's original terminology it should be called the projective twistor space). We can understand its geometry by writing the equation of an α-plane in the homogeneous form

$$x^{AA'} \pi_{A'} = \omega^A. \tag{7.2.14}$$

Any solution of this equation can be translated according to

$$x^{AA'} \longrightarrow x^{AA'} + \kappa^A \pi^{A'},$$

where κ^A is arbitrary, so the space of solutions is indeed an α-plane in $M_\mathbb{C}$.

The spinors $(\omega^A, \pi_{A'})$ are homogeneous coordinates on $\mathcal{PT}$. They are determined by an α-plane up to the equivalence $(\omega^A, \pi_{A'}) \sim (c\omega^A, c\pi_{A'})$, where $c \in \mathbb{C}^*$. Conversely each pair of spinors (ω, π) with $\pi_{A'} \neq (0, 0)$ determines an α-plane. The twistor space $\mathcal{PT}$ is the complex manifold[1] $\mathbb{CP}^3 - \mathbb{CP}^1$. It fibres over $\mathbb{CP}^1$ by $(\omega^A, \pi_{A'}) \longrightarrow \pi_{A'}$.

An alternate interpretation of (7.2.14) is to fix $x^{AA'}$. This determines ω^A as a pair of linear functions of $\pi_{A'}$, that is, a projective line on $\mathcal{PT}$. This line is a holomorphic section of the fibration $\mathcal{PT} \to \mathbb{CP}^1$.

The outlined twistor correspondence can be summarized as follows (Figure 7.1):

Points $\longleftrightarrow$ holomorphic sections of $\mathcal{PT} \longrightarrow \mathbb{CP}^1$

α-planes $\longleftrightarrow$ points

Two points lie on the same α-plane $\longleftrightarrow$ two holomorphic sections intersect at a point

[1] If one considers the complexified and compactified Minkowski space then the twistor space becomes $\mathbb{CP}^3$ with the additional $\mathbb{CP}^1$ worth of α-planes at infinity. See [132] for details.

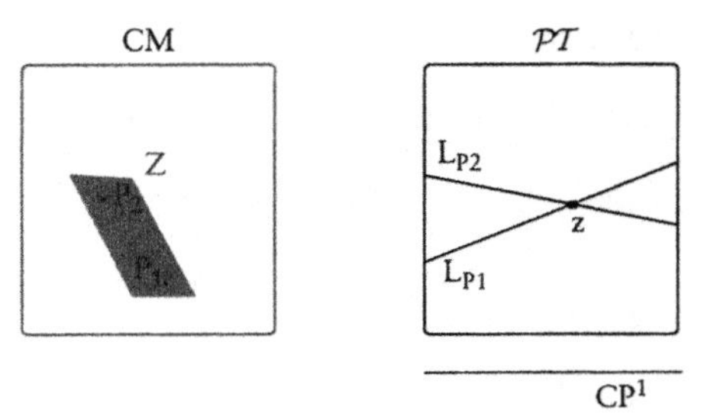

Figure 7.1 *Twistor correspondence*

Another way of defining $\mathcal{PT}$ is by the double fibration:

$$\mathbb{C}^4 \xleftarrow{r} \mathcal{F} \xrightarrow{q} \mathcal{PT}. \tag{7.2.15}$$

The correspondence space $\mathcal{F} = \mathbb{C}^4 \times \mathbb{CP}^1$ has a natural fibration over $\mathbb{C}^4$, and the projection $q : \mathcal{F} \longrightarrow \mathcal{PT}$ is a quotient of $\mathcal{F}$ by the two-dimensional distribution (7.1.8) of vectors tangent to an α-plane. Set $\partial_{AA'} = \partial/\partial x^{AA'}$. In concrete terms this distribution is spanned by $\delta_A = \pi^{A'}\partial_{AA'}$, or

$$\delta_0 = l = \partial_{\tilde{z}} - \lambda\partial_w, \quad \delta_1 = m = \partial_{\tilde{w}} - \lambda\partial_z$$

and for each λ it generates translations along α-plane. Objects on $\mathcal{F}$ which are Lie-derived along δ_A descend to $\mathcal{PT}$. For example,

$$w + \lambda\tilde{z}, \quad z + \lambda\tilde{w}, \quad \text{and} \quad \lambda$$

are twistor functions on $\mathcal{F}$. They are independent as

$$d(w + \lambda\tilde{z}) \wedge d(z + \lambda\tilde{w}) \wedge d\lambda = \omega(\lambda) \wedge d\lambda \neq 0$$

and so they give local holomorphic coordinates on $\mathcal{PT}$ in an open set containing $\lambda = 0$. Note that $\omega(\lambda)$ is the SD two-form defined in (7.1.9).

- To single out the Euclidean reality conditions consider an anti-holomorphic involution $\sigma : \mathcal{PT} \to \mathcal{PT}$ given by

$$\sigma(Z^0, Z^1, Z^2, Z^3) = (\overline{Z^1}, -\overline{Z^0}, \overline{Z^3}, -\overline{Z^2}), \tag{7.2.16}$$

  where $(Z^0, Z^1, Z^2, Z^3) = (\omega^A, \pi_{A'})$. Therefore $\sigma^2 = -\text{Id}$.
  For any point $Z \in \mathcal{PT}$ the line L_p joining Z to $\sigma(Z)$ is called a *real line.* The real lines do not intersect in $\mathcal{PT}$ so $\mathcal{PT}$ is fibred by such lines, and the quotient space (the space of all real lines) is $\mathbb{R}^4$ (or S^4 if the line $Z^2 = Z^3 = 0$ is included). See [186] for more details.
- There are two neutral slices as explained in Section 7.1:

  1. Neutral slice (a)

$$\sigma(Z^\alpha) = \overline{Z^\alpha}, \quad \alpha = 0, \ldots, 3$$

2. Neutral slice (**b**)

$$\sigma(Z^0, Z^1, Z^2, Z^3) = (\overline{Z^1}, \overline{Z^0}, \overline{Z^3}, \overline{Z^2})$$

In both cases $\sigma^2 = \text{Id}$ and there exist fixed points of σ on $\mathcal{PT}$. They correspond to real α-planes in $M_{\mathbb{C}}$ that intersect the real slice in real two-planes. There is an S^1 worth of real α-planes through each real point.

7.2.4 Penrose–Ward correspondence

The following result is a far-reaching consequence of Proposition 7.1.1 which stated that ASDYM fields vanish on α-planes. Both the formulation and the proof of the result rely heavily on the complex analysis machinery presented in Appendix B.

Theorem 7.2.2 (Ward [169]) *There is a one-to-one correspondence between*

1. *Gauge equivalence classes of ASD connections on $M_{\mathbb{C}}$ with the gauge group $G = GL(n, \mathbb{C})$*
2. *Holomorphic rank n vector bundles E over twistor space $\mathcal{PT}$ which are trivial on each degree one section of $\mathcal{PT} \to \mathbb{CP}^1$*

Proof Let A be an ASD connection . Therefore the pair of linear PDEs $LV = MV = 0$, where L and M are given by (7.1.6), is integrable. This assigns an n-dimensional vector space to each α-plane $Z \subset \mathbb{C}^4$, and so to a point $Z \in \mathcal{PT}$. It is the fibre of a holomorphic vector bundle E. The bundle E is trivial on each section, since we can identify fibres of $E|_{L_p}$ at Z_1, Z_2 because covariantly constant vector fields at α-planes Z_1, Z_2 coincide at a common point $p \in \mathbb{C}^4$. In concrete terms the patching matrix and its splitting are given by the path-ordered integral

$$F_{01} = \mathcal{P}\exp\left(\int_{Q(Z)}^{P(Z)} A_\mu dx^\mu\right)$$

where $P(Z)$ and $Q(Z)$ are unique points on the α-plane Z such that $x^{A1'} = 0$ on $P(Z)$ and $x^{A0'} = 0$ on $Q(Z)$. These points vary holomorphically with $(\omega^A, \pi_{A'})$ and the integral is taken over any contour. The choice of contour does not matter since the gauge field is flat when restricted to the α-plane Z. The transition function when restricted to L_p splits as in (B3), where

$$\widetilde{H} = \mathcal{P}\exp\left(\int_x^{P(Z)} A_\mu dx^\mu\right), \quad H = \mathcal{P}\exp\left(\int_x^{Q(Z)} A_\mu dx^\mu\right), \quad \text{and}$$

$$F = F(x^{AA'}\pi_{A'}, \pi^{A'}).$$

Conversely, assume that we are given a holomorphic vector bundle over $\mathcal{PT}$ which is trivial on each section. Since $E|_{L_p}$ is trivial, and $L_p \cong \mathbb{CP}^1$, Theorem B.2.5 gives

$$E|_{L_p} = \mathcal{O} \oplus \mathcal{O} \oplus \cdots \oplus \mathcal{O} \tag{7.2.17}$$

and

$$\Gamma(L_p, E|_{L_p}) = \mathbb{C}^n.$$

This gives us a holomorphic rank n vector bundle $\hat{E} \to \mathbb{C}^4$. We shall give a concrete method of constructing an ASD connection on this bundle. Let $\mu : E \to \mathcal{PT}$, and let

$$U = \{(\omega^A, \pi_{A'}), \pi_{1'} \neq 0\}, \quad \tilde{U} = \{(\omega^A, \pi_{A'}), \pi_{0'} \neq 0\}, \quad \text{and} \quad \lambda = \pi_{0'}/\pi_{1'}$$

be the covering of $\mathcal{PT}$. Let

$$\chi : \mu^{-1}(U) \to U \times \mathbb{C}^n \quad \text{and} \quad \tilde{\chi} : \mu^{-1}(\tilde{U}) \to \tilde{U} \times \mathbb{C}^n$$

be local trivializations of E, and let $F_{01} = F = \tilde{\chi} \circ \chi^{-1} : \mathbb{C}^n \to \mathbb{C}^n$ be a holomorphic patching matrix defined on $U \cap \tilde{U}$.

1. Restrict F to L_p and pull it back to $\mathcal{F}$. This is achieved by substituting $\omega^A = x^{AA'}\pi_{A'}$ in $F(\omega^A, \pi_{A'})$, so that $\pi^{A'}\partial_{AA'}F = 0$. Define the fibres of $\hat{E} \to M_{\mathbb{C}}$ as holomorphic sections of E restricted to L_p. The 'triviality on sections' condition implies (7.2.17) and so F is homogeneous of degree 0 in $\pi_{A'}$. Therefore each fibre of $\hat{E}$ is isomorphic to $\mathbb{C}^n$.
2. $E|_{L_p}$ is trivial, so Lemma B.2.2 implies that the patching matrix can be split
$$F = \tilde{H}H^{-1}, \tag{7.2.18}$$
where H and $\tilde{H}$ are holomorphic for $\pi_{A'}$ in $\mu^{-1}(U \cap L_p)$ and $\mu^{-1}(\tilde{U} \cap L_p)$, respectively. We note that $H, \tilde{H}$ do not descend to the twistor space. The patching relation $\tilde{V} = FV$, where $\tilde{V}$ and V are column vectors whose components depend on coordinates of $\tilde{U}$ and U, implies that $V = H\xi$ and $\tilde{V} = \tilde{H}\xi$, where ξ is a constant vector.
3. Note that $\pi^{A'}\partial_{AA}F = 0$ implies
$$\tilde{H}^{-1}\pi^{A'}\partial_{AA'}\tilde{H} = H^{-1}\pi^{A'}\partial_{AA'}H. \tag{7.2.19}$$
Both sides are homogeneous of degree one, and holomorphic, so by the Liouville theorem B.2.4, they must be linear in $\pi^{A'}$, and equal to $\pi^{A}A_{AA'}$ for some $A_{AA'}(x^\mu)$.
4. Now we show that $A_{AA'}$ is ASD. Operating on
$$\pi^{A'}A_{AA'} = H^{-1}\pi^{A'}\partial_{AA'}H$$

with $\pi^{C'}\partial^B_{C'}$ leads to

$$\partial^B_{(C'}A_{A')B} + A^B{}_{(C'}A_{A')B} = 0,$$

which is the spinor version $[D_{A(A'}, D_{B')B}] = 0$ of the ASDYM. One can now deduce that H^{-1} and $\tilde{H}^{-1}$ are in the kernel of the Lax pair (7.1.6):

$$L_A H^{-1} = \delta_A H^{-1} + H^{-1}(\delta_A H)H^{-1} = 0,$$

where $L_A = \pi^{A'}(\partial_{AA'} + A_{AA'})$.

□

In practice, constructing A form H and $\tilde{H}$ can be simplified by the following:

Lemma 7.2.3 *Let* $B = H|_{\lambda=0}$, $\tilde{B} = \tilde{H}|_{\lambda=\infty}$. *Then*

$$A = B^{-1}\tilde{\partial}B + \tilde{B}^{-1}\partial\tilde{B}, \tag{7.2.20}$$

where $\partial = dw \otimes \partial_w + dz \otimes \partial_z$ *and* $\tilde{\partial} = d\tilde{w} \otimes \partial_{\tilde{w}} + d\tilde{z} \otimes \partial_{\tilde{z}}$.

Proof This follows from evaluating the LHS and the RHS of (7.2.19) at $\pi^{A'} = (1, 0)$ and $\pi^{A'} = (0, 1)$, respectively. □

Remarks

1. We have constructed a trivial vector bundle $\hat{E}$ on the space-time $M_{\mathbb{C}}$ with a connection satisfying a local PDE from a holomorphic vector bundle (with no connection) over the twistor space. All the information sits in the patching matrix F. Equations (6.4.23) appear as the integrability conditions for the existence of E.
2. The splitting $F = \tilde{H}H^{-1}$ (known as the Riemann–Hilbert problem or the Birkhoff factorization problem – compare Section 3.3.1) is the hardest part of this approach (and others) to integrable PDEs.
3. All complex, Euclidean, and real analytic neutral ASDYM fields, including instantons, can be obtained from the construction.
4. To obtain real solutions on $\mathbb{R}^4$ with the gauge group $G = SU(n)$ the bundle must be compatible with the involution (7.2.16). This comes down to $\det F = 1$, and

$$F^*(Z) = F[\sigma(Z)],$$

where $*$ denotes the Hermitian conjugation and F is the patching matrix. In Section 6.4 we explained how the instanton solutions of ASDYM extend from $\mathbb{R}^4$ to S^4. The corresponding vector bundles extend from $\mathcal{PT}$ to $\mathbb{CP}^3$. The holomorphic vector bundles over $\mathbb{CP}^3$ have been extensively studied by algebraic geometers. All such bundles (and thus the instantons) can be generated by the monad construction [9].

- **Example. ASD Maxwell equations.** Take $G = GL(1, \mathbb{C})$, therefore $E = L$ is a line bundle with $c_1(L) = 0$ (as L restricts to a trivial bundle $\mathcal{O}$ on each section), and $F = F(\omega^A, \lambda)$ is a nowhere vanishing function holomorphic on $U \cap \tilde{U}$. Put

$$F = e^f, \quad H = e^h, \quad \text{and} \quad \tilde{H} = e^{\tilde{h}}.$$

The nonlinear splitting (B3) can now be done additively as in (B8):

$$f = \tilde{h} - h,$$

where f, h, and $\tilde{h}$ are homogeneous of degree 0. Choose a point $[\iota] \in \mathbb{CP}^1$. Then

$$h = \frac{1}{2\pi i}\oint_\Gamma \frac{\pi \cdot \iota}{(\pi \cdot \rho)(\iota \cdot \rho)} f(x^{AA'}\rho_{A'}, \rho)\rho \cdot d\rho \quad \text{and}$$

$$\tilde{h} = \frac{1}{2\pi i}\oint_{\tilde{\Gamma}} \frac{\pi \cdot \iota}{(\pi \cdot \rho)(\iota \cdot \rho)} f(x^{AA'}\rho_{A'}, \rho)\rho \cdot d\rho,$$

where ρ are homogeneous coordinates on $\mathbb{CP}^1$, and $\rho \cdot d\rho = d\zeta$ in affine coordinates. Now

$$\begin{aligned}\pi^{A'} A_{AA'} &= H^{-1}\pi^{A'}\partial_{AA'}H = \pi^{A'}\partial_{AA'}h \\ &= \frac{1}{2\pi i}\oint_\Gamma \frac{\pi \cdot \iota}{\rho \cdot \iota}\frac{\partial f}{\partial \omega^A}\rho \cdot d\rho,\end{aligned}$$

where we have made the replacement

$$\frac{\partial}{\partial x^{AA'}} \longrightarrow \rho_{A'}\frac{\partial}{\partial \omega^A}$$

under the integral sign. The choice of the spinor $\iota_{A'}$ is a gauge choice. Finally

$$A_{BB'} = \frac{1}{2\pi i}\oint_\Gamma \frac{\iota_{B'}}{(\iota_{C'}\rho^{C'})}\frac{\partial f}{\partial \omega^B}\rho_{D'}d\rho^{D'}. \tag{7.2.21}$$

The ASD Maxwell equations in double null coordinates are

$$\partial_w A_z - \partial_z A_w = 0, \quad \partial_{\tilde{w}} A_{\tilde{z}} - \partial_{\tilde{z}} A_{\tilde{w}} = 0, \quad \text{and}$$

$$\partial_z A_{\tilde{z}} - \partial_{\tilde{z}} A_z - \partial_w A_{\tilde{w}} + \partial_{\tilde{w}} A_w = 0.$$

The first two equations can be interpreted as integrability conditions for the existence of $u(x^\mu)$, $v(x^\nu)$ such that

$$A = \partial_w u\, dw + \partial_z u\, dz + \partial_{\tilde{w}} v\, d\tilde{w} + \partial_{\tilde{z}} v\, d\tilde{z}.$$

Making a gauge transformation $A \longrightarrow A - du$ (i.e. setting $A_w = A_z = 0$), and redefining v reduces the ASD Maxwell equations (the third equation) to the wave equation

$$\frac{\partial^2 v}{\partial z \partial \tilde{z}} - \frac{\partial^2 v}{\partial w \partial \tilde{w}} = 0 \tag{7.2.22}$$

and

$$A_{AA'} = l_{A'} o^{B'} \partial_{AB'} v. \tag{7.2.23}$$

Proposition 7.2.4 *All solutions to the holomorphic wave equation* (7.2.22) *are given by*

$$v = \frac{1}{2\pi i} \oint_{\Gamma} \hat{f}(w + \lambda \tilde{z}, z + \lambda \tilde{w}, \lambda) d\lambda, \tag{7.2.24}$$

where the twistor function $\hat{f}$ is holomorphic in its arguments on the intersection $U \cap \tilde{U}$ and, when expressed in homogeneous coordinates $(\omega^A, \rho_{A'})$ with $\rho_{0'}/\rho_{1'} = \lambda$, it is homogeneous of degree -2 in $\rho_{A'}$.

Proof Let v be given by (7.2.24). Differentiating under the integral sign gives (7.2.22) with

$$\hat{f} = \frac{1}{(\rho \cdot o)(\rho \cdot \iota)} f.$$

Conversely, given a solution to (7.2.22) construct the Maxwell field (7.2.23). Comparing (7.2.23) with (7.2.21) yields (7.2.24) with $d\lambda = \rho_{D'} d\rho^{D'}$. □

The formula (7.2.24) can be applied outside the realm of ASD, as any reality conditions can be imposed. In particular setting

$$z = \frac{x^1 + x^4}{\sqrt{2}}, \quad \tilde{z} = \frac{x^1 - x^4}{\sqrt{2}}, \quad w = -\frac{x^2 + i x^3}{\sqrt{2}}, \quad \text{and} \quad \tilde{w} = \overline{w}$$

give solutions to the Lorentzian wave equations

$$\frac{\partial^2 v}{\partial (x^1)^2} - \frac{\partial^2 v}{\partial (x^2)^2} - \frac{\partial^2 v}{\partial (x^3)^2} - \frac{\partial^2 v}{\partial (x^4)^2} = 0.$$

Proposition 7.2.4 is an example of the Penrose transform which gives a general correspondence between solutions to the zero-rest-mass equations on complexified Minkowski space and cohomology classes on twistor space:

- All solutions to the spin $-n/2$ massless field equation

  $$\partial_A{}^{A_1'} \phi_{A_1' A_2' \cdots A_n'} = 0$$

  are given by

  $$\phi_{A_1' A_2' \cdots A_n'} = \frac{1}{2\pi i} \oint_{\Gamma} \pi_{A_1'} \pi_{A_2'} \cdots \pi_{A_n'} f \, \pi \cdot d\pi,$$

where $f \in H^1(\mathbb{CP}^1, \mathcal{O}(-n-2))$.

- All solutions to the spin $n/2$ equation

$$\partial_{A'}{}^{A_1} \psi_{A_1 A_2 \cdots A_n} = 0$$

are given by

$$\psi_{A_1 A_2 \cdots A_n} = \frac{1}{2\pi i} \oint_\Gamma \frac{\partial^n f}{\partial \omega^{A_1} \partial \omega^{A_2} \cdots \partial \omega^{A_n}} \pi \cdot d\pi,$$

where $f \in H^1(\mathbb{CP}^1, \mathcal{O}(n-2))$.

The details of the Penrose correspondence are presented in [12, 132, 175, 186]. In our next example we shall return to the ASDYM equations.

- **Example. Atiyah–Ward ansatz for SL(2, $\mathbb{C}$) ASDYM.** One way to construct holomorphic vector bundles is to produce extensions of line bundles, which comes down to using upper triangular matrices as patching functions.

  Let E be a rank-two holomorphic vector bundle over $\mathcal{PT}$ which arises as an extension of a line bundle $L_1 \to \mathcal{PT}$ by another line bundle $L_2 \to \mathcal{PT}$:

$$0 \longrightarrow L_1 \longrightarrow E \longrightarrow L_2 \longrightarrow 0. \tag{7.2.25}$$

Let us assume $\det F = 1$ so that the gauge group is $SL(2, \mathbb{C})$. This implies

$$L_1 = L \otimes \mathcal{O}(-k) \quad \text{and} \quad L_2 = L^* \otimes \mathcal{O}(k),$$

where L is the line bundle from the ASD Maxwell example, and $\mathcal{O}(k)$ is the pull-back of a line bundle from $\mathbb{CP}^1$ to $\mathcal{PT}$. Therefore

$$F = \begin{pmatrix} \lambda^k e^f & \Omega \\ 0 & \lambda^{-k} e^{-f} \end{pmatrix}, \tag{7.2.26}$$

where Ω and f are holomorphic, and homogeneous of degree 0, and

$$\Omega \in H^1(\mathcal{PT}, \mathcal{O}(L_1 \otimes L_2^{-1})) = H^1(\mathcal{PT}, L^2 \otimes \mathcal{O}(2k))$$

classifies all extensions. The resulting solution to (6.4.23) depends on two arbitrary functions of three variables. The relatively simple form of F corresponds to very complicated (as k increases) formulae for A. Let us work out the details if $k = 1$, $f = 0$. Let

$$\Omega = \sum_{-\infty}^{\infty} \phi_i \lambda^i = \Omega_- + \phi_0 + \Omega_+.$$

Since $\delta_A \Omega = 0$ we have the recursion relations

$$\partial_{\tilde{z}} \phi_{i+1} = \partial_w \phi_i \quad \text{and} \quad \partial_{\tilde{w}} \phi_{i+1} = \partial_z \phi_i, \tag{7.2.27}$$

from which it follows by cross-differentiating that each $\phi_i(x^\mu)$ satisfies the scalar wave equation (7.2.22).

One can take the splitting matrices to be

$$H = \frac{1}{\sqrt{\phi_0}} \begin{pmatrix} -\lambda^{-1}\Omega_+ & -(\phi_0 + \Omega_+) \\ 1 & \lambda \end{pmatrix} \quad \text{and} \quad \tilde{H} = \frac{1}{\sqrt{\phi_0}} \begin{pmatrix} \phi_0 + \Omega_- & \lambda\Omega_- \\ \lambda^{-1} & 1 \end{pmatrix},$$

and use the Lemma 7.2.3. This implies

$$B = \frac{1}{\sqrt{\phi_0}} \begin{pmatrix} -\phi_1 & -\phi_0 \\ 1 & 0 \end{pmatrix} \quad \text{and} \quad \tilde{B} = \frac{1}{\sqrt{\phi_0}} \begin{pmatrix} \phi_0 & \phi_{-1} \\ 0 & 1 \end{pmatrix},$$

and finally

$$A = \frac{1}{2\phi} \begin{pmatrix} (\partial - \tilde{\partial})\phi & 2(\phi_{\tilde{z}} dw + \phi_{\tilde{w}} dz) \\ 2(\phi_w d\tilde{z} + \phi_z d\tilde{w}) & (\tilde{\partial} - \partial)\phi \end{pmatrix}, \tag{7.2.28}$$

where $\phi = \phi_0$ is any solution to the wave equation. This solution is invariant under the Euclidean reality conditions. If ϕ is a harmonic function on $\mathbb{R}^4$ we recovered the anzatz (6.4.26) with $\rho = \phi$.

If $k > 1$ then A is given in terms of a solution to the linear zero-rest-mass field equations with higher helicity. In general one can show

Theorem 7.2.5 (Atiyah–Ward [8]) *Every $SU(2)$ ASDYM instanton (with the boundary conditions (6.4.21)) over $\mathbb{R}^4$ arises from some ansatz (7.2.26).*

There is a relationship between the Atiyah–Ward ansatz and the dressing method described in Section 3.3.2. See [154] for details.

Exercises

1. Show that the two-forms

   $$\omega_1 = dw \wedge dz, \quad \omega_2 = dw \wedge d\tilde{w} - dz \wedge d\tilde{z}, \quad \text{and} \quad \omega_3 = d\tilde{z} \wedge d\tilde{w}$$

   span the space of SD two-forms in $\mathbb{C}^4$, where

   $$ds^2 = 2(dz d\tilde{z} - dw d\tilde{w}) \quad \text{and} \quad \text{vol} = dw \wedge d\tilde{w} \wedge dz \wedge d\tilde{z}.$$

   Show that a two-form F is ASD iff $F \wedge \omega_i = 0$.
2. Show that in four dimensions $**F = \pm F$ where the sign depends on the signature of the metric on $\mathbb{R}^4$.

   Show that in the $U(1)$ theory $F \to *F$ interchanges the electric and magnetic fields with factors of ± 1 or $\pm i$.

3. Use the Lax pair formulation of ASDYM to
   (a) Deduce the existence of a gauge such that $A = A_{\tilde{w}} d\tilde{w} + A_{\tilde{z}} d\tilde{z}$
   (b) Deduce the existence of a $\mathfrak{g}$-valued function $K = K(w, z, \tilde{w}, \tilde{z})$ such that
   $$A_{\tilde{w}} = \partial_z K \quad \text{and} \quad A_{\tilde{z}} = \partial_w K.$$
   (c) Reduce the ASDYM equations to a single second-order PDE
   $$\partial_z \partial_{\tilde{z}} K - \partial_w \partial_{\tilde{w}} K + [\partial_w K, \partial_z K] = 0.$$

   What is the residual gauge freedom in K?
4. Show that
   - $\xi_A \rho^A = 0$ iff ξ_A, ρ_A are proportional.
   - A vector in $\mathbb{C}^4$ is null iff $V^{AA'} = \lambda^A \xi^{A'}$ for some spinors $\lambda^A, \xi^{A'}$.
   - $\tau_{AB} = \tau_{(AB)} + \frac{1}{2}\tau\varepsilon_{AB}$ where $(\ldots)$ denotes symmetrization, and τ should be determined.
   - The ASDYM equations are of the form
   $$\partial_{B(B'} A_{C')}{}^B + A_{B(B'} A_{C')}{}^B = 0$$
   and are equivalent to $[L_0, L_1] = 0$, where $L_B = \pi^{B'}(\partial_{BB'} + A_{BB'})$ for some constant spinor $\pi^{B'}$.
5. Show that the factorization of the patching matrix $F = \tilde{H}H^{-1}$ in the Penrose–Ward correspondence is unique up to multiplication of H and $\tilde{H}$ on the right by a non-singular matrix g depending on the space-time coordinates, but not λ. Show that different choices of factorization give gauge-equivalent connections.
6. Let the patching matrix for a rank-two Ward bundle over the twistor space $\mathcal{PT}$ be given by
   $$F = \begin{pmatrix} 1 & f \\ 0 & 1 \end{pmatrix},$$
   where $f = f(\omega^A, \pi_{A'})$ is en element of $H^1(\mathcal{PT}, \mathcal{O})$. Find the YM potential A in terms of the ASD electromagnetic field generated by f.

8 Symmetry reductions and the integrable chiral model

8.1 Reductions to integrable equations

Most integrable systems arise as symmetry reductions of ASDYM, where the potential one-form is assumed not to depend on one or more coordinates on the space-time. The resulting system will admit a (reduced) Lax pair with a spectral parameter as well as a twistor correspondence, and thus will be integrable. This leads to a classification of those integrable systems that can be obtained by reduction from the ASDYM equations as well as a unification of the theory of these equations by reduction of the corresponding theory for the ASDYM equations. The programme of reducing the ASDYM equations to various integrable equations has been proposed and initiated by Ward [171] and fully implemented in the monograph [118].

The general scheme and classification of reductions involves the following choices:

1. A subgroup H of the complex conformal group $PGL(4, \mathbb{C})$.
   The complex conformal group consists of linear transformations ρ of $M_{\mathbb{C}}$ such that

$$\rho^*(ds^2) = \Omega^2 ds^2 \quad \text{and} \quad \rho^*(\text{vol}) = \Omega^4(\text{vol})$$

   for some $\Omega : M_{\mathbb{C}} \to \mathbb{C}$. It is a symmetry group of ASDYM, as conformal transformations map ASD two-forms to ASD two-forms, and therefore preserve equations (6.4.23). It has 15 generators which are the conformal Killing vectors, that is, solutions to

$$\partial_{(\mu} K_{\nu)} = \frac{1}{4}\eta_{\mu\nu}\partial_\rho K^\rho.$$

   These generators are given by

$$K_\mu = T_\mu + L_{\mu\nu}x^\nu + Rx_\mu + x^\nu x_\nu S_\mu - 2S_\nu x^\nu x_\mu, \tag{8.1.1}$$

   where the constant coefficients T_μ, $L_{\mu\nu} = -L_{\nu\mu}$, R, and S_μ label translations, Lorentz rotations, dilatations, and special conformal transformations, respectively. It can be shown (see e.g. [118]) that the conformal group of complexified Minkowski space is isomorphic to the projective general linear group PGL(4, $\mathbb{C}$).

The reduction is performed by assuming that the components of A do not depend on one or more variables which parameterize the orbits of some set of generators of the conformal group. The chosen generators must form a Lie sub-algebra of the conformal algebra, as otherwise the reduction would not be consistent. These generators integrate to a subgroup $H \subset PGL(4, \mathbb{C})$.

2. A real section.

   To obtain hyperbolic equations in lower dimensions we need to work with ASDYM in the neutral signature. For elliptic reductions one chooses the Euclidean reality conditions. Once the choice is made, the reductions are partially classified by rank and signature of the metric tensor on $M_{\mathbb{C}}$ restricted to the space of orbits of H.
3. The gauge group G.
4. Canonical forms of the Higgs fields.

   Any generator of $X \in H$ will correspond to a Higgs field:

$$\Phi = X \lrcorner A.$$

The gauge transformation $g \in G$ is also invariant in the sense that $X(g) = 0$, so (6.1.1) reduces to

$$\Phi \longrightarrow \Phi' = g\Phi g^{-1}.$$

This transformation can be used to put Φ into a canonical form which depends on the Jordan normal form.

Here we are assuming that the infinitesimal action of H on $M_{\mathbb{C}}$ is free, and the invariant gauges exist. If the action is not free (e.g if $H = SO(3, \mathbb{C})$) the invariant gauges do not have to exist, which leads to additional complications. See [118] for the full discussion.

Below we shall give several examples of symmetry reductions. In all cases H will be an abelian subgroup of the conformal group generated by translations in $M_{\mathbb{C}}$ and we will not need to worry about non-trivial lifts of H to the YM bundle and non-invariant gauges.

- **Example.** Let us impose the Euclidean reality conditions and consider a reduction of ASDYM by a one-dimensional group of translations. The simplest way to impose a symmetry is to drop the dependence on x^4. In this case one must also restrict the gauge transformations so that they too do not depend on x^4 and this implies that the component

$$-\Phi = \partial_4 \lrcorner A$$

transforms homogeneously under gauge transformations $\Phi \to g\Phi g^{-1}$. We have

$$F_{i4} = -\partial_i \Phi - [A_i, \Phi] = -D_i \Phi,$$

and the ASDYM equations (6.4.23) reduce down to

$$\frac{1}{2}\varepsilon_{ijk} F_{jk} = D_i \Phi$$

which are the Bogomolny equations (6.3.16). One could now guess that the solutions to the Bogomolny equations with boundary conditions (6.3.11) give rise to solutions of ASDYM with finite action by

$$A = -\Phi(x^j)dx^4 + A_i(x^j)dx^i.$$

This is not the case. The resulting YM potential is constant in the x^4 directions so the (6.3.11) does not hold.

In the next three examples we shall choose the neutral reality conditions where all coordinates $(z, \tilde{z}, w, \tilde{w})$ are real. The fourth example uses Euclidean reality conditions.

- **Example.** Consider the $SU(2)$ ASDYM in neutral signature and choose a gauge $A_{\tilde{z}} = 0$. Let $T_\alpha, \alpha = 1, 2, 3$ be 2×2 constant matrices such that

  $$[T_\alpha, T_\beta] = -\epsilon_{\alpha\beta\gamma} T_\gamma.$$

  Then ASDYM equations are solved by the ansatz

  $$A_w = 2\cos\phi\, T_1 + 2\sin\phi\, T_2, \quad A_{\tilde{w}} = 2T_1, \quad \text{and} \quad A_z = \partial_z \phi\, T_3$$

  provided that $\phi = \phi(z, \tilde{z})$ satisfies

  $$\phi_{z\tilde{z}} + 4\sin\phi = 0$$

  which is the Sine-Gordon equation.
- **Example.** In Section 3.3.3 we have obtained a zero-curvature representation of the KdV equation given by (3.3.21). We shall extend this representation to the Lax pair of ASDYM in neutral signature with two symmetries, exactly one of which is null. First we need to get rid of the quadratic term in λ from the KdV Lax pair. This is easily done, as $[\partial_x - U_L, \partial_t - V_A] = 0$ is equivalent to

  $$[\partial_x - U_L, \partial_t - V_A - 4\lambda(\partial_x - U_L)] = 0.$$

  For convenience we replace λ by $\lambda/4$ in this last expression so that it takes the form

  $$[\partial_x - A + \lambda B, \partial_t + C + \lambda(-\partial_x + D)] = 0$$

where (A, B, C, D) are matrices given by

$$A = \begin{pmatrix} 0 & 1 \\ u & 0 \end{pmatrix}, \quad B = \begin{pmatrix} 0 & 0 \\ 1/4 & 0 \end{pmatrix},$$

$$C = \begin{pmatrix} u_x & -2u \\ -2u^2 + u_{xx} & -u_x \end{pmatrix}, \quad \text{and} \quad D = \begin{pmatrix} 0 & 0 \\ u/2 & 0 \end{pmatrix}, \tag{8.1.2}$$

where $u = u(x, t)$. This can be extended to the ASDYM Lax pair. Introduce two auxiliary coordinates (p, q) and set

$$L = \partial_p + \partial_x - A + \lambda(\partial_q + B) \quad \text{and} \quad M = \partial_t + C + \lambda(\partial_p - \partial_x + D). \tag{8.1.3}$$

Then the KdV equation (2.1.1) is equivalent to $[L, M] = 0$. But the Lax pair (8.1.3) is in the form (7.1.6), and so it gives rise to a solution to the ASDYM equations, where the underlying metric is

$$ds^2 = dp^2 - dx^2 - 4dqdt.$$

The solution admits two translational symmetries: $\partial/\partial q$, which is null, and $\partial/\partial p$. The Higgs field B corresponding to the null symmetry is nilpotent, which is a gauge-invariant property. This fact was used by Mason and Sparling to establish a converse to this construction

Proposition 8.1.1 (Mason–Sparling [117]) *Any solution to the ASDYM equations with the gauge group* $SL(2, \mathbb{R})$ *and invariant under two translations exactly one of which is null, and such that the Higgs field corresponding to the null translation is nilpotent is gauge equivalent to* (8.1.2).

The proof given in [117] comes down to exploring the gauge freedom in reduced ASDYM. First a gauge is chosen so that the Higgs field B is given by a constant nilpotent matrix, and then it is shown that the ASDYM equations and the residual gauge freedom imply the form (8.1.2) for (A, C, D). Note that the gauge choices made by Mason and Sparing are different than the ones we use in (8.1.2).

- **Example.** This example deals with a reduction of $SL(3, \mathbb{R})$ ASDYM. We shall require that the connection possesses two commuting translational symmetries X_1, X_2 which in our coordinates are $X_1 = \partial_w$ and $X_2 = \partial_{\tilde{w}}$. Direct calculation shows [58] that the ASDYM equations are solved by the following ansatz for Higgs fields $Q = A_w$ and $P = A_{\tilde{w}}$, and gauge fields A_z and $A_{\tilde{z}}$:

$$A_w = \begin{pmatrix} 0 & 0 & 0 \\ 0 & 0 & 0 \\ e^{\phi} & 0 & 0 \end{pmatrix}, \quad A_{\tilde{w}} = \begin{pmatrix} 0 & 0 & 1 \\ 0 & 0 & 0 \\ 0 & 0 & 0 \end{pmatrix},$$

$$A_z = \begin{pmatrix} \phi_z & 0 & 0 \\ 1 & -\phi_z & 0 \\ 0 & 1 & 0 \end{pmatrix}, \quad \text{and} \quad A_{\tilde{z}} = \begin{pmatrix} 0 & e^{-2\phi} & 0 \\ 0 & 0 & e^{\phi} \\ 0 & 0 & 0 \end{pmatrix}, \tag{8.1.4}$$

iff $\phi(z, \tilde{z})$ satisfies the Tzitzéica equation [164]

$$\phi_{z\tilde{z}} = e^{\phi} - e^{-2\phi}. \tag{8.1.5}$$

This equation first arose in a study of surfaces in $\mathbb{R}^3$ for which the ratio of the negative Gaussian curvature to the fourth power of a distance from a tangent plane to some fixed point is a constant. Tzitzéica has shown [164] that if z and $\tilde{z}$ are coordinates on such a surface in which the second fundamental form is off-diagonal, then there exists a real function $\phi(z, \tilde{z})$ such that the Peterson–Codazzi equations reduce to (8.1.5).

This reduction can also be achieved in a gauge-invariant manner and the analogue of Proposition 8.1.1 can be established.

Proposition 8.1.2 [58] *Let A be a solution of the $SL(3, \mathbb{R})$ ASDYM equations in the neutral signature invariant under two non-null translations X_1, X_2 such that the metric on the space of orbits of these translations has signature $(+-)$. Then the coordinates can be chosen so that $X_1 = \partial_w$ and $X_2 = \partial_{\tilde{w}}$, and $(Q = A_w, P = A_{\tilde{w}}, A_z, A_{\tilde{z}})$ can be transformed into (8.1.4) by a gauge and coordinate transformation iff the following conditions hold:*

1. *P and Q have minimal polynomial t^2, with $Tr(PQ) \neq 0$.*
2. *$Tr\left[(D_z P)^2\right] = 0 = Tr\left[(D_{\tilde{z}} Q)^2\right]$ and $Tr\left[(D_z P)^2 (D_{\tilde{z}} Q)^2\right] > 0$.*
3. *$Tr\left[(PQ)^4 + (PQ)^2 (D_z P)(D_{\tilde{z}} Q) - PQ(D_z P)QP(D_{\tilde{z}} Q)\right] = 0$.*

The details are more complicated than in the KdV case. This is because there are more normal forms in the $SL(3)$ case, and additional gauge-invariant conditions (2) and (3) need to be imposed to select the normal forms leading to the Tzitzéica equation. Dropping the condition (3) leads to the $\mathbb{Z}_3$ reduction of the two-dimensional Toda chain [119]. See [58] for details.

- **Example.** By imposing three translational symmetries one can reduce ASDYM to an ODE. Choose the Euclidean reality condition, and assume that the YM potential is independent on x^1, x^2, and x^3.

  Select a gauge $A_4 = 0$, and set $A_j = \Phi_j$, where the Higgs fields Φ_j are real $\mathfrak{g}$-valued functions of $x^4 = t$. The ASDYM equations reduce to the Nahm equations

$$\dot{\Phi}_1 = [\Phi_2, \Phi_3], \quad \dot{\Phi}_2 = [\Phi_3, \Phi_1], \quad \text{and} \quad \dot{\Phi}_3 = [\Phi_1, \Phi_2]. \tag{8.1.6}$$

These equations admit a Lax representation which comes from taking a linear combination of L and M in (7.1.6). Let

$$A(\lambda) = (\Phi_1 + i\Phi_2) + 2\Phi_3\lambda - (\Phi_1 - i\Phi_2)\lambda^2.$$

Then

$$\begin{aligned}\dot{A} &= [\Phi_2 - i\Phi_1, \Phi_3] + 2[\Phi_1, \Phi_2]\lambda - [\Phi_2 + i\Phi_1, \Phi_3]\lambda^2 \\ &= [A, -i\Phi_3 + i(\Phi_1 - i\Phi_2)\lambda] \\ &= [A, B], \quad \text{where} \quad B = -i\Phi_3 + i(\Phi_1 - i\Phi_2)\lambda. \end{aligned} \tag{8.1.7}$$

This representation reveals the existence of many conserved quantities for (8.1.6). We have

$$\frac{d}{dt}\mathrm{Tr}(A^p) = \mathrm{Tr}(p[A, B]A^{p-1}) = p\mathrm{Tr}(ABA^{p-1} - BA^p) = 0$$

by the cyclic property of trace. Therefore all the coefficients of the polynomials $\mathrm{Tr}[A(\lambda)^p]$ for all p are conserved (which implies that the whole spectrum of $A(\lambda)$ is constant in t).

Taking $G = SU(2)$ and setting $\Phi_k(t) = i\sigma_k w_k(t)$, where σ_k are the Pauli matrices reduces the problem to the Euler equations

$$\dot{w}_1 = w_2w_3, \quad \dot{w}_2 = w_1w_3, \quad \text{and} \quad \dot{w}_3 = w_1w_2,$$

or

$$(\dot{w}_3)^2 = (w_3^2 - C_1)(w_3^2 - C_2)$$

(where C_1 and C_2 are constants) which is solvable by Weierstrass elliptic function.

8.2 Integrable chiral model

We have seen that symmetry reductions of the ASDYM equations in neutral signature can lead to hyperbolic equations in lower dimensions, where one can make direct contact with time evolution of moving solitons. In this section we shall analyse one such reduction in detail.

Consider the ASDYM equations in neutral signature invariant under the action of one-parameter group of non-null translations. The underlying metric on $\mathbb{R}^{2,2}$ is of the form (7.1.1) where all the coordinates are real. Let us assume, without loss of generality, that the metric on the three-dimensional space of orbits has signature (2, 1). Therefore we can choose the coordinates such that the Killing vector is $K = \partial/\partial\tau$ and

$$z = \frac{x-\tau}{\sqrt{2}}, \quad \tilde{z} = \frac{x+\tau}{\sqrt{2}}, \quad w = \frac{t+y}{\sqrt{2}}, \quad \text{and} \quad \tilde{w} = \frac{t-y}{\sqrt{2}}$$

where (t, x, y, τ) are all real. With these choices the Higgs field is $\Phi = \partial_\tau \lrcorner A = (A_{\tilde{z}} - A_z)/\sqrt{2}$ and the ASDYM equations become

$$D_x\Phi = F_{yt}, \quad D_y\Phi = F_{tx}, \quad \text{and} \quad D_t\Phi = F_{yx},$$

or

$$D\Phi = *F, \tag{8.2.8}$$

where $*$ is the Hodge operator on $\mathbb{R}^{2,1}$ taken with respect to the Minkowski metric

$$ds^2 = \eta_{\mu\nu}dx^\mu dx^\nu = -dt^2 + dx^2 + dy^2.$$

The first-order Yang–Mills–Higgs system (8.2.8) was introduced by Ward [173]. The unknowns $(A = A_t dt + A_x dx + A_y dy, \Phi)$ are a one-form and a function which depend on local coordinates $x^\mu = (t, x, y)$, and take values in the Lie algebra of gauge group $G = U(n)$. They are subject to gauge transformations

$$A \longrightarrow gAg^{-1} - dg\, g^{-1}, \quad \Phi \longrightarrow g\Phi g^{-1}, \quad \text{and} \quad g = g(x, y, t) \in U(n). \tag{8.2.9}$$

The system (8.2.8) formally resembles the BPS equations (6.3.16) arising in the study of non-abelian monopoles. We shall however see that the Lorentzian reality conditions dramatically change the overall structure and the behaviour of solutions.

There are not any known examples of Lorentz-invariant integrable equations admitting time-dependent soliton solutions in 2 + 1 dimensions, but the system (8.2.8) almost does the job: it is Lorentz invariant and in the next section moving soliton solutions will be written down explicitly. It cannot however be regarded as a genuine soliton system, because the energy functional associated to the Lagrangian density

$$\mathcal{L} = \frac{1}{2}\mathrm{Tr}(F_{\mu\nu}F^{\mu\nu}) - \mathrm{Tr}(D_\mu\Phi D^\mu\Phi) \tag{8.2.10}$$

is not positive-definite and its density vanishes on all solutions to (8.2.8). Note that $\mathcal{L}$ cannot be regarded as a Lagrangian for (8.2.8) as the system (8.2.8) is first order and the Euler–Lagrange equation arising from $\mathcal{L}$ are second order. There is however a connection: the second-order Euler–Lagrange equations are satisfied by solutions to the first-order system (8.2.8).

There exists a different positive functional associated to (8.2.8). To see it, note that the equations (8.2.8) arise as the integrability conditions for an overdetermined system of linear Lax equations:

$$L_0\Psi = 0 \quad \text{and} \quad L_1\Psi = 0, \quad \text{where} \quad L_0 = D_y + D_t - \lambda(D_x + \Phi) \quad \text{and}$$

$$L_1 = D_x - \Phi - \lambda(D_t - D_y), \tag{8.2.11}$$

which is the reduction of the ASDYM Lax pair (7.1.6). The extended solution Ψ is a $GL(n, \mathbb{C})$-valued function of x^μ and a complex parameter $\lambda \in \mathbb{CP}^1$, which satisfies the unitary reality condition

$$\Psi(x^\mu, \bar{\lambda})^* \Psi(x^\mu, \lambda) = 1. \tag{8.2.12}$$

The matrix Ψ is also subject to gauge transformation $\Psi \longrightarrow g\Psi$. The integrability conditions for (8.2.11) imply the existence of a gauge $A_t = A_y$ and $A_x = -\Phi$, and a matrix $J : \mathbb{R}^3 \longrightarrow U(n)$ such that

$$A_t = A_y = \frac{1}{2}J^{-1}(J_t + J_y) \quad \text{and} \quad A_x = -\Phi = \frac{1}{2}J^{-1}J_x,$$

where $J_\mu = \partial_\mu J$. With this gauge choice the equations (8.2.8) become

$$(J^{-1}J_t)_t - (J^{-1}J_x)_x - (J^{-1}J_y)_y - [J^{-1}J_t, J^{-1}J_y] = 0. \tag{8.2.13}$$

A positive-definite conserved energy functional can now be introduced by

$$E = \int_{\mathbb{R}^2} \mathcal{E} dxdy, \tag{8.2.14}$$

where the energy density is given by

$$\mathcal{E} = -\frac{1}{2}\mathrm{Tr}\left[(J^{-1}J_t)^2 + (J^{-1}J_x)^2 + (J^{-1}J_y)^2\right]. \tag{8.2.15}$$

This came at the price of losing the full Lorentz invariance since the commutator term fixes a space-like direction. If we rewrite (8.2.13) as

$$(\eta^{\mu\nu} + V_\alpha \varepsilon^{\alpha\mu\nu})\partial_\mu(J^{-1}\partial_\nu J) = 0$$

then the fixed direction is given by[1] a space-like vector $V = \partial/\partial x$. This breaks the symmetry to $SO(1, 1)$. In general the finiteness of E is ensured by imposing the boundary condition (valid for all t)

$$J = J_0 + J_1(\varphi)r^{-1} + O(r^{-2}) \quad \text{as} \quad r \longrightarrow \infty, \quad x + iy = re^{i\varphi} \tag{8.2.16}$$

[1] Manakov and Zakharov [110] studied a closely related system of equations where the unit vector V was taken to be time-like. The resulting equations do not appear to have a positive-definite energy functional, and no static solutions can exist globally on $\mathbb{R}^2$.

where J_0 is a constant matrix, and so for a fixed value of t the matrix J extends to a map from S^2 (the conformal compactification of $\mathbb{R}^2$) to $U(n)$.

The equation (8.2.13) is known as the Ward model or the integrable chiral model [173, 176, 177]. The ordinary chiral model in (2+1) dimensions

$$\eta^{\mu\nu}\partial_\mu(J^{-1}\partial_\nu J) = 0$$

has $V = 0$. It is fully Lorentz–invariant, but it lacks integrability, and explicit time-dependent solutions (other than trivial Lorentz boosts of static solutions) cannot be constructed. In the remainder of this chapter we shall study the solutions and properties of (8.2.13).

We shall finish this section by presenting another gauge choice leading to a different potential formulation of (8.2.8). Choose the familiar gauge $A_y = A_t$, $A_x = -\Phi$. The vanishing of the term proportional to λ in the compatibility conditions (8.2.11) implies the existence of $K : \mathbb{R}^{2,1} \longrightarrow \mathfrak{u}(n)$ such that

$$A_y = A_t = \frac{1}{2}K_x \quad \text{and} \quad A_x = -\Phi = \frac{1}{2}(K_t - K_y),$$

where $K_\mu = \partial_\mu K$. The zeroth-order term in the compatibility conditions now yields

$$K_{tt} - K_{xx} - K_{yy} + [K_x, K_t - K_y] = 0. \tag{8.2.17}$$

The relation between $K \in \mathfrak{u}(n)$ and $J \in U(n)$ is

$$K_x = J^{-1}(J_t + J_y) \quad \text{and} \quad K_t - K_y = J^{-1}J_x,$$

and exhibits a duality between the two formulations: the compatibility condition $K_{xt} - K_{xy} = K_{tx} - K_{yx}$ yields the field equation (8.2.13). The K-equation (8.2.17) admits a Lagrangian formulation with the Lagrangian density

$$-\mathrm{Tr}\left\{\frac{1}{2}((K_t)^2 - (K_x)^2 - (K_y)^2) - \frac{1}{3}K[K_x, K_t - K_y]\right\}.$$

The Lagrangian formulation of (8.2.13) is more complicated – we shall present it in Section 8.2.2.

8.2.1 Soliton solutions

One method [173] of constructing explicit solutions is based on the associated linear problem (8.2.11). Let $\Psi(x^\mu, \lambda)$ be the fundamental matrix solution to the Lax pair (8.2.11) and let $u = (t + y)/2$, $v = (t - y)/2$. Then

$$A_u - \lambda(A_x + \Phi) = (-\partial_u\Psi + \lambda\partial_x\Psi)\Psi^{-1} \quad \text{and}$$

$$A_x - \Phi - \lambda A_v = (-\partial_x\Psi + \lambda\partial_v\Psi)\Psi^{-1}, \tag{8.2.18}$$

and in the gauge leading to (8.2.13) we have $A_v = A_x + \Phi = 0$. Thus in this gauge given a solution Ψ to the linear system (8.2.11) one can construct a solution to (8.2.13) by

$$J(x^\mu) = \Psi^{-1}(x^\mu, \lambda = 0) \tag{8.2.19}$$

and all solutions to (8.2.13) arise from some Ψ. This can be an effective method of finding solutions (also known as the 'Riemann problem with zeros'), if we know $\Psi(x^\mu, \lambda)$ in the first place. One class of solutions can be obtained by assuming that

$$\Psi = 1 + \sum_{k=1}^{m} \frac{N_k(x, y, t)}{\lambda - \mu_k}, \quad \text{where} \quad \mu_k = \text{const.} \tag{8.2.20}$$

Let us restrict to the case where $G = SU(2)$. The unitarity condition (8.2.12) implies rank $N_k = 1$. Thus $(N_k)_{\alpha\beta} = n^k_\alpha m^k_\beta$. Demanding that the RHS of (8.2.18) is independent of λ (like the LHS) yields $N_k = N_k(\omega_k)$, where

$$\omega_k = u\mu_k^2 + x\mu_k + v.$$

It also follows that

$$n^k_\alpha = -\sum_{l=1}^{m} (\Gamma^{-1})^{kl} \overline{m}^l_\alpha$$

where the $m \times m$ matrix Γ is given by

$$\Gamma^{kl} = \sum_{\alpha=1}^{2} (\overline{\mu}_k - \mu_l)^{-1} \overline{m}_\alpha{}^k m_\alpha{}^l.$$

We can use the homogeneity of the extended solution in m^k_α to rescale m^k_α and set $m^k_\alpha = (1, f_k(\omega_k))$. Finally, dividing Ψ by the square root of its determinant to achieve $\det \Psi = 1$ yields

$$(J^{-1})_{\alpha\beta} = \chi^{-1/2} \left[\delta_{\alpha\beta} + \sum_{k,l} \mu_k^{-1} (\Gamma^{-1})^{kl} \overline{m}_\alpha{}^l m^k_\beta \right], \quad \text{where} \quad \chi = \prod_{k=1}^{m} \frac{\overline{\mu}_k}{\mu_k}. \tag{8.2.21}$$

The soliton solutions correspond to rational functions $f_k(\omega_k)$.

- **Example.** The solution (8.2.21) with $m = 1$ and $\mu_1 = \mu = |\mu| e^{i\phi}$ is given by

$$J_1 = \frac{1}{1 + |f|^2} \begin{pmatrix} e^{i\phi} + e^{-i\phi}|f|^2 & 2i \sin\phi \, f \\ 2i \sin\phi \, \overline{f} & e^{-i\phi} + e^{i\phi}|f|^2 \end{pmatrix}, \tag{8.2.22}$$

where $f = f(u\mu^2 + x\mu + v)$ is a holomorphic, rational function.

To obtain the static solution put $\mu = i$ which gives

$$J = \frac{i}{1+|f|^2}\begin{pmatrix} 1-|f|^2 & 2f \\ 2\overline{f} & |f|^2-1 \end{pmatrix}, \tag{8.2.23}$$

where the holomorphic function f is rational in $z = x + iy$ and $f(z) \longrightarrow 1$ as $|z| \longrightarrow \infty$. All such maps are classified by integer winding numbers N with values in $\pi_2(S^2) = \mathbb{Z}$. This integer is precisely the degree of f: for a given N, f is of the form

$$f(z) = \frac{p(z)}{r(z)} = \frac{(z-p^1)\cdots(z-p^N)}{(z-r^1)\cdots(z-r^N)}. \tag{8.2.24}$$

The N static lumps are positioned at $(p^1, \ldots, p^N)$, as the maxima of $\mathcal{E}$ occur at these points. For $\mu \neq \pm i$ there is time dependence, and $m > 1$ corresponds to m solitons moving with different velocities which however do not scatter.

Allowing Ψ to have poles of order higher than one gives solutions which exhibit soliton scattering. Explicit time-dependent solutions corresponding to scattering can be obtained by choosing $\mu_1 = i + \varepsilon$, $\mu_2 = i - \varepsilon$ in (8.2.20) with $m = 2$ and taking the limit $\varepsilon \to 0$. This yields [178]

$$J_2 = \left(1 - \frac{2q_1{}^* \otimes q_1}{||q_1||^2}\right)\left(1 - \frac{2q_2{}^* \otimes q_2}{||q_2||^2}\right), \tag{8.2.25}$$

where

$$q_1 = (1, f), \quad q_2 = (1+|f|^2)(1, f) - 2i(tf' + h)(\overline{f}, -1), \tag{8.2.26}$$

and f and h are rational functions of $z = x + iy$. In [178] 90° scattering is illustrated by choosing $f = z$ and $h = z^2$. The positions of the solitons correspond to the maxima of the energy density which in this case is given by

$$\mathcal{E} = 128\frac{3x^2 + 3y^2 - 10tx^2 + 10ty^2 + t^2x^2 + t^2y^2 - 4t^2 - 3x^4 - 6x^2y^2 - 3y^4 + 1}{\left(4t^2 + 8tx^2 - 8ty^2 + 5x^4 + 10x^2y^2 + 5y^4 + 1 + 2x^2 + 2y^2\right)^2}.$$

The following series of plots of this energy density demonstrates soliton scattering. Two solitons approach along the x-axis, collide by forming a ring, and finally move away scattered by 90° along the y-axis.

Two-soliton scattering. Energy density at times $t = -1, -0.2, 0, 0.2$, and 1.

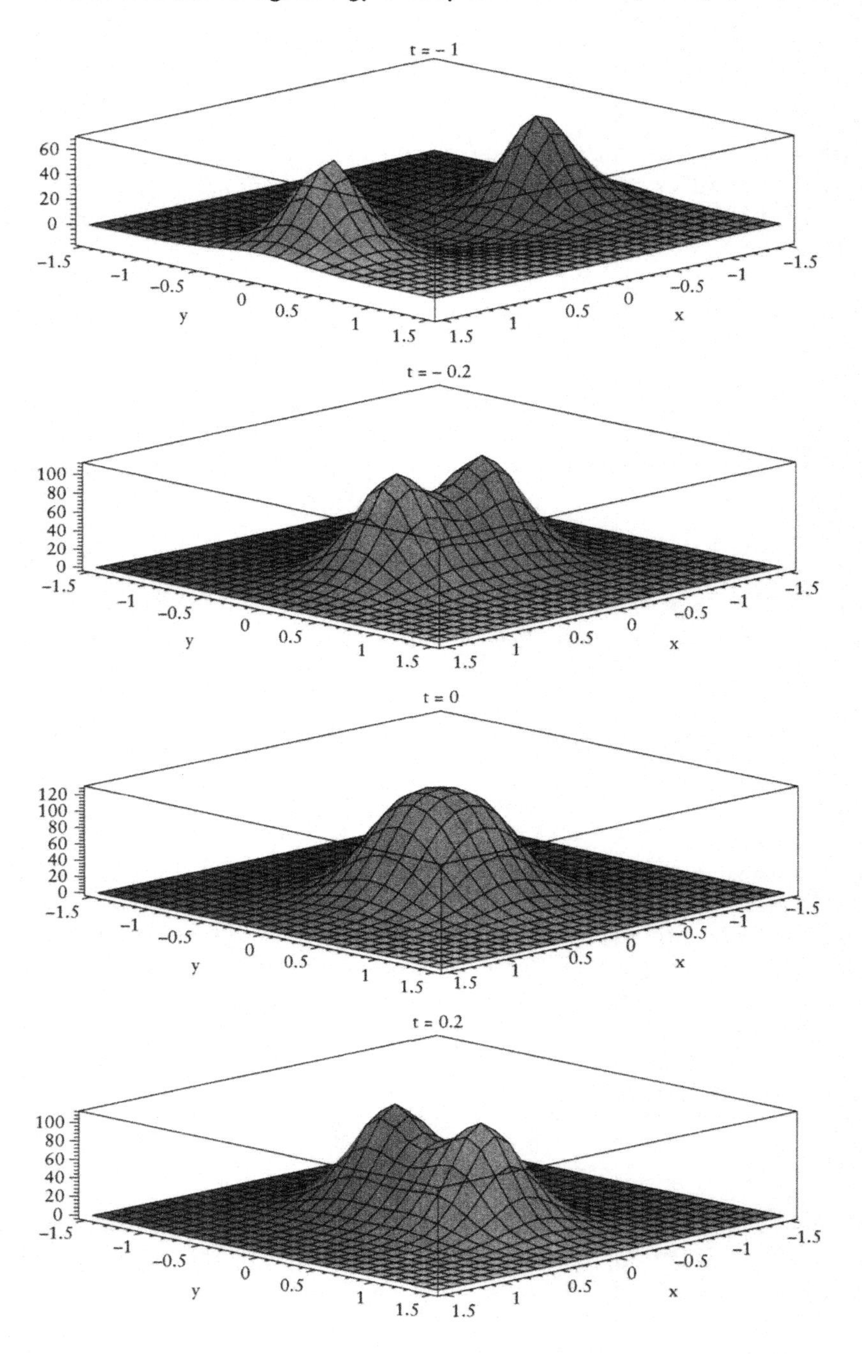

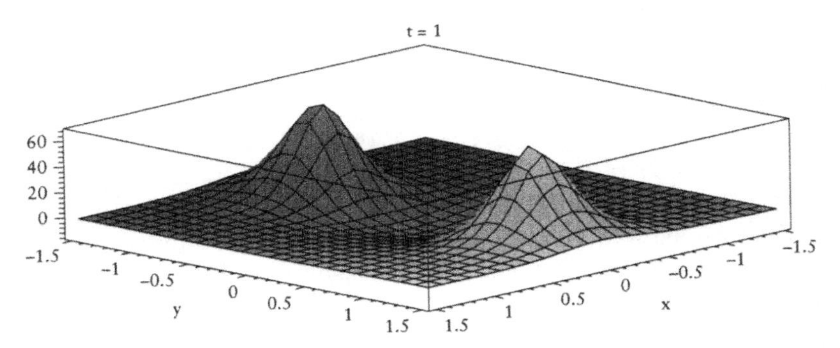

More complicated examples were considered in [35, 86].

8.2.1.1 *'Trivial-scattering' boundary condition*

The last example fits into a class of time-dependent solutions which depend on finite number of parameters. This finiteness of the energy alone does not pick this finite-dimensional family and one needs stronger boundary conditions which we shall now describe.

Let us restrict Ψ from $\mathbb{R}^{2,1} \times \mathbb{CP}^1$ to the space-like plane $t = 0$. We shall also restrict the spectral parameter to lie in the real equator $S^1 \subset \mathbb{CP}^1$ parameterized by θ:

$$\Psi(t, x, y, \lambda) \longrightarrow \psi(0, x, y, \theta) := \Psi(x, y, -\cot\frac{\theta}{2}), \tag{8.2.27}$$

where now $\psi : \mathbb{R}^2 \times S^1 \longrightarrow U(n)$ and we made the change of variable for real $\lambda = -\cot(\frac{\theta}{2})$. Note that ψ satisfies

$$(u^\mu D_\mu - \Phi)\psi = 0, \tag{8.2.28}$$

where the operator annihilating ψ is the spatial part of the Lax pair (8.2.11), given by

$$\frac{\lambda L_0 + L_1}{1 + \lambda^2} = u^\mu D_\mu - \Phi, \quad \text{where}$$

$$\mathbf{u} = \left(0, \frac{1 - \lambda^2}{1 + \lambda^2}, \frac{2\lambda}{1 + \lambda^2}\right) = (0, -\cos\theta, -\sin\theta)$$

is a unit vector tangent to the $t = 0$ plane.

Definition 8.2.1 *The matrix J satisfies the 'trivial-scattering' boundary condition* [4, 179] *if*

$$\psi(x, y, \theta) \longrightarrow \psi_0(\theta) \quad as \quad r \longrightarrow \infty, \tag{8.2.29}$$

where $\psi_0(\theta)$ is a $U(n)$-valued function on S^1.

We shall now demonstrate that if J satisfies the trivial-scattering boundary condition then ψ extends to a map from S^3 to $U(n)$. First note that (8.2.29) implies the existence of the limit of ψ at spatial infinity for all values of θ, while the finite-energy boundary condition (8.2.16) only implies the limit at $\theta = \pi$. Thus the condition (8.2.29) extends the domain of ψ to $S^2 \times S^1$. However (8.2.29) is also a sufficient condition for ψ to extend to the suspension $SS^2 = S^3$ of S^2. This can be seen as follows. The domain $S^2 \times S^1$ can be considered as $S^2 \times [0, 1]$ with $\{0\}$ and $\{1\}$ identified. The *suspension* SX of a manifold X is the quotient space [21]

$$SX = ([0, 1] \times X)/((\{0\} \times X) \cup (\{1\} \times X)).$$

This definition is compatible with spheres in the sense that $SS^d = S^{d+1}$.

Now the only condition ψ needs to fulfil for the suspension is an equivalence relation between all the points in $S^2 \times \{0\}$, since such relation for $S^2 \times \{1\}$ will follow from the identification of $\{0\}$ and $\{1\}$. This equivalence can be achieved by choosing a gauge

$$\psi(x, y, 0) = 1. \tag{8.2.30}$$

Therefore ψ extends to a map from $SS^2 = S^3$ to $U(n)$ if it satisfies the zero-scattering boundary condition. In addition, after fixing the gauge (8.2.30), there is still some residual freedom in ψ given by

$$\psi \longrightarrow \psi K, \tag{8.2.31}$$

where $K = K(x, y, \theta) \in U(n)$ is annihilated by $u^\mu \partial_\mu$. Setting $K = [\psi_0(\theta)]^{-1}$ results in

$$\psi(\{\infty\}, \theta) = 1. \tag{8.2.32}$$

The gauge (8.2.32) picks a base point $\{x_0 = \infty\} \in S^2$, and this implies that the trivial-scattering condition is also sufficient for ψ to extend to the reduced suspension of S^2, given by

$$S_{\text{red}} S^2 = ([0, 1] \times S^2)/((\{0\} \times S^2) \cup (\{1\} \times S^2) \cup ([0, 1] \times \{x_0\})).$$

This is also homeomorphic to S^3. The idea of (reduced) suspension is illustrated in (Figure 8.1).

Now let us justify the term 'trivial scattering' in (8.2.29). Consider equation (8.2.28) and restrict it to a line $(x(\sigma), y(\sigma)) = (x_0 - \sigma \cos\theta, y_0 - \sigma \sin\theta)$, $\sigma \in \mathbb{R}$ on the $t = 0$ plane. The equation (8.2.28) becomes an ODE describing the propagation of

$$\psi = \psi(x_0 - \sigma \cos\theta, y_0 - \sigma \sin\theta, \theta)$$

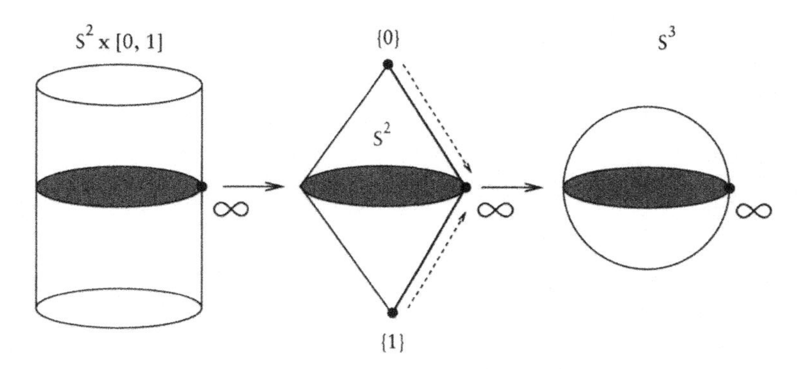

Figure 8.1 *Suspension and reduced suspension*

along the oriented line through (x_0, y_0) in $\mathbb{R}^2$. We can choose a gauge such that

$$\lim_{\sigma \to -\infty} \psi = \mathbf{1},$$

and define the scattering matrix $S : TS^1 \to U(n)$ on the space of oriented lines in $\mathbb{R}^2$ as

$$S = \lim_{\sigma \to \infty} \psi. \tag{8.2.33}$$

The trivial-scattering condition (8.2.29) then implies this matrix is trivial,

$$S = \mathbf{1}. \tag{8.2.34}$$

As we have explained, the boundary conditions (8.2.16) and (8.2.29) imply that for each value of θ the function ψ extends to the one-point compactification S^2 of $\mathbb{R}^2$. The straight lines on the plane are then replaced by the great circles, and in this context the trivial-scattering condition implies that the differential operator $u^\mu D_\mu - \Phi$ has trivial monodromy along the compactification $S^1 = \mathbb{R} \cup \{\infty\}$ of a straight line parameterized by σ.

8.2.1.2 *Time-dependent unitons*

A general class of solutions to (8.2.13) which satisfy the trivial-scattering boundary condition (8.2.29) is given by the so-called *time-dependent unitons*

$$J(x, y, t) = M_1 M_2 \cdots M_m, \tag{8.2.35}$$

where the unitary matrices M_k, $k = 1, \ldots, m$, are given by

$$M_k = i \left[\mathbf{1} - \left(1 - \frac{\mu}{\bar{\mu}} \right) R_k \right] \quad \text{and} \quad R_k \equiv \frac{q_k^* \otimes q_k}{||q_k||^2}. \tag{8.2.36}$$

Here $\mu \in \mathbb{C} \backslash \mathbb{R}$ is a non-real constant and $q_k = (1, f_{k1}, \ldots, f_{k(N-1)}) \in \mathbb{C}^N$, with $k = 1, \ldots, m$, are vectors whose components $f_{kj} = f_{kj}(x^\mu) \in \mathbb{C}$ are smooth functions which tend to a constant at spatial infinity.[2] The terminology 'trivial scattering' is rather confusing as solitons described by (8.2.35) do physically scatter in $\mathbb{R}^2$ if $m > 1$.

If $m = 1$ then q_1 is holomorphic and rational in

$$\omega = \mu x + \frac{1}{2}\mu^2(t+y) + \frac{1}{2}(t-y)$$

and J is a generalization of (8.2.22) to the case $n > 2$. If $m > 1$ q_1 is still holomorphic and rational in ω, but $q_2, q_3, \ldots$ are not holomorphic. For $m > 1$ the Bäcklund transformations [35, 87] can be used to determine the q's recursively as we will show next.

The extended solutions corresponding to the uniton solutions (8.2.35) factorize into the m-uniton factors

$$\Psi = G_m G_{m-1} \cdots G_1, \quad \text{where}$$

$$G_k = -i\left(1 - \frac{\overline{\mu} - \mu}{\lambda - \mu} R_k\right) \in GL(n, \mathbb{C}) \quad \text{and} \quad R_k = \frac{q_k^* \otimes q_k}{||q_k||^2}. \tag{8.2.37}$$

The exact form of q_k's is determined from the Lax pair (8.2.11) by demanding that the expressions

$$[\partial_x \Psi - \lambda(\partial_t - \partial_y)\Psi]\Psi^{-1} \quad \text{and} \quad [(\partial_t + \partial_y)\Psi - \lambda \partial_x \Psi]\Psi^{-1} \tag{8.2.38}$$

are independent of λ.

Proposition 8.2.2 [35, 87] *Let $J : \mathbb{R}^{2,1} \to U(n)$ be a solution to the integrable chiral model* (8.2.13), *and let*

$$M = i\left[1 - \left(1 - \frac{\mu}{\overline{\mu}}\right) R\right] \quad \textit{and} \quad R \equiv \frac{q^* \otimes q}{||q||^2}.$$

Then $\tilde{J} = JM$ is another solution to (8.2.13) *if the Grasmannian projector R satisfies a pair of first-order Bäcklund relations*

$$RL_0|_{\lambda=\mu}(1 - R) = 0 \quad \textit{and} \quad RL_1|_{\lambda=\mu}(1 - R) = 0, \tag{8.2.39}$$

where $L_0|_{\lambda=\mu}$ and $L_1|_{\lambda=\mu}$ are the Lax operators (8.2.11) *evaluated at $\lambda = \mu$.*

Proof Let Ψ be an extended solution to the Lax pair (8.2.11) corresponding to J which satisfies (8.2.13). Set

$$\tilde{\Psi} = G\Psi = -i\left(1 - \frac{\overline{\mu} - \mu}{\lambda - \mu} R\right)\Psi, \quad \text{so that} \quad \tilde{J} = \tilde{\Psi}^{-1}|_{\lambda=0} = JM. \tag{8.2.40}$$

[2] The matrix R_k is a Hermitian projection satisfying $(R_k)^2 = R_k$, and the corresponding M_k is a Grassmanian embedding of $\mathbb{CP}^{n-1}$ into $U(n)$. A more general class of unitons can be obtained from the complex Grassmanian embeddings of $\mathrm{Gr}(K, n)$ into the unitary group. For μ pure imaginary, a complex K-plane $V \subset \mathbb{C}^n$ corresponds to a unitary transformation $i(\pi_V - \pi_{V^\perp})$, where π_V denotes the Hermitian orthogonal projection onto V. The formula (8.2.36) with $\mu = i$ corresponds to $K = 1$ where $\mathrm{Gr}(1, n) = \mathbb{CP}^{n-1}$. See Appendix A for discussion of Grassmanians and their homotopy.

The matrix $\widetilde{\Psi}$ will be an extended solution if expressions (8.2.38) with Ψ replaced by $\widetilde{\Psi}$ are independent of λ. Set $\delta_0 = \partial_u - \lambda\partial_x$ and consider the relation

$$(\delta_0\widetilde{\Psi})\,\widetilde{\Psi}^{-1} = (\delta_0 G)\,G^{-1} - GA_0G^{-1}$$

where $A_0 = J^{-1}\partial_u J$. The unitarity condition (8.2.12) holds as $G(\lambda)G^*(\bar{\lambda}) = 1$. We use it to find G^{-1} and equate the residue of the above expression at $\lambda = \mu$ to zero. This gives

$$(\delta_0 R)(1 - R) - RA_0(1 - R) = 0.$$

The identity $R(1 - R) = 0$ gives $(\delta_0 R)(1 - R) = -R\delta_0(1 - R)$ and finally

$$R(\partial_u - \mu\partial_x + J^{-1}J_u)(1 - R) = 0$$

which is the first relation in (8.2.39). The second relation arises in the same way with δ_0 replaced by $\delta_1 = \partial_x - \lambda\partial_v$ and A_0 replaced by $A_1 = J^{-1}\partial_x J$. The overdetermined Bäcklund relations (8.2.39) are compatible as J is a solution to (8.2.13). □

8.2.2 Lagrangian formulation

The Lagrangian formulation of (8.2.13) contains the Wess–Zumino–Witten (WZW) term [19, 87]. This involves an extended field $\hat{J}$ defined in the interior of a cylinder which has the 2+1 Minkowski space-time as one of its boundary components

$$\hat{J} : \mathbb{R}^{2+1} \times [0, 1] \longrightarrow U(n)$$

such that $\hat{J}(x^\mu, 0)$ is a constant group element, which we take to be the identity $1 \in U(n)$, and $\hat{J}(x^\mu, 1) = J(x^\mu)$.

The equation (8.2.13) can be derived as a stationarity condition for the action functional

$$\begin{aligned} S &= S_C + S_M, \\ S_C &= -\frac{1}{2}\int_{[t_1,t_2]\times\mathbb{R}^2} \mathrm{Tr}\,(\mathbb{J} \wedge *\mathbb{J}), \qquad (8.2.41) \\ S_M &= \frac{1}{3}\int_{[t_1,t_2]\times\mathbb{R}^2\times[0,1]} \mathrm{Tr}\left(\hat{\mathbb{J}} \wedge \hat{\mathbb{J}} \wedge \hat{\mathbb{J}} \wedge \mathbb{V}\right). \end{aligned}$$

Here $*$ is a Hodge star of $\eta_{\mu\nu}$ and

$$\mathbb{J} = J^{-1}J_\mu\, dx^\mu \quad \text{and} \quad \hat{\mathbb{J}} = \hat{J}^{-1}\hat{J}_p\, dx^p, \quad \text{where} \quad p = 0, 1, 2, 3 \equiv t, x, y, \rho,$$

are $\mathfrak{u}(n)$-valued one-forms on $\mathbb{R}^{2,1}$ and $\mathbb{R}^{2,1} \times [0, 1]$, respectively and $\mathbb{V} = 1\,dx$ is a constant one-form on $\mathbb{R}^{2,1} \times [0, 1]$. We make the assumption that the extension $\hat{J}$ is of the form

$$\hat{J}(x^\mu, \rho) = \mathcal{F}(J(x^\mu), \rho) \tag{8.2.42}$$

for some smooth function $\mathcal{F} : U(n) \times [0, 1] \longrightarrow U(n)$. The WZW term S_M in the action is topological in the sense that its integrand does not depend on the metric on $\mathbb{R}^{2,1}$.

Following [184] we can obtain a more geometric picture by regarding the domain of $\hat{J}$ as $B \times \mathbb{R}$, where B is a ball in $\mathbb{R}^3$ with the boundary $\partial B = S^2$, and rewriting S_M as

$$S_M = \int_{[t_1,t_2]\times B} \hat{J}^*(T) \wedge V, \quad \text{where} \quad V = dx.$$

Here T is the preferred three-form [19] on $U(n)$ in the third cohomology group given by $T = \mathrm{Tr}[(\phi^{-1}d\phi)^3]$ for $\phi \in U(n)$. This three-form coincides with torsion of a flat connection ∇ on $U(n)$ which parallel propagates left-invariant vector fields, that is,

$$T(X, Y, Z) = h(\nabla_X Y - \nabla_Y X - [X, Y],\ Z),$$

where $h = -\mathrm{Tr}(\phi^{-1}d\phi\, \phi^{-1}d\phi)$ is the metric on $U(n)$ given in terms of the Maurer–Cartan one-form (this definition makes sense for any matrix Lie group – see Appendix A). The torsion three-form T can be pulled back to B. It is closed, so $T = d\beta$, where β is a two-form on G which can be defined only locally. Stokes' theorem now yields

$$\begin{aligned} S_M &= \int_{[t_1,t_2]\times B} d[\hat{J}^*(\beta) \wedge V] \\ &= \frac{1}{2}\int_{S^2\times[t_1,t_2]} (\varepsilon^{\mu\nu\alpha} V_\alpha)\beta_{ij}(\phi)\partial_\mu \phi^i \partial_\nu \phi^j\, dxdydt, \end{aligned}$$

where $\phi^i = \phi^i(x^\mu)$ are local coordinates on the group (e.g. the components of the matrix J). In the above derivation we have neglected the boundary component $(t_1 \times B) \cup (t_2 \times B)$, as variations of the corresponding integrals vanish identically.

Let us make some comments about the extensions of J used in the variational principle. In general any J can be extended, as the obstruction group $\pi_2[U(n)]$ vanishes. In the case of time-dependent uniton solutions (8.2.35) to (8.2.13) we can be more explicit. All unitons factorize as $J = \prod_k M_k$ into a finite number of time-dependent unitons of the form $M_k = i[\mathbf{1} - (1 - e^{2i\phi_k})R_k]$, where $R_k = R_k(t, x, y)$ are Hermitian projectors, and the real constants ϕ_k are the phases of the poles on the spectral plane. Any of these factors can be extended by

$$M_k \longrightarrow \hat{M}_k = i[\mathbf{1} - (1 - e^{2i\rho\phi_k})R_k], \tag{8.2.43}$$

thus giving the extension $\hat{J} = \prod_k \hat{M}_k$.

8.2.2.1 *Noether charges*

Time translational invariance of the action S gives rise to the conservation of the energy functional with density (8.2.15) which is the same as for the ordinary chiral model in 2+1 dimensions with $V = 0$. Ward [173] has observed that the y-momentum for (8.2.13) also coincides with the expression derived from the ordinary chiral model, but the x-momentum of the ordinary chiral model is not conserved by the time evolution (8.2.13) of the initial data. Here we shall follow [56] and construct the x-momentum using the WZW Lagrangian (8.2.41) written in terms of the torsion on $U(n)$.

The Lagrangian density takes the form

$$\mathcal{L} = -\frac{1}{2}\eta^{\mu\nu}\partial_\mu\phi^i\partial_\nu\phi^j h_{ij}(\phi) + \frac{1}{2}V_\alpha\varepsilon^{\alpha\mu\nu}\beta_{ij}(\phi)\partial_\mu\phi^i\partial_\nu\phi^j,$$

where h is the metric on the group and β is a local two-form potential for the totally anti-symmetric torsion [176]. The conserved Noether energy–momentum tensor is

$$T_{\mu\nu} = \eta_{\mu\nu}\mathcal{L} - \frac{\partial\mathcal{L}}{\partial(\partial^\mu\phi^j)}\partial_\nu\phi^j.$$

The energy corresponding to T_{00} is given by (8.2.14), and the momentum densities are

$$\mathcal{P}_y = T_{02} = -\mathrm{Tr}\,(J^{-1}J_tJ^{-1}J_y) \quad \text{and}$$

$$\mathcal{P}_x = T_{01} = -\mathrm{Tr}\,(J^{-1}J_tJ^{-1}J_x) - \beta_{ij}\partial_x\phi^i\partial_y\phi^j. \tag{8.2.44}$$

The additional term in the conserved x-momentum $P_x = \int_{\mathbb{R}^2}\mathcal{P}_x dxdy$ does not depend on the choice of β, since for fixed t

$$\Theta := \int_{\mathbb{R}^2}\beta_{ij}(\phi)\partial_x\phi^i\partial_y\phi^j dxdy = \int_{\mathbb{R}^2} J^*\beta. \tag{8.2.45}$$

This expression does not change under the transformation $\beta \to \beta + d\alpha$ because $\int_{\mathbb{R}^2} d(J^*\alpha) = 0$ as a consequence of the boundary condition (8.2.16). We can choose the extension

$$\hat{J} = \cos(\pi\rho/2)1 + \sin(\pi\rho/2)J$$

to find the additional term Θ using the identity

$$\int_{\mathbb{R}^2}\beta_{ij}\partial_x\phi^i\partial_y\phi^j \mathrm{d}x\mathrm{d}y = \int_{\mathbb{R}^2}\int_0^1 \mathrm{Tr}\left(\hat{J}^{-1}\hat{J}_\rho\left[\hat{J}^{-1}\hat{J}_y, \hat{J}^{-1}\hat{J}_x\right]\right) d\rho dxdy, \tag{8.2.46}$$

which follows from calculating $\mathcal{P}_x$ in terms of $\hat{J}$ directly from (8.2.41).

- **Example.** Consider the time-dependent one-soliton solution generalizing (8.2.22) to the case of $G = U(n)$:

$$J = i\left[1 - \left(1 - \frac{\mu}{\bar{\mu}}\right)R\right].$$

Here $\mu \in \mathbb{C}/\mathbb{R}$ is a non-real constant, $R = q^* \otimes q/||q||^2$ is the Grassmanian projection (A11) and the components of $q : \mathbb{R}^{2,1} \to \mathbb{C}^n$ are holomorphic and rational in $\omega = x + \frac{\mu}{2}(t+y) + \frac{\mu^{-1}}{2}(t-y)$. In this case the additional term Θ is proportional to the topological charge (A12) which is itself a constant of motion as the time evolution is continuous.

8.2.3 Energy quantization of time-dependent unitons

The fact that the allowed energy levels of some physical systems can take only discrete values has been well known since the the early days of quantum theory. The hydrogen atom and the harmonic oscillator are two well-known examples. In these two cases the boundary conditions imposed on the wave function imply discrete spectra of the Hamiltonians. The reasons are therefore global.

The quantization of energy can also occur at the classical level in non-linear field theories if the energy of a smooth-field configuration is finite. The reasons are again global: The potential energy of static-soliton solutions in the Bogomolny limit of certain field theories must be proportional to integer homotopy classes of smooth maps. The details depend on the model: In pure gauge theories the energy of solitons satisfying the Bogomolny equations is given by one of the Chern numbers of the curvature. The equalities of the BPS bounds (6.3.15) or (6.4.24) are examples of this mechanism. In scalar (2+1)-dimensional sigma models, allowed energies of Bogomolny solitons are given by elements of $\pi_2(\Sigma)$, where the manifold Σ is the target space. The example is given by equality in Proposition 5.4.1.

The situation is different for moving solitons: The total energy is the sum of kinetic and potential terms, and the Bogomolny bound is not saturated. One expects that the moving (non-periodic) solitons will have continuous energy. Attempts to construct theories with quantized total energy based on compactifying the time direction are physically unacceptable, as they lead to paradoxes related to the existence of closed time-like curves.

In this section we shall follow [55] and demonstrate that the energy of (8.2.35) is quantized in the rest frame,[3] and given by the third homotopy class of the extended solution to (8.2.13). Restricting the extended solution Ψ to a space-like plane in $\mathbb{R}^3$ and an equator in the Riemann sphere of the spectral parameter gives a map ψ, whose domain is $\mathbb{R}^2 \times S^1$. If J is an m-uniton solution (8.2.35), the corresponding extended solution satisfies stronger boundary conditions which promote ψ to a map $S^3 \longrightarrow U(n)$ as we have explained in Section 8.2.1. We will prove

[3] The model (8.2.13) is $SO(1,1)$ invariant, and the Lorentz boosts correspond to rescaling μ by a real number. The rest frame corresponds to $|\mu| = 1$, when the y-component of the momentum vanishes. The $SO(1,1)$-invariant generalization of (8.2.47) is given in [55].

Theorem 8.2.3 *[55] The total energy of the m-uniton solution* (8.2.35) *with* $\mu = |\mu|e^{i\phi}$ *is quantized and equal to*

$$E_{(m)} = 4\pi \left(\frac{1 + |\mu|^2}{|\mu|}\right) |\sin\phi| \, [\psi], \tag{8.2.47}$$

where for any fixed value of t the map $\psi : S^3 \longrightarrow U(n)$ *is given by*

$$\psi = \prod_{k=m}^{1} \left[1 + \frac{\overline{\mu} - \mu}{\mu + \cot\left(\frac{\theta}{2}\right)} R_k\right], \qquad \theta \in [0, 2\pi], \tag{8.2.48}$$

and

$$[\psi] = \frac{1}{24\pi^2} \int_{S^3} Tr \, [(\psi^{-1} d\psi)^3] \tag{8.2.49}$$

takes values in $\pi_3[U(n)] = \mathbb{Z}$.

In Section 8.2.1 we have explained that we can regard ψ as a map from S^3 to $U(n)$. All such maps are characterized by their homotopy type [21] given by (8.2.49). The element $[\psi] \in \pi_3[U(n)]$ is invariant under continuous deformations of ψ.

The restricted map ψ (8.2.27) corresponding to (8.2.37) is given by

$$\psi = g_m g_{m-1} \cdots g_1, \quad \text{where} \quad g_k = 1 + \frac{\bar{\mu} - \mu}{\mu + \cot\left(\frac{\theta}{2}\right)} R_k \in U(n), \tag{8.2.50}$$

where $\lambda = -\cot\left(\frac{\theta}{2}\right) \in S^1 \subset \mathbb{CP}^1$ as before and all the maps are restricted to the $t = 0$ plane (we have removed the irrelevant constant factor $(-i)$ from each map g_k).

Each element g_k has the limit at spatial infinity for all values of θ:

$$g_k(x, y, \theta) \longrightarrow g_{0k}(\theta) = 1 + \frac{\overline{\mu} - \mu}{\mu + \cot\left(\frac{\theta}{2}\right)} R_{0k}, \quad \text{as} \quad x^2 + y^2 \longrightarrow \infty.$$

The existence of the limit at spatial infinity $R_{0k} = \lim_{r\to\infty} R_k(x, y) = \text{const}$ is guaranteed by the finite-energy condition (8.2.16). Hence ψ (8.2.50) satisfies the trivial-scattering condition (8.2.29) and extends to a map from S^3 to $U(n)$. The scattering matrix (8.2.33) is $S = 1$. Note, however, that the g_k's and ψ in (8.2.50) only extend to the ordinary suspension of S^2. One needs to perform the transformation (8.2.31) with $K = \prod_{k=1}^{m} g_{0k}^{-1}$ for ψ to extend to the reduced suspension of S^2. We shall use ψ as in (8.2.50), because the transformation (8.2.31) does not contribute to the degree and $[K(\theta)\psi] = [\psi]$.

The proof of Theorem 8.2.3 relies on the following result.

Proposition 8.2.4 *The third homotopy class of* ψ *is given by the formula*

$$[\psi] = \pm\frac{i}{2\pi}\int_{\mathbb{R}^2}\sum_{k=1}^{m} Tr(R_k[\partial_x R_k, \partial_y R_k])dxdy \quad \begin{cases} 0<\phi<\pi \\ \pi<\phi<2\pi, \end{cases} \tag{8.2.51}$$

where $\mu = |\mu|e^{i\phi}$.

Proof The recursive application of (A9) implies that

$$[\psi] = \sum_{k=1}^{m}[g_k].$$

Using (8.2.49), with $z = x + iy$,

$$\begin{aligned}[g_k] &= \frac{1}{8\pi^2}\int_{S^1\times\mathbb{R}^2}\mathrm{Tr}(g_k^{-1}\partial_\theta g_k\,[g_k^{-1}\partial_z g_k, g_k^{-1}\partial_{\bar z} g_k])\, d\theta\wedge dz\wedge d\bar z \\ &= \frac{1}{16\pi^2}\mathcal{I}(\mu)\int_{\mathbb{R}^2}\mathrm{Tr}(R_k[\partial_z R_k, \partial_{\bar z} R_k])\, dz\wedge d\bar z,\end{aligned}$$

where

$$\begin{aligned}\mathcal{I}(\mu) &= \int_0^{2\pi}\frac{(\bar\mu-\mu)^3\sin^2\left(\frac{\theta}{2}\right)}{\left[|\mu|^2+(1-|\mu|^2)\cos^2\left(\frac{\theta}{2}\right)+(\mu+\bar\mu)\cos\left(\frac{\theta}{2}\right)\sin\left(\frac{\theta}{2}\right)\right]^2}d\theta \\ &= \pm 8\pi i \quad \begin{cases} 0<\phi<\pi \\ \pi<\phi<2\pi. \end{cases}\end{aligned}$$

Hence, changing to the (x, y) coordinates, we obtain

$$[g_k] = \pm\frac{i}{2\pi}\int_{\mathbb{R}^2}\mathrm{Tr}(R_k[\partial_x R_k, \partial_y R_k])dxdy \quad \begin{cases} 0<\phi<\pi \\ \pi<\phi<2\pi. \end{cases} \tag{8.2.52}$$

Therefore, the third homotopy class of ψ is given by (8.2.51). □

The proof of Theorem (8.2.3) makes use of the above proposition and a recursive Bäcklund procedure (8.2.39) of adding unitons to a given solution.

Proof of Theorem 8.2.3 We first consider a solution of the form $\tilde J = JM$, where J is an arbitrary solution of (8.2.13) and M is given by (8.2.36). Noting that M is unitary and writing it in terms of R, the difference between the energy densities (8.2.15) of $\tilde J$ and J is given by

$$\Delta\mathcal{E} \equiv \tilde{\mathcal{E}} - \mathcal{E} = \sum_a \mathrm{Tr}\,[\kappa\bar\kappa R_a R_a R + \kappa(1-\bar\kappa R)J^{-1}J_a R_a], \tag{8.2.53}$$

where a stands for (t, x, y), $R_a = \partial_a R$, $\tilde{\mathcal{E}}$ and $\mathcal{E}$ are the energy densities of $\tilde J$ and J, respectively, and $\kappa = (1 - \frac{\mu}{\bar\mu})$.

The Bäcklund relations (8.2.39) can be rewritten as

$$R[R_t - J^{-1}J_t(1-R)] = B \qquad (8.2.54)$$
$$RR_t = C,$$

where

$$B = (\mu R_x - R_y + RJ^{-1}J_y)(1-R) \quad \text{and}$$
$$C = \frac{1}{\mu}\left[(\mu R_y + R_x - RJ^{-1}J_x)(1-R)\right].$$

Multiplying these Bäcklund relations and their Hermitian conjugates yields the following identities:

$$\mathrm{Tr}(R_t R_t R) = \mathrm{Tr}(CC^*),$$
$$\mathrm{Tr}(J^{-1}J_t R_t) = \mathrm{Tr}(CB^* - BC^*), \quad \text{and}$$
$$\mathrm{Tr}(RJ^{-1}J_t R_t) = \mathrm{Tr}[(C-B)C^*]. \qquad (8.2.55)$$

The terms involving time derivatives in (8.2.53) are of the form $R_t R_t R$, $J^{-1}J_t R_t$, and $RJ^{-1}J_t R_t$, which, by (8.2.55), can be written in terms of the spatial derivatives only. Thus by direct substitution and some rearrangements (8.2.53) becomes

$$\Delta\mathcal{E} = -\frac{\kappa}{\mu}\mathrm{Tr}\left[(1+|\mu|^2)R[R_x, R_y] + \mathcal{T}\right],$$

where $\mathcal{T} = \partial_x(RJ^{-1}J_y) - \partial_y(RJ^{-1}J_x)$ gives no contribution to the difference in the energy functionals of $\tilde{J}$ and J. This is because

$$\begin{aligned}\mathrm{Tr}\int_{\mathbb{R}^2}\mathcal{T}dx\wedge dy &= \lim_{r\to\infty}\int_{D_r} d[\mathrm{Tr}(RJ^{-1}dJ)]\\ &= \lim_{r\to\infty}\oint_{C_r}\mathrm{Tr}(RJ^{-1}dJ) = \mathrm{Tr}\left[\lim_{r\to\infty}\oint_{C_r}(J\,R)^*dJ\right]\\ &\le \lim_{r\to\infty}\left(\mathrm{Tr}\left\{\frac{[(J\,R)_0]^*}{r}[J_1(\varphi=2\pi) - J_1(\varphi=0)]\right\}\right.\\ &\quad\left. +2\pi r\left\{\frac{|c_2|}{r^2}+\frac{|c_3|}{r^3}+\cdots\right\}\right) = 0,\end{aligned}$$

by Stokes' theorem, where C_r denotes the circle enclosing the disc D_r of radius r, φ is a coordinate on C_r, and $|c_i|$ is the bound of $\mathrm{Tr}[(J\,R)_i^*\partial_\varphi J]$, $i = 1, 2, \ldots$. We have used the boundary condition

$$\lim_{r\to\infty} J\,R = (J\,R)_0 + (J\,R)_1(\varphi)r^{-1} + O(r^{-2}), \qquad (8.2.56)$$

which follows from (8.2.16) for $\tilde{J} = JM$, and the fact that integrands are continuous on the circle and hence bounded. Since $(JR)_0$ is a constant matrix, the first term in the series is a total derivative.

So far we have only used the assumption that J is a solution of (8.2.13), but not that it has to be a uniton solution defined by (8.2.35). Therefore, we have a more general result for the total energy of a solution of the form $\tilde{J} = JM$, where J is an arbitrary solution to the integrable chiral equation. Let $\tilde{E}$ and E be the total energies of $\tilde{J}$ and J, respectively, then

$$\tilde{E} = E + \frac{(\mu - \bar{\mu})(1 + |\mu|^2)}{|\mu|^2} \int_{\mathbb{R}^2} \mathrm{Tr}(R[R_x, R_y])dxdy. \tag{8.2.57}$$

From this, the explicit expression for the total energy of an m-uniton solution (8.2.35) follows. First, consider a one-uniton solution $J_{(1)} = M_1$. It can be written as $J_{(1)} = J_{(0)}M_1$, where the constant matrix $J_{(0)}$, which satisfies (8.2.13) trivially, is chosen to be the identity matrix. Then, from (8.2.57), the total energy of a one-uniton solution is given by

$$E_{(1)} = \frac{(\mu - \bar{\mu})(1 + |\mu|^2)}{|\mu|^2} \int_{\mathbb{R}^2} \mathrm{Tr}(R_1[\partial_x R_1, \partial_y R_1])dxdy. \tag{8.2.58}$$

Therefore, using (8.2.57), we show by induction that the total energy of an m-uniton solution (8.2.35) is given by

$$\begin{aligned} E_{(m)} &= \frac{(\mu - \bar{\mu})(1 + |\mu|^2)}{|\mu|^2} \sum_{k=1}^{n} \int_{\mathbb{R}^2} \mathrm{Tr}(R_k[\partial_x R_k, \partial_y R_k])dxdy \qquad (8.2.59) \\ &= \pm 4\pi \left(\frac{1 + |\mu|^2}{|\mu|} \right) \sin\phi \, [\psi] \quad \begin{cases} 0 < \phi < \pi \\ \pi < \phi < 2\pi, \end{cases} \end{aligned}$$

where $\mu = |\mu|e^{i\phi}$, and we have used (8.2.51). □

The formula (8.2.52) reveals another topological interpretation of the energy quantization which is useful in practical calculations. Consider the group element (8.2.50) with the index k dropped. The Grassmanian projector R in (8.2.37) corresponds to a smooth map from the compactified space to the complex projective space $R : S^2 \longrightarrow \mathbb{CP}^{n-1}$. The homotopy group $\pi_2(\mathbb{CP}^{n-1}) = \mathbb{Z}$ is non-trivial and the degree of R is obtained by evaluating the homology class on a standard generator for $H^2(\mathbb{CP}^{n-1})$ represented in a map $q = (1, f_1, \ldots, f_{n-1})$ by the Kähler form (A13) (see Appendix A). We conclude that the energy is proportional to the sum of the topological degrees (A12) of the Grassmanian projectors involved in the definition of unitons.

- **Example.** Consider the $SU(2)$ case, where the third homotopy class is equal to the topological degree and set $\mu = i$. The uniton factors are of the form

$$M_k = \frac{i}{1 + |f_k|^2} \begin{pmatrix} |f_k|^2 - 1 & -2f_k \\ -2\overline{f_k} & 1 - |f_k|^2 \end{pmatrix}.$$

n = 1 – In the one-uniton case $\partial_t M_1 = 0$ and M_1 is given by (8.2.36) with $f_1 = f_1(z)$ a rational function of some fixed degree N. The energy density is

$$\mathcal{E}_1 = \frac{8|f_1'|^2}{(1 + |f_1|^2)^2} = -i\mathrm{Tr}\,(M_1[\partial_z M_1, \partial_{\bar{z}} M_1])$$

and $E = 8\pi\,\deg(g_1)$ in agreement with (8.2.47). In this case g_1 is the suspension of a rational map $f_1 : \mathbb{CP}^1 \longrightarrow \mathbb{CP}^1$ and $\deg(g_1) = \deg(f_1)$.

n = 2 – In the two-uniton case M_1 and M_2 are given by (8.2.36) with $\mu = i$ and (q_1, q_2) of the form (8.2.26). Define $k = 2(tf' + h)$. The total energy density is

$$\mathcal{E} = \frac{8|(1 + |f|^2)k' - 2k\overline{f}f'|^2 + 16|kf'|^2 + 16(1 + |f|^2)^2|f'|^2}{[|k|^2 + (1 + |f|^2)^2]^2} \qquad (8.2.60)$$

and

$$E = \int_{\mathbb{R}_2} \mathcal{E}\,dxdy = 8\pi[\deg(g_1) + \deg(g_2)]$$

for all t. The quantization of energy in this case was first observed by Ioanidou and Manton in [88], where it was shown that $E = 8\pi N$ where generically $N = 2\deg f + \deg h$. However, $N = \max(2\deg f, \deg h)$ if both f and h are polynomials. The formula (8.2.47) is valid for all pairs (f, h).

8.2.4 Moduli space dynamics

We have seen that the integrability of equations (8.2.8), or equivalently of (8.2.13), allows a construction of explicit static and also time-dependent solutions. In this section we choose a different route [54, 56] and construct slow-moving solitons using a modification of the geodesic approximation [113] which may involve a background magnetic field in the moduli space of static solutions. This will allow a comparison between exact solutions and the solutions obtained in the moduli space approximation.

The argument is based on the analogy with a particle in $\mathbb{R}^{n+1}$ moving in a potential U and coupled to a magnetic vector potential $\mathbf{A}(\mathbf{q})$; the

Lagrangian is

$$L = \frac{1}{2}|\dot{\mathbf{q}}|^2 + \mathbf{A} \cdot \dot{\mathbf{q}} - U(\mathbf{q}),$$

where $U : \mathbb{R}^{n+1} \to \mathbb{R}$ is a potential whose minimum value is 0. The equilibrium positions are on a submanifold $X \subset \mathbb{R}^{n+1}$ given by $U = 0$. If the kinetic energy of the particle is small and the initial velocity is tangent to X, the exact motion will be approximated by motion on X with the Lagrangian L' given by a restriction of L to X:

$$L' = \frac{1}{2}h_{jk}\dot{\gamma}_j\dot{\gamma}_k + A_j\dot{\gamma}_j.$$

Here, the γ's are local coordinates on X, and the metric h and the one-form $A_j d\gamma_j$ are induced on X from the Euclidean inner product and the magnetic vector potential $\mathbf{A}$, respectively. In the absence of the magnetic term we expect the true motion to have small oscillations in the direction transverse to X, with the approximation becoming exact at the limit of zero initial velocity. The presence of a magnetic force may in some cases balance the contribution from a centrifugal force so that the oscillations do not occur, and the exact motion is confined to X.

The dynamics of finite-energy solutions to (8.2.13) will be put in this framework with $\mathbb{R}^{n+1}$ replaced by an infinite-dimensional configuration space of the field J, and X replaced by the the moduli space $\mathcal{M}$ of rational maps from $\mathbb{CP}^1$ to $\mathbb{CP}^{n-1}$ the important point being that the static solutions to (8.2.13) obtained from such maps give the absolute minimum of the potential energy in a given topological class. The time-dependent solutions to (8.2.13) with small total energy (hence small potential energy) above the absolute minimum will be approximated by a sequence of static states, that is, a motion in $\mathcal{M}$. This comes down to three steps:

1. Construct finite-dimensional families of static solutions to (8.2.13) with finite energy.
2. Allow time-dependence of the parameters, and read off the metric h and the magnetic one-form A on the moduli space from the Lagrangian (8.2.41) for J. Investigate whether A has a non-vanishing or vanishing magnetic two-form $F = dA$. Some of the parameters may have to be fixed to ensure that this metric is complete, and all tangent vectors have finite length.
3. The geodesic motion, possibly with magnetic forcing, should then approximate the slow (non-relativistic) motion of rational map, or lump solutions to (8.2.13).

8.2.4.1 *Static solutions*

All static solutions to (8.2.13) are the chiral fields on $\mathbb{R}^2$, that is, solutions to

$$\overline{\partial_z}(J^{-1}\partial_z J) + \partial_z(J^{-1}\overline{\partial_z}J) = 0, \tag{8.2.61}$$

where $z = x + iy$ and $\partial_z = \partial/\partial z$. In the case of $SU(2)$ the static finite-energy solutions are given by (8.2.23) and $\mathcal{M} = \mathcal{M}_N$ is the moduli space of rational maps from $\mathbb{CP}^1$ to itself with degree N.

All finite-energy static solutions to (8.2.13) can be factorized in terms of maps $R_{(k)}$ of $\mathbb{R}^2$ into Grassmanian manifolds [166, 193]:

$$J = K_0(\mathbf{1} - 2R_{(1)})(\mathbf{1} - 2R_{(2)}) \cdots (\mathbf{1} - 2R_{(m)}), \tag{8.2.62}$$

where K_0 is a constant unitary matrix, $R_{(1)}$ is holomorphic, and $m \leq n - 1$ is the so-called uniton number.

The family of static solutions in moduli space construction should minimize the energy for a given value of topological charge. These instanton (or anti-instanton) solutions are given by

$$J = K_0(\mathbf{1} - 2R), \tag{8.2.63}$$

where R is given by (A11) and f_l, $l = 1, \ldots, (n-1)$ are rational holomorphic (respectively anti-holomorphic) functions of $z = x + iy$ given by

$$f_l = \frac{p_l(z)}{r_l(z)} = \frac{(z - p_{l,1}) \cdots (z - p_{l,N_l})}{(z - r_{l,1}) \cdots (z - r_{l,N_l})}, \qquad l = 1, \cdots, n-1. \tag{8.2.64}$$

Here $N_l = k_{\text{alg}}(f_l)$ is the algebraic degree of f_l. The numbers $\gamma = (\text{Re}(p), \text{Im}(p), \text{Re}(r), \text{Im}(r))$ are real coordinates on a finite-dimensional moduli space $\mathcal{M}_N \subset \mathcal{M}$.

The finiteness of the energy requires the base condition to be imposed. We therefore need to fix the limit of each f_l at spatial infinity. We have taken this limit to be equal to one for all functions f_l.

One-uniton solutions (8.2.63) correspond to R being an (anti-)instanton solution, which at the level of the Grassmanian model minimizes the value of energy in its topological sector. For such solutions the energy is proportional to the topological charge N of the Grassmanian projector (given by the formula (A12)). This is also true for the potential energy of the chiral field J, defined in (8.2.14), since in the case of (8.2.63) it is equal to the energy of R. The non-instanton solutions corresponding to $m > 1$ in (8.2.62) are unstable in the space of all Grassmanian embedding, and so are not suitable from the moduli-space perspective. Therefore[4] we shall concentrate on instantons, in

[4] Note that (8.2.65) holds for any smooth map $R : S^2 \longrightarrow \mathbb{CP}^{n-1}$ with N being the homotopy class under the standard isomorphism $\pi_2(\mathbb{CP}^{n-1}) = \mathbb{Z}$. To see this (e.g. [21]) consider the homology

which case [193]

$$N = \max_l N_l. \tag{8.2.65}$$

The boundary conditions (8.2.16) imply that the finite-energy static solutions to (8.2.13) are maps from S^2 (conformal compactification of $\mathbb{R}^2$) into $U(n)$. In the moduli-space approximation we choose a class of such solutions which are homotopic as maps of S^2 into $U(n)$ and all have the same value of potential energy. Ideally every such map ought to provide minimum of the potential energy. This is the case on the level of the Grassmanian models for constructions which involve (anti-)instanton solutions. For chiral models one can show that all finite-energy static solutions are saddle points of the potential energy functional [134]. This raises a question about stability of the approximate solutions.

To summarize, for a given value of the topological charge, all solutions in the class (8.2.63) can be described by a finite set of parameters which are the positions of zeros and poles of holomorphic functions. To ensure finite values of kinetic energy we needed to impose the base condition on the solutions by fixing their value at spatial infinity. Then the parameters define a map on the resulting moduli space.

8.2.4.2 *Metric and magnetic field*

Next we allow the parameters to depend on time and so time-dependent approximate solutions correspond to paths in the moduli space. Let us denote the solutions contributing to the moduli space by $J(\gamma; x, y)$, where γ denote real parameters in (8.2.64). Approximate time-dependent solutions are then of the form $J(\gamma(t); x, y)$ and time differentiation gives

$$J_t = J_j \dot{\gamma}^j, \quad j = 1, \ldots, \dim(\mathcal{M}), \quad \text{where} \quad J_j = \frac{\partial J}{\partial \gamma^j}. \tag{8.2.66}$$

The dynamics is governed by the action obtained as a restriction of (8.2.41) to the moduli space

$$S_{\mathcal{M}} = \int_{t_1}^{t_2} \left(\frac{1}{2} h_{jk} \dot{\gamma}^j \dot{\gamma}^k + A_j \dot{\gamma}^j \right) dt. \tag{8.2.67}$$

group $H_2(\mathbb{CP}^{n-1})$. This is isomorphic to $\mathbb{Z}$. If $R : S^2 \longrightarrow \mathbb{CP}^{n-1}$ is a map from the compactified space to $\mathbb{CP}^{n-1}$, representing a homology class $R_*[S^2]$, we obtain the corresponding integer by evaluating $R_*[S^2]$ on a standard generator for $H^2(\mathbb{CP}^{n-1})$ represented by the Kähler form $\mathbf{\Omega}$. In terms of differential forms, evaluating a cohomology class on a homology class just means integrating, so the evaluation of $R_*[S^2]$ on $\mathbf{\Omega}$ is given by the RHS of (A12). Now consider the Hurewicz homomorphism from $\pi_2(\mathbb{CP}^{n-1})$ to $H_2(\mathbb{CP}^{n-1})$ sending the homotopy class of $R : S^2 \longrightarrow \mathbb{CP}^{n-1}$ to $R_*[S^2]$, where $[S^2] \in H_2(S^2)$ is the fundamental class. The projective space $\mathbb{CP}^{n-1}$ is simply connected, so this is an isomorphism $\pi_2(\mathbb{CP}^{n-1}) = H_2(\mathbb{CP}^{n-1}) = \mathbb{Z}$.

The metric term can be obtained from the kinetic energy form (8.2.14) by using (8.2.66):

$$T = \frac{1}{2} h_{jk} \dot{\gamma}^j \dot{\gamma}^k, \quad \text{where} \quad h_{jk} = -\int \mathrm{Tr}\left(J^{-1} J_j J^{-1} J_k\right) dx dy, \tag{8.2.68}$$

and the magnetic term can similarly be obtained from the WZW term, which can be rewritten by the cyclic property of the trace as

$$\begin{aligned} S_{\mathrm{M}} &= \int_{t_1}^{t_2} \int_{\mathbb{R}^2} \int_0^1 \mathrm{Tr}([\hat{J}^{-1}\hat{J}_t, \hat{J}^{-1}\hat{J}_y]\hat{J}^{-1}\hat{J}_\rho) d\rho dx dy dt \\ &= \int_{t_1}^{t_2} A_j \dot{\gamma}^j dt, \end{aligned}$$

where

$$A_j = \int_{\mathbb{R}^2} \int_0^1 \mathrm{Tr}([\hat{J}^{-1}\hat{J}_j, \hat{J}^{-1}\hat{J}_y]\hat{J}^{-1}\hat{J}_\rho) d\rho dx dy. \tag{8.2.69}$$

Then $A = A_j d\gamma^j$ is the *magnetic one-form* on the moduli space. We shall now prove the following

Proposition 8.2.5 [56] *The magnetic field* (8.2.70) *vanishes on moduli spaces constructed from embeddings* (8.2.63) *of Grassmanian solutions.*

Proof One property of the WZW term is that its variation does not depend on the particular choice of extension $\hat{J}$. We consider the variations restricted to the moduli space $\delta J = J_i \delta\gamma^i$, and find

$$\begin{aligned} \delta S_{\mathrm{M}} &= \int_{t_1}^{t_2} \int_{\mathbb{R}^2} \mathrm{Tr}\left(J^{-1} J_y \left[J^{-1}\delta J, J^{-1} J_t\right]\right) dx dy dt, \\ &= -\int_{t_1}^{t_2} \int_{\mathbb{R}^2} \mathrm{Tr}\left(J^{-1} J_y \left[J^{-1} J_i, J^{-1} J_j\right]\right) dx dy \, \dot{\gamma}^j \delta\gamma^i dt. \end{aligned}$$

Comparing this expression with the variation of (8.2.69)

$$\delta S_{\mathrm{M}} = \int_{t_1}^{t_2} F_{ij} \dot{\gamma}^j \delta\gamma^i \, dt, \quad F_{ij} = \partial_i A_j - \partial_j A_i$$

gives

$$F_{ij} = -\int_{\mathbb{R}^2} \mathrm{Tr}\left(J^{-1} J_y \left[J^{-1} J_i, J^{-1} J_j\right]\right) dx dy, \tag{8.2.70}$$

where $F = \frac{1}{2} F_{ij}(\gamma) d\gamma^i \wedge d\gamma^j$ is the the magnetic field. We can see that although the magnetic one-form A in general depends on the choice of the extension $\hat{J}$, its exterior derivative F does not. Changing the extension merely corresponds to a gauge transformation of A.

Note that the potential energy term has not been included in the effective action (8.2.67). The potential is proportional to the topological charge (A12), and does not contribute to the effective equations of motion.

Let us now consider a Grassmanian projector R depending smoothly on some set of variables, which we shall denote by a, b, and c. From idempotency and the Leibniz rule we deduce

$$R_a = R_a R + R R_a,$$

and

$$\begin{aligned} R_a R_b R_c &= R R_a R_b R_c + R_a R R_b R_c = R R_a R_b R_c + R_a R_b R_c - R_a R_b R R_c \\ &= R R_a R_b R_c + R_a R_b R_c - R_a R_b R_c + R_a R_b R_c R = R R_a R_b R_c + R_a R_b R_c R. \end{aligned}$$

Taking the trace of the above expression gives

$$\mathrm{Tr}(R_a R_b R_c) = \mathrm{Tr}(2 R R_a R_b R_c). \tag{8.2.71}$$

If J is given by (8.2.63) then

$$\mathrm{Tr}\left(J^{-1}J_y\,[J^{-1}J_i, J^{-1}J_j]\right) \sim \mathrm{Tr}\left[(1-2R)R_y\,[R_i, R_j]\right], \tag{8.2.72}$$

where we have assumed that K_0 does not depend on parameters γ on the moduli space to ensure the finiteness of the kinetic energy. The RHS of (8.2.72) vanishes because of (8.2.71), which in turn implies the vanishing of the magnetic field (8.2.70). □

We defined the metric (8.2.68) as a restriction of the kinetic energy to $\mathcal{M}_N$. Its completeness is equivalent to the requirement that the kinetic energy is finite along all curves in $\mathcal{M}_N$. Although the base condition was necessary to ensure finite kinetic energies, it appears not to be sufficient as the metric is complete only on leaves of appropriate foliation of $\mathcal{M}_N$ [172, 153] and we need to restrict the dynamics to these leaves. These restrictions are assumed to hold and we will use the symbol $\mathcal{M}_N$ to denote some particular leaf.

The moduli-space metric can be obtained explicitly from (8.2.68) by using (8.2.63), (A11), and (8.2.64). This metric is Kähler with respect to the natural complex structure induced by map (8.2.64), with the Kähler potential

$$\Omega_b = 8\int_{\mathbb{R}^2} \ln \sum_{l=1}^{n-1}(|p_l|^2 + |r_l|^2)dxdy. \tag{8.2.73}$$

- **Example.** Consider the case $G = SU(2)$ where the static finite-energy solutions are of the form (8.2.23).

  In the moduli-space approximation J stays on the equatorial $S^2 \subset S^3$, and the lumps are located where J departs from its asymptotic value (8.2.16). In these regions the energy density of (8.2.14) attains its local maxima. The velocities of the lumps are the velocities of these local maxima.

The charge-one solution is given by

$$f = \alpha + \frac{\beta}{z + \gamma}, \tag{8.2.74}$$

and we need to fix α and β in order for (8.2.68) to be well defined. Choosing $\alpha = 0, \beta = 1, \gamma = \gamma(t)$, and setting $\gamma(t) = r(t)e^{i\theta(t)}$ we find the metric and the one-form

$$h = 8\pi(dr^2 + r^2 d\theta^2) \quad \text{and} \quad A = 4\pi^2 d(r\cos\theta).$$

Therefore the metric is flat, and the motion is along straight lines, $\gamma(t) = -vt$, because $dA = 0$ does not contribute to the Euler–Lagrange equations. The energy density is approximated by

$$\mathcal{E} = (1 + |z - vt|^2)^{-2}.$$

Next we look at the charge-two case:[5]

$$f = \alpha + \frac{\beta z + \gamma}{z^2 + \delta z + \kappa}. \tag{8.2.75}$$

The corresponding metric was constructed by Ward [172]. The parameters α and β have to be fixed to ensure finiteness of kinetic energy, and δ can be set to 0 by exploiting the translational invariance of (8.2.23). Moreover the Möbius transformations can be used to ensure $\alpha = 0, \beta \in \mathbb{R}$, and here Ward makes an additional choice $\beta = 0$. The resulting metric is therefore defined on four-dimensional leaves of a foliation of $\mathcal{M}_2$, with local coordinates $(\gamma, \overline{\gamma}, \kappa, \overline{\kappa})$. The Kähler potential is given by

$$\Omega_h = -4\pi|\kappa| + \pi|\gamma| \int_0^{\pi/2} \sqrt{1 + |\kappa/\gamma|^2 \sin^2\theta}\, d\theta.$$

The structure is invariant under the torus action and a homothety

$$\gamma \to e^{i\tau_1}\gamma, \quad \kappa \to e^{i\tau_2}\kappa, \quad \text{and} \quad |\gamma|^2 + |\kappa|^2 \to \tau_3(|\gamma|^2 + |\kappa|^2).$$

In the next two examples we shall compare the moduli-space motion with the limiting cases of exact soliton solutions [54]

- **Example.** Consider the one-soliton solution (8.2.22). The energy density

$$\mathcal{E} = 2\sin^2\phi \frac{(1 + |\mu|^2)^2 |f'|^2}{|\mu|^2 (1 + |f|^2)^2}$$

[5] We remark that the boundary condition for f in these examples is $f(z) \longrightarrow 0$ as $|z| \longrightarrow \infty$, and hence $J_0 = i\sigma^3$. The conclusion $F = 0$ does not depend on these boundary conditions.

has local maxima which give the locations $\{(x_a, y_a), a = 1, \dots, N\}$ of N lumps. The velocities $(\dot{x}_a, \dot{y}_a) = (-2|\mu|\cos\phi/(1 + |\mu|^2), (1 - |\mu|^2)/(1 + |\mu|^2))$ are the same for each lump so (8.2.22) should be regarded as a one-soliton solution. To make contact with the moduli-space approximation write

$$J_1 = \cos\phi \mathbf{1} + i\mathbf{a} \cdot \boldsymbol{\sigma}$$

to reveal that $\cos\phi$ measures the deviation of J_1 from the unit sphere in the Lie algebra $\mathfrak{su}(2)$. If J is initially tangent to the space of static solutions, then $\cos\phi = 0$, and we can set $\mu = i(1 + \varepsilon)$, where $\varepsilon \in \mathbb{R}$. The solution is of the form (8.2.23), but f is rational in

$$\omega = z + \varepsilon(z + it) + \frac{\varepsilon^2}{2}\left(\frac{z - \bar{z}}{2} + it\right),$$

so

$$f(\omega) = \frac{(z - Q_1)\cdots(z - Q_N)}{(z - Q_{N+1})\cdots(z - Q_{2N})},$$

where the Q's are linear functions of $(\varepsilon^2\bar{z}, \varepsilon t)$. The (squared) velocity is

$$\mathcal{V}^2 = 1 - 4(1 + \varepsilon)^2/\left[1 + (1 + \varepsilon)^2\right]^2,$$

so in the non-relativistic limit (which underlies the moduli-space approximation) we regard ε as small. Therefore the Q's depend only on t, and they all move at velocity ε. Setting $N = 1$, we recover the charge-one solution (8.2.74). More generally we find that J is given by (8.2.23) with

$$f = f_2(z) + tf_1(z),$$

where $f_2 = f(\omega)|_{\varepsilon=0}$ and $f_1 = \partial f/\partial\varepsilon|_{\varepsilon=0}$ are rational functions of z.

- **Example.** In Theorem 8.2.3 we have demonstrated that the total rest-frame energy of the solution (8.2.25) is quantized in units of 8π. Solutions to (8.2.13) obtained in the moduli-space approximation have energies close to their potential energy as their kinetic energy is small. We should therefore expect that some of these approximate solutions arise from (8.2.25) by a limiting procedure.

  Consider (8.2.25) and for convenience set $f = -if_1/2$ and $h = -if_2/2$. To demonstrate how the limiting procedure is achieved first observe that solutions to (8.2.13) are defined up to a multiple by a constant element of $SU(2)$. The static solution (8.2.23) with $f = f_2$ arises from (8.2.25) by using this freedom and setting $f_1 = 0$:

$$\begin{pmatrix} i & 0 \\ 0 & -i \end{pmatrix} J_2|_{f_1=0} = \frac{i}{1 + |f_2|^2}\begin{pmatrix} 1 - |f_2|^2 & 2f_2 \\ 2\bar{f_2} & |f_2|^2 - 1 \end{pmatrix} = J_{\text{static}}.$$

Moreover the energy density of (8.2.25) has maxima where $f = f_2 + tf_1' = 0$. The lumps are located at the zeros $z_a = z_a(t)$, $a = 1, \dots, \deg f$ of f and the squared velocity of each lump is

$$\mathcal{V}_a^2 = \left.\frac{|f_1'|^2}{|f_2' + tf_1''|^2}\right|_{z=z_a},$$

so that $|f_1'|^2$ is small in the non-relativistic limit. Therefore $|f_1|$ is also small as we choose J_2 to be tangent to the space of static solutions at $t = 0$. Keeping only the linear terms in f_1 in (8.2.25) yields

$$\begin{pmatrix} i & 0 \\ 0 & -i \end{pmatrix} J_2 =$$

$$\frac{i}{1+|f|^2}\begin{pmatrix} 1-|f|^2 - i(f_2\overline{f_1} + \overline{f_2}f_1) & 2f \\ 2\overline{f} & |f|^2 - 1 - i(f_2\overline{f_1} + \overline{f_2}f_1) \end{pmatrix}.$$

The term $(f_2\overline{f_1} + \overline{f_2}f_1)$ can be dropped by rescaling the coordinates $x^\mu \to x^\mu/\varepsilon$.

Comparing the resulting expression with (8.2.23) will give a motion on the moduli space of static solutions if f_2, tf_1', and $(f_1)^2$ lie in the common space of rational maps of degree deg f_2. To achieve this, we therefore take

$$f_1 = \frac{p(z)}{q(z)} \quad \text{and} \quad f_2 = \frac{r(z)}{q(z)^2},$$

where r is of degree $2N$ and p and q are of degree at most N. The total energy is equal to 8π deg f_2. The resulting motion on the moduli space of static solutions of charge deg f_2 is given by (8.2.23) with

$$f(z,t) = \frac{r + t(p'q - pq')}{q^2}. \tag{8.2.76}$$

This motion is restricted to a geodesic submanifold as the parameters in the denominator of f are fixed. In particular, setting $q = 1$, we can take $f_2(z)$ to be a polynomial of degree $2N$ and $f_1(z)$ to be a polynomial of degree at most N.

8.2.5 Mini-twistors

The geometric interpretation of the Lax representation (8.2.11) is the following: For any fixed pair of real numbers (μ, λ) the plane

$$\mu = v + x\lambda + u\lambda^2 \tag{8.2.77}$$

is null with respect to the Minkowski metric $h = dx^2 - 4dudv$ on $M = \mathbb{R}^{2,1}$, and conversely all null planes can be put in this form if one allows $\lambda = \infty$. The two vector fields

$$\delta_0 = \partial_u - \lambda\partial_x \quad \text{and} \quad \delta_1 = \partial_x - \lambda\partial_v \tag{8.2.78}$$

span this null plane. Thus the Lax equations (8.2.11) imply that the generalized connection (A, Φ) is flat on null planes. This underlies the twistor approach [174, 177], where one works in a complexified Minkowski space $\mathbb{C}^3$, and interprets (μ, λ) as coordinates in a patch of the mini-twistor space $\mathcal{Z} = T\mathbb{CP}^1$, with $\mu \in \mathbb{C}$ being a coordinate on the fibres and $\lambda \in \mathbb{CP}^1$ being an affine coordinate on the base.

In this section we shall study the geometry of the mini-twistor space, and its compactification in some detail. It is convenient to make use of the spinor formalism based on the isomorphism

$$TM = \mathbb{S} \odot \mathbb{S},$$

where $\mathbb{S}$ is rank-two real vector bundle (spin bundle) over the (2+1)-dimensional Minkowski space. The fibre coordinates of this bundle are denoted by[6] (π_0, π_1). Rearrange the space-time coordinates (u, x, v) as a symmetric two-spinor

$$x^{AB} := \begin{pmatrix} u & x/2 \\ x/2 & v \end{pmatrix},$$

such that the space-time metric and the volume form are

$$h = -2dx_{AB}dx^{AB} \text{ and } \mathrm{vol}_h = dx_A{}^B \wedge dx_C{}^A \wedge dx_B{}^C.$$

As usual, two-dimensional spinor indices are raised and lowered with the symplectic form ε_{AB}, such that $\varepsilon_{01} = 1$.

The mini-twistor space of $\mathbb{R}^{2,1}$ is the two-dimensional complex manifold $\mathcal{Z} = T\mathbb{CP}^1$ which is the total space of the line bundle $\mathcal{O}(2)$ of Chern class 2 over $\mathbb{CP}^1$. Points of $\mathcal{Z}$ correspond to null two-planes in $\mathbb{R}^{2,1}$ via the incidence relation

$$x^{AB}\pi_A\pi_B = \omega. \tag{8.2.79}$$

Here (ω, π_0, π_1) are homogeneous coordinates on $\mathcal{Z}$: $(\omega, \pi_A) \sim (c^2\omega, c\pi_A)$, where $c \in \mathbb{C}^*$. In the affine coordinates $\lambda := \pi_0/\pi_1$ and $\mu := \omega/(\pi_1)^2$ equation (8.2.79) gives (8.2.77).

First fix (ω, π_A). If (μ, λ) are both real then (8.2.79) defines a null plane in $\mathbb{R}^{2,1}$. If both μ and λ are complex then the solution to (8.2.79) is a time-like curve in $\mathbb{R}^{2,1}$. We shall say that this curve is oriented to the future if $\mathrm{Im}(\lambda) > 0$

[6] Strictly speaking the spinor indices used in this section should be primed if one views the integrable chiral model as a reduction of ASDYM. However only one type of indices will be used, so using unprimed indices should not lead to misunderstandings.

and to the past otherwise. If λ is real and μ is complex then (8.2.79) has no solutions for finite x^{AB}.

An alternate interpretation of (8.2.79) is to fix x^{AB}. This determines ω as a function of π_A, that is, a section of $\mathcal{O}(2) \to \mathbb{CP}^1$ when factored out by the relation $(\omega, \pi_A) \sim (c^2\omega, c\pi_A)$. These are embedded rational curves with self-intersection number 2, as infinitesimally perturbed curve $\mu + \delta\mu$ with $\delta\mu = \delta v + \delta x\lambda + \delta u\lambda^2$ intersects (8.2.77) at two points.

Two rational curves L_1 and L_2 (corresponding to $p_1 = (u_1, x_1, v_1)$ and $p_2 = (u_2, x_2, v_2)$, respectively) intersect at two points

$$\lambda_{1,2} = \frac{2R_2 \mp \sqrt{h(R,R)}}{2R_1}, \quad \text{where} \quad R_i := (u_1 - u_2, x_1 - x_2, v_1 - v_2).$$

Therefore the incidence of curves in $\mathcal{Z}$ encodes the causal structure of $\mathbb{R}^{2+1}$ in the following sense: L_1 and L_2 intersect at (a) one point, (b) two real points, and (c) two complex points conjugates of each other, iff p_1, p_2 are (a) null separated, (b) space-like separated, and (c) time-like separated.

Examining the relevant cohomology groups (see formula (B5)) shows that the moduli space of curves with self-intersection number 2 (and so the normal bundle $\mathcal{O}(2)$) in $\mathcal{Z}$ is $\mathbb{C}^3$. The real space-time $\mathbb{R}^{2,1}$ arises as the moduli space of curves that are invariant under the conjugation $(\omega, \pi_A) \mapsto (\bar{\omega}, \bar{\pi}_A)$, which corresponds to real x^{AB}.

The correspondence space is

$$\mathcal{F} = \mathbb{C}^3 \times \mathbb{CP}^1 = \{(p, Z) \in \mathbb{C}^3 \times \mathcal{Z} | Z \in L_p\}.$$

By definition, it inherits fibrations over both $\mathbb{C}^3$ and $\mathcal{Z}$, and the fibration of $\mathcal{F} = \mathbb{C}^3 \times \mathbb{CP}^1$ over $\mathcal{Z}$ has fibres spanned by the distribution $\delta_A = \pi^B \partial_{AB}$, where $\partial_{AB} x^{CD} = 1/2(\varepsilon_A^C \varepsilon_B^D + \varepsilon_B^C \varepsilon_A^D)$. In the affine coordinates where $\pi^A = (1, -\lambda)$ this distribution is given by (8.2.78) where we have ignored the constant factor π_1.

8.2.5.1 *Ward correspondence*

There is a one-to-one correspondence between the gauge equivalence classes of complex solutions to (8.2.8) in complexified Minkowski space, and holomorphic vector bundles over the mini-twistor space $T\mathbb{CP}^1$ which are trivial on the holomorphic sections of $T\mathbb{CP}^1 \to \mathbb{CP}^1$. The proof is entirely analogous to that of Theorem 7.2.2. Here we shall concentrate on the backward construction, where the additional subtlety arises due to the choice of Ward gauge leading to the integrable chiral model (8.2.13).

Let F be a patching matrix for a vector bundle over $T\mathbb{CP}^1$. Restrict F to a section (8.2.79) where the bundle is trivial, and therefore F can be split

$$F = \tilde{H}H^{-1},$$

where H and $\tilde{H}$ are holomorphic in π^A around $\pi^A = o^A = (1, 0)$ and $\pi^A = \iota^A = (0, 1)$, respectively. As a consequence of $\delta_A F = 0$, the splitting matrices satisfy

$$H^{-1}\delta_A H = \tilde{H}^{-1}\delta_A \tilde{H} = \pi^B \Phi_{AB}, \tag{8.2.80}$$

where

$$\Phi_{AB} = \Phi_{(AB)} + \varepsilon_{AB}\Phi$$

gives a one-form $A = \Phi_{AB}dx^{AB}$ and a scalar field Φ on the complexified Minkowski space, that is,

$$\Phi_{AB} = \begin{pmatrix} A_u & A_x + \Phi \\ A_x - \Phi & A_v \end{pmatrix}.$$

Let

$$h := H(x^\mu, \pi^A = o^A) \quad \text{and} \quad \tilde{h} := \tilde{H}(x^\mu, \pi^A = \iota^A)$$

so that

$$\Phi_{A0} = h^{-1}\partial_{A0}h \quad \text{and} \quad \Phi_{A1} = \tilde{h}^{-1}\partial_{A1}\tilde{h}.$$

The splitting matrices are defined up to a multiple by the inverse of a non-singular matrix $g = g(x^\mu)$ independent of π^A:

$$H \to Hg^{-1} \text{ and } \tilde{H} \to \tilde{H}g^{-1}.$$

This corresponds to the gauge transformation (8.2.9) of Φ_{AB}.

We choose g such that $\tilde{h} = \mathbf{1}$ so

$$\Phi_{A1} = \iota^A \Phi_{AB} = 0$$

and

$$\Phi_{AB} = -\iota_B o^C h^{-1}\partial_{AC}h,$$

that is,

$$A_x + \Phi = A_v = 0.$$

This is the Ward frame with $J(x^\mu) = h$. The Higgs field is given by

$$\Phi = \frac{1}{2}\varepsilon^{AB}\Phi_{AB} = V^\mu J^{-1}\partial_\mu J,$$

where $V = o^A\iota^B\partial_{AB} = \partial_x$ is the unit space-like vector which breaks the Lorentz invariance. The Lax pair (8.2.11) becomes

$$L_A = \delta_A + H^{-1}\delta_A H,$$

where $\delta_A = \pi^B \partial_{AB}$, so that

$$L_A(H^{-1}) = -H^{-1}(\delta_A H)H^{-1} + H^{-1}(\delta_A H)H^{-1} = 0$$

and $\Psi = H^{-1}$ is a solution to the Lax equations regular around $\lambda = 0$. In the Ward frame

$$J(x^\mu) = \Psi^{-1}(x^\mu, \lambda = 0)$$

is a solution to (8.2.13). Let us show explicitly that (8.2.13) holds. Differentiating both sides of (8.2.80) yields

$$\delta^A(H^{-1}\delta_A H) = -(H^{-1}\delta^A H)(H^{-1}\delta_A H)$$

which holds for all π if

$$D_{A(C}\Phi^A{}_{B)} = 0$$

where $D_{AC} = \partial_{AC} + \Phi_{AC}$. This is the spinor form of the Yang–Mills–Higgs system (8.2.8). In the Ward gauge the system reduces to

$$\partial^A{}_1 \Phi_{A0} = 0$$

which is (8.2.13).

8.2.5.2 *Abelian case and the wave equation*

In the abelian case $j = \log(J)$ satisfies the three-dimensional wave equation

$$\frac{\partial^2 j}{\partial x^2} - \frac{\partial^2 j}{\partial u \partial v} = 0.$$

Repeating the steps leading to (7.2.24) we find the integral formula

$$j = \oint_\Gamma \frac{f(\omega, \rho)}{(\rho \cdot \iota)(\rho \cdot o)} \rho \cdot d\rho,$$

where Γ is a real contour in a rational curve $\omega = x^{AB}\pi_A\pi_B$ and f is a cohomology class homogeneous of degree zero in ρ giving rise to a patching function $F = e^f$. This solution is gauge dependent as the integrand depends on a choice of the normalized spin dyad (o, ι). The abelian Higgs field also satisfies the wave equation and is given by a gauge-independent formula

$$\Phi = \oint_\Gamma \frac{\partial f}{\partial \omega} \rho \cdot d\rho.$$

8.2.5.3 *Compactified mini-twistor space*

The solutions to (8.2.8) which satisfy the trivial scattering boundary condition (8.2.29) extend to a compactification of the mini-twistor space. In this section we shall fill in the details in [179] and describe this compactification and its corresponding compactified space-time $\overline{M} = \mathbb{RP}^3$.

Let us first notice that the non-compact mini-twistor space $\mathcal{Z} = T\mathbb{CP}^1$ is biholomorphic to the quadratic cone in $\mathbb{CP}^3$ minus the vertex. Indeed, let $[Z_0 : Z_1 : Z_2 : Z_3]$ be homogeneous coordinates in $\mathbb{CP}^3$, and let T be a cone

$$(Z_1)^2 + (Z_2)^2 + (Z_3)^2 = 0.$$

The map from the mini-twistor space $T\mathbb{CP}^1$ to the cone is given by

$$(\lambda, \mu) \to [-2\mu, 1 - \lambda^2, -2\lambda, -i(1 + \lambda^2)] \quad \text{in the patch } \lambda \neq 0 \quad \text{and}$$
$$(\tilde{\lambda}, \tilde{\mu}) \to [2\tilde{\mu}, \tilde{\lambda}^2 - 1, 2\tilde{\lambda}, i(1 + \tilde{\lambda}^2)] \quad \text{in the patch } \tilde{\lambda} \neq 0,$$

where on the overlap

$$\tilde{\lambda} = \frac{1}{\lambda} \text{ and } \tilde{\mu} = \frac{\mu}{\lambda^2}.$$

In the patch of $\mathbb{CP}^3$ where $Z_0 \neq 0$ the cone T is

$$(z_1)^2 + (z_2)^2 + (z_3)^2 = 0, \tag{8.2.81}$$

where $z_i = Z_i/Z_0$ are inhomogeneous coordinates on $\mathbb{CP}^3$. Now consider a plane $P \subset \mathbb{CP}^3$ which omits the vertex $[1 : 0 : 0 : 0]$. Planes in $\mathbb{CP}^3$ are of the form $P^\alpha Z_\alpha = 0$, so we need $P^0 \neq 0$, and thus the plane is

$$1 + z_1 x^1 + z_2 x^2 + z_3 x^3 = 0, \tag{8.2.82}$$

where $x^i = P^i/P^0$. Those planes correspond to points in complexified Minkowski space $M_\mathbb{C} = \mathbb{C}^3$, that is,

$$P \leftrightarrow p = (x^1, x^2, x^3) \in M_\mathbb{C}.$$

The real planes (with P^α real) are parameterized by points in the real Minkowski space $M = \mathbb{R}^{2,1}$.

The conic sections in T corresponding to points in $M_\mathbb{C}$ are given by the locus of (8.2.81) and (8.2.82). Two planes P, P' intersect in a line, which intersects a cone in two points Z, Z' and all conics (8.2.82) through Z, Z' correspond to geodesics in $M_\mathbb{C}$ joining p and p'. In the special case if the planes touch on the cone the geodesic is null (Figure 8.2). Now consider a compactification of $M_\mathbb{C}$ by including the planes through the vertex, that is, allowing $P^0 = 0$. The space $\overline{M}_\mathbb{C}$ of such real planes has homogeneous coordinates

$$[P^0 : P^1 : P^2 : P^3]$$

so $\overline{M}_\mathbb{C} = \mathbb{CP}^3$. The 'added planes' are of the form $0 + P^1 Z_1 + P^2 Z_2 + P^3 Z_3 = 0$, and are parameterized by $[P^1 : P^2 : P^3]$ which is $\hat{\mathcal{J}} = \mathbb{CP}^2$. Therefore

$$\overline{M}_\mathbb{C} = \mathbb{C}^3 \cup \mathbb{CP}^2.$$

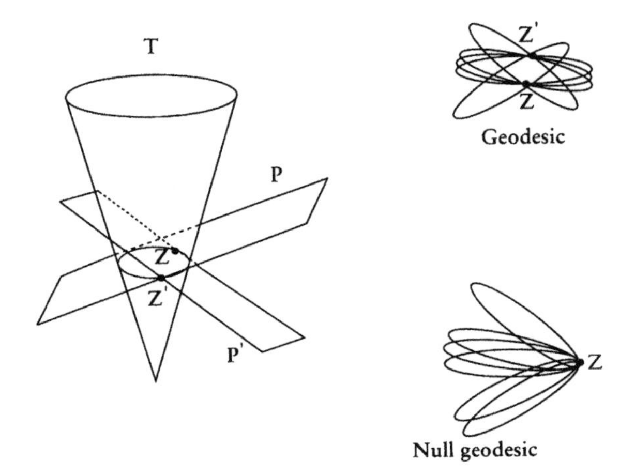

Figure 8.2 *A geodesic joining two points in the complexified Minkowski space corresponds to all conic sections intersecting at two points. The null geodesics arise as a limiting case*

These additional planes intersect the cone in the generator lines which pass through the vertex. To construct the compactified mini-twistor space $\tilde{T}$ blow up the vertex and replace it by a line $L_\infty = \mathbb{CP}^1$ (Figure 8.3). The resulting compact complex two-fold is called the Hirzebruch surface and denoted $\overline{\mathcal{O}(2)}$ to stress that each fibre of $\mathcal{O}(2) \to \mathbb{CP}^1$ has been compactified. The lines in $\tilde{T}$ corresponding to points on $\hat{\mathcal{J}}$ intersect L_∞. From our discussion it is clear that there is a one-to-one correspondence between the following geometric objects in T, $\overline{\mathcal{O}}(2)$, and $\overline{M}$:

1. A generator λ of a cone $T \subset \mathbb{CP}^3$
2. A pencil of planes through that generator
3. One-parameter family of null planes in the space time with the same normal $\pi^A\pi^B$, where $\lambda = \pi_0/\pi_1$
4. A fibre λ of $\overline{\mathcal{O}}(2) \to \mathbb{CP}^1$

After the blow-up all the $\mathbb{CP}^1$–worth of generators become disjoint, and correspond to points on L_∞, therefore a point on L_∞ corresponds, via (2), to a line

$$l_\lambda \in \hat{\mathcal{J}} = \mathbb{CP}^2 \subset \mathbb{CP}^3$$

(we recall that $\hat{\mathcal{J}}$ is the infinity of $\overline{M}_\mathbb{C}$ corresponding to the vertex $\mathcal{J}$ of the cone). This is however not a one-to-one correspondence. The planes through the vertex are not disjoint after the blow-up.

There is a unique plane through any two generators λ_1 and λ_2 of the cone. This plane corresponds to the intersection point L_{12} of two lines l_{λ_1} and l_{λ_2}

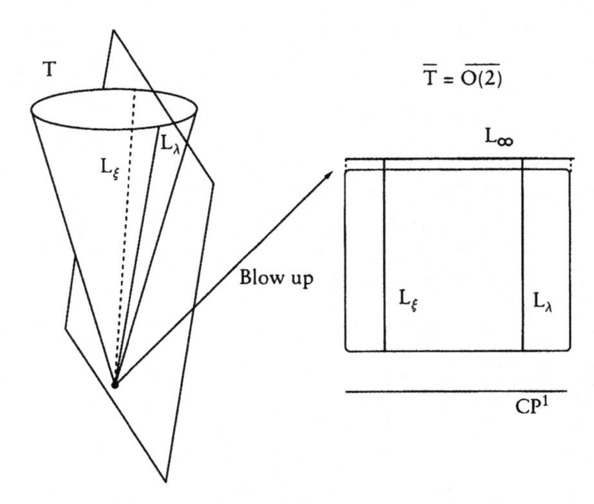

Figure 8.3 *Blow-up of the vertex of the cone. Planes thorough the vertex correspond to points at infinity in* $\overline{M}$

in $\hat{\mathcal{J}}$. Any point in $\hat{\mathcal{J}}$ corresponds to two or one fibres in that way, as any generator also corresponds to a point, because it determines a unique tangent plane to a cone.

Any finite point $p \in \overline{M}_{\mathbb{C}} = \mathbb{CP}^{3*}$ corresponds to a section L_p of $\overline{\mathcal{O}(2)}$ which intersects the line L_{12} at two points (respectively, one point for a point in $\hat{\mathcal{J}}$). The set of all sections of $\overline{\mathcal{O}(2)}$ through these two points is a geodesic (respectively, a null geodesic) joining p to L_{12}. This geodesic is compactified to a projective line in the space-time by inclusion of L_{12}.

All these structures are compatible with the real structure corresponding to real planes in $\mathbb{CP}^3$. Thus the compactified Minkowski space is

$$\overline{M} = \mathbb{R}^3 \cup \mathbb{RP}^2.$$

8.2.5.4 *Initial value problem*

Let Π be a non-null plane in the complexified Minkowski space $\mathbb{C}^3$. To any such plane we can associate an involution $\sigma_\Pi : \mathcal{Z} \to \mathcal{Z}$ defined as follows. Given a null plane Z in $\mathbb{C}^3$ we define $\sigma_\Pi(Z)$ to be a null plane such that the pair of planes $(Z, \sigma_\Pi(Z))$ intersect Π in the same line. Take Π to be a space-like plane $t = 0$ with a normal vector $\mathrm{t} = t^{AB}\partial_{AB}$. Then the involution σ acts on homogeneous coordinates on the twistor space by

$$\sigma(\omega, \pi^A) = (-\omega, t^A{}_B\pi^B),$$

or, in the inhomogeneous form,

$$\sigma(\mu,\lambda)=\left(-\frac{\mu}{\lambda^2},-\frac{1}{\lambda}\right).$$

The fixed points of this involution have $\lambda=\pm i$. The σ-invariant sections of $T\mathbb{CP}^1\to\mathbb{CP}^1$ correspond to points in the $t=0$ plane in Minkowski space. The involution σ commutes with the real structure

$$\tau:(\omega,\pi_A)=(\overline{\omega},\overline{\pi_A}).$$

These two involutions fix a real space-like plane in $\mathbb{R}^{2,1}$.

Now consider the compactified picture. The space-like plane $t=0$ intersects $\hat{\mathcal{J}}$ in $\mathbb{CP}^1{}_\infty$. Any point on L_∞ corresponds to a point on $\mathbb{CP}^1{}_\infty$. To see it consider the generator λ of the cone, and draw the unique plane through λ and (the generator corresponding to) $-1/\lambda$. These two generators are swapped by the involution τ, and so the point corresponding to this plane belongs to $\mathbb{CP}^1{}_\infty$. However this is not a bijection, as any point in $\mathbb{CP}^1{}_\infty$ corresponds to two generators, and so two points of L_∞ (the special case is $\lambda=\pm i$).

Exercises

1. Show that the generators of the conformal group are given by (8.1.1).
2. Use the zero-curvature representation (3.3.12) to show that the NLS equation is a symmetry reduction of ASDYM.
   [Hint: Proceed by analogy with the calculation leading to KdV Lax pair (8.1.3).]
3. Find the energy functional corresponding to the Lagrangian (8.2.10) and show that the energy density vanishes on solutions to (8.2.8).
4. Use the Bäcklund relations (8.2.39) to obtain the two-soliton solution (8.2.25) starting from the one-soliton (8.2.23).
5. Verify that the WZW action (8.2.41) gives rise to the integrable chiral equation (8.2.13).
6. Show that a conformal Killing vector $K=K^{AA'}\partial_{AA'}$ on the complexified four-dimensional Minkowski space maps α-planes to α-planes and gives rise to a holomorphic vector field on the corresponding twistor space given by

$$\mathcal{K}=K^{AA'}\pi_{A'}\frac{\partial}{\partial\omega^A}+\phi_{A'B'}\pi^{B'}\frac{\partial}{\partial\pi_{A'}}$$

   for some $\phi_{A'B'}$ which should be determined.

Describe the infinitesimal conformal group action on the space of SD two-forms in terms of $\phi_{A'B'}$.

7. Assume that K in the previous question is a non-null translation so that $\phi_{A'B'} = 0$. Obtain the mini-twistor space $T\mathbb{CP}^1$ as a quotient of the twistor space $\mathcal{PT} = \mathbb{CP}^3 - \mathbb{CP}^1$ by $\mathcal{K}$. Conversely, show that $\mathcal{PT}$ is a holomorphic bundle over $T\mathbb{CP}^1$ and determine the transition functions for this bundle.

9 Gravitational instantons

Gravitational instantons are solutions to the four-dimensional Einstein equations in Riemannian signature which give complete metrics and asymptotically 'look-like' flat space, in a sense to be made precise in the remainder of this chapter. Gravitational instantons cannot usually be analytically continued to Lorentzian signature. They are nevertheless physically relevant as they can allow a semi-classical description of the as-yet-unknown theory of quantum gravity. Our discussion in this chapter assumes familiarity with basic general relativity, but we shall summarize the standard definitions in Section 9.2.

9.1 Examples of gravitational instantons

- **Example. Euclidean Schwarzchild solution.** Our first example will be an analytic continuation of a well-known solution describing a static black hole. Replacing t by $i\tau$ in the Schwarzchild metric gives the following Riemannian metric:

$$g = \left(1 - \frac{2m}{r}\right)^{-1} dr^2 + \left(1 - \frac{2m}{r}\right) d\tau^2 + r^2(d\theta^2 + \sin^2\theta d\phi^2). \qquad (9.1.1)$$

The metric g is an analytic continuation of a Ricci-flat metric, therefore g itself is Ricci-flat and there is no need to verify the Einstein equations.

The Schwarzchild metric in the Lorentzian signature is singular at $r = 0$, but in the Riemannian setting this singularity can be removed by considering the range $2m \leq r < \infty$. The apparent singularity at $r = 2m$ can also be avoided at the price of allowing τ to be periodic. To see this define

$$\rho = 4m\sqrt{1 - \frac{2m}{r}}.$$

Expanding the metric near $\rho = 0$, which corresponds to $r = 2m$, gives $dr \sim (\rho/4m)d\rho$ and

$$g \sim d\rho^2 + \frac{\rho^2}{16m^2}d\tau^2 + 4m^2(d\theta^2 + \sin^2\theta d\phi^2).$$

If τ is taken to be periodic with a period of $8\pi m$ then the term $d\rho^2 + \frac{\rho^2}{16m^2}d\tau^2$ describes the flat metric on $\mathbb{R}^2$ and the four-dimensional metric g is regular.

For the next two examples we need to introduce left-invariant one-forms σ_i, $i = 1, 2, 3$, on the group manifold $SU(2)$ such that

$$d\sigma_1 + \sigma_2 \wedge \sigma_3 = 0, \quad d\sigma_2 + \sigma_3 \wedge \sigma_1 = 0, \quad \text{and} \quad d\sigma_3 + \sigma_1 \wedge \sigma_2 = 0. \tag{9.1.2}$$

These one-forms can be represented in terms of Euler angles by

$$\sigma_1 + i\sigma_2 = e^{-i\psi}(d\theta + i\sin\theta d\phi) \quad \text{and} \quad \sigma_3 = d\psi + \cos\theta d\phi,$$

where to cover $SU(2) = S^3$ we require the ranges

$$0 \leq \theta \leq \pi, \quad 0 \leq \phi \leq 2\pi, \quad \text{and} \quad 0 \leq \psi \leq 4\pi.$$

In terms of the left-invariant one-forms the flat metric on $\mathbb{R}^4$ is given by

$$g_{\text{flat}} = d\rho^2 + \frac{1}{4}\rho^2\left(\sigma_1^2 + \sigma_2^2 + \sigma_3^2\right),$$

where $\rho \geq 0$.

The following two examples do not arise as analytic continuations of Lorentzian solutions. Their Riemann tensor is ASD which makes them analogous to the ASD instantons in YM theory studied in Section 6.4. We shall postpone the discussion of the ASD condition to the next section and here only give the expressions for the metrics. We shall demonstrate that the expressions give rise to regular metrics but leave the verification of the Ricci-flat condition to Section 9.4 where a convenient formalism is developed.

- **Example. Eguchi–Hanson metric.** The Eguchi–Hanson metric [60, 61] is given by

$$g = \left(1 - \frac{a^4}{r^4}\right)^{-1} dr^2 + \frac{1}{4}r^2\left(1 - \frac{a^4}{r^4}\right)\sigma_3^2 + \frac{1}{4}r^2\left(\sigma_1^2 + \sigma_2^2\right). \tag{9.1.3}$$

The apparent singularity at $r = a$ is removed by allowing

$$r > a, \quad 0 \leq \psi \leq 2\pi, \quad 0 \leq \phi \leq 2\pi, \quad \text{and} \quad 0 \leq \theta \leq \pi.$$

To see that the metric is complete set $\rho^2 = r^2\left[1-(a/r)^4\right]$. Expanding the metric near $r=a$ and fixing (θ,ϕ) gives

$$g \sim \frac{1}{4}\left(d\rho^2 + \rho^2 d\psi^2\right).$$

In the standard spherical polar coordinates ψ has a period 4π on $SU(2)$. In our case the period of ψ is 2π to achieve regularity. Therefore the surfaces of constant r are real projective planes defined by identifying the antipodal points on the sphere, $\mathbb{RP}^3 = S^3/\mathbb{Z}^2$. At large values of r the metric looks like $\mathbb{R}^4/\mathbb{Z}^2$ rather than Euclidean space. The Eguchi–Hanson metric is an example of the asymptotically locally euclidean (ALE) manifold.

The Eguchi–Hanson example motivates the following definition:

Definition 9.1.1 *A complete regular four-dimensional Riemannian manifold (M,g) which solves the Einstein equations (possibly with cosmological constant) is called ALE if it approaches $\mathbb{R}^4/\Gamma$ at infinity, where Γ is a discrete subgroup of $SO(4)$.*

If Γ contains only the identity then asymptotically g is Euclidean and Ricci-flat. The positive action theorem [183] implies that any such g is a flat metric on $\mathbb{R}^4$. The Eguchi–Hanson metric corresponds to Γ generated by a single reflection.

The following example also has ASD Riemann tensor, but its asymptotic behaviour is rather different from that of the Eguchi–Hanson metric.

- **Example. Taub-NUT gravitational instanton.** Consider the metric [77]

$$g = \frac{1}{4}\frac{r+m}{r-m}dr^2 + m^2\frac{r-m}{r+m}\sigma_3^2 + \frac{1}{4}(r^2-m^2)\left(\sigma_1^2+\sigma_2^2\right). \tag{9.1.4}$$

At $r=m$ the three-sphere of constant r collapses to a point – an example of a NUT singularity (see below). The change of variables $r = m+\rho^2/(2m)$ gives near $r=m$

$$g \sim d\rho^2 + \frac{\rho^2}{4}\left(\sigma_1^2+\sigma_2^2+\sigma_3^2\right),$$

where we set $r+m \sim 2m$ in all numerators. Thus the metric extends smoothly over this point, and therefore is defined on the whole of $\mathbb{R}^4$. This metric is not ALE as the coefficient of σ_3^2 approaches a constant as $r\to\infty$ while the coefficient of $(\sigma_1^2+\sigma_2^2)$ grows like r^2. Thus the circle generated by σ_3 separates from the three-sphere at large distances. The infinity has a topology of the S^1 bundle over S^2. Examining the transition functions shows that this is in fact the one-monopole bundle considered in Section 6.2.1. This kind of behaviour is referred to as asymptotic local flatness. It implies flatness in the three-dimensional sense. The fourth (imaginary time) dimension is periodic.

Definition 9.1.2 *A complete regular four-dimensional Riemannian manifold (M, g) which solves the Einstein equations (possibly with cosmological constant) is called ALF (asymptotically locally flat) if it approaches S^1 bundle over S^2 at infinity.*

The case when the asymptotic bundle is globally $S^1 \times S^2$ corresponds to asymptotically flat metrics. The Euclidean Schwarzchild solution (9.1.1) is one example. The ALE and ALF manifolds belong to the class of gravitational instantons because their curvature is concentrated in a finite region of a space-time.

We shall end this section by discussing the removable singularities of the metrics (9.1.3) and (9.1.4) in a more general context. Consider a general metric of the form

$$g = dr^2 + a_1(r)\sigma_1^2 + a_2(r)\sigma_2^2 + a_3(r)\sigma_3^2.$$

This metric is regular if the functions $a_i(r)$ are regular and non-vanishing. It can however be regular even if some of the functions vanish, as the apparent singularity may result from choosing a singular coordinate system. Following Gibbons and Hawking [72] we consider two types of removable singularities at $r = 0$.

- The metric has a removable NUT singularity if

  $$a_1(r)^2 = a_2(r)^2 = a_3(r)^2 = \frac{r^2}{4} + O(r^3)$$

  near $r = 0$. This singularity may be removed by using the Cartesian coordinate system near the origin. The singularity $r = m$ in the Taub-NUT metric (9.1.4) is an example of a NUT singularity.
- The metric has a bolt singularity if

  $$a_1(r)^2 = a_2(r)^2 = \text{const} \quad \text{and} \quad a_3(r)^2 = n^2r^2, \quad \text{where} \quad n \in \mathbb{Z}$$

  up to the higher order terms in r near $r = 0$. Let us set the constant to one without the loss of generality. Thus near the bolt singularity the metric is of the form

  $$g \sim dr^2 + \frac{n^2r^2}{4}(d\psi + \cos\theta d\phi)^2 + \frac{1}{4}(d\theta^2 + \sin^2\theta d\phi^2)$$

  up to the higher order terms. This can be made regular if ψ is a periodic coordinate with the adjusted range

  $$0 \leq \psi \leq \frac{4\pi}{n}$$

  as then the singularity for fixed values of (θ, ϕ) just arises from using the plane polar coordinates (r, ψ) and can be removed by using Cartesian coordinates on $\mathbb{R}^2$. Thus at the bolt singularity $r = 0$ the three-dimensional orbits of the $SU(2)$ action collapse to a two-sphere S^2 with constant radius. The $r = a$ singularity in the Eguchi–Hanson metric (9.1.3) is an example of a bolt.

9.2 Anti-self-duality in Riemannian geometry

Let (M, g) be an n-dimensional Riemannian manifold. Thus there exists a non-degenerate symmetric tensor $g_{\mu\nu} = g_{\mu\nu}(x)$ on M giving rise to a line element

$$g = g_{\mu\nu}\, dx^\mu \otimes dx^\nu, \qquad \mu, \nu = 1, \dots, n.$$

A linear connection ∇ on M is a map sending a pair of vector fields (X, Y) on M to a vector field $\nabla_X Y$ such that for any vector fields X, Y, Z and any function f on M we have

- $\nabla_X(fY) = f\nabla_X Y + X(f)Y$
- $\nabla_X(f) = X(f)$
- $\nabla_X(Y + Z) = \nabla_X(Y) + \nabla_X(Z)$
- $\nabla_{X+fY} Z = \nabla_X Z + f\nabla_Y Z$

If $\partial_\mu = \partial/\partial x^\mu$ is a coordinate basis for vector fields we can define the Christoffel symbols by

$$\nabla_{\partial_\mu}\partial_\nu = \Gamma^\lambda_{\nu\mu}\partial_\lambda.$$

Moreover we define $\nabla_\mu := \nabla_{\partial_\mu}$. The Levi-Civita connection ∇ is the unique connection on TM that is torsion-free, that is,

$$[\nabla_\mu, \nabla_\nu] f = 0$$

for any function f on M, and preserves the metric, that is, $\nabla_\mu g_{\nu\lambda} = 0$. From now on we shall assume that ∇ is the Levi-Civita connection.

Expanding the expressions $\nabla_X V$ where $X = X^\mu \partial_\mu$ and $V = V^\mu \partial_\mu$ using the properties of ∇ we find that the covariant derivative of a vector V^μ and a form V_μ are given by

$$\nabla_\mu V^\nu = \partial_\mu V^\nu + \Gamma^\nu_{\mu\lambda} V^\lambda \quad \text{and} \quad \nabla_\mu V_\nu = \partial_\mu V_\nu - \Gamma^\lambda_{\mu\nu} V_\lambda,$$

respectively, where the Christoffel symbols $\Gamma^\lambda_{\mu\nu}$ of the Levi-Civita connection are uniquely determined by the metric

$$\Gamma^\gamma_{\mu\nu} = \frac{1}{2} g^{\gamma\delta}\left(\frac{\partial g_{\mu\delta}}{\partial x^\nu} + \frac{\partial g_{\nu\delta}}{\partial x^\mu} - \frac{\partial g_{\mu\nu}}{\partial x^\delta}\right). \tag{9.2.5}$$

The Riemann curvature tensor $R_{\mu\nu\gamma}{}^{\delta}$ is defined by

$$(\nabla_\mu \nabla_\nu - \nabla_\nu \nabla_\mu) V^\delta = R_{\mu\nu\gamma}{}^{\delta} V^\gamma, \tag{9.2.6}$$

where V^μ is an arbitrary vector. The symmetric Ricci tensor $R_{\mu\nu}$ and the Ricci scalar R are defined as

$$R_{\mu\nu} = R_{\mu\gamma\nu}{}^{\gamma} \quad \text{and} \quad R = R_{\mu\nu} g^{\mu\nu}.$$

The definition (9.2.6) of the Riemann tensor implies the following algebraic identities:

$$R_{\mu[\nu\gamma\delta]} = 0, \quad R_{\mu\nu\gamma\delta} = R_{\gamma\delta\mu\nu} = -R_{\nu\mu\gamma\delta}, \quad \text{and} \quad R_{\mu\nu} = R_{\nu\mu},$$

as well as the differential Bianchi identity

$$\nabla_{[\mu} R_{\nu\gamma]\delta\epsilon} = 0.$$

An n-dimensional Riemannian manifold is called Einstein if

$$R_{\mu\nu} = \frac{R}{n} g_{\mu\nu}. \tag{9.2.7}$$

In the Einstein case for $n > 2$ the Bianchi identity implies that the Ricci scalar is a constant. The quantity R/n is often referred to as the cosmological constant.

An alternative way to present the Riemannian metric is to use a moving frame (also called a tetrad in four dimensions) and write

$$g = \eta_{ab} \mathrm{e}^a \otimes \mathrm{e}^b, \quad a, b = 1, \ldots, n,$$

where $\eta_{ab} = \mathrm{diag}(1, 1, \ldots, 1)$ is a constant matrix and $\mathrm{e}^a = \mathrm{e}^a{}_\mu dx^\mu$ are one-forms such that

$$g_{\mu\nu} = \eta_{ab} \mathrm{e}^a_\mu \mathrm{e}^b_\nu.$$

The Latin indices $a, b, c, \ldots$ are raised and lowered using the constant matrix η_{ab} and its inverse, and the Greek indices are manipulated with $g_{\mu\nu}$. In the moving-frame formalism the connection Γ^a_b and the curvature R^a_b are $\mathfrak{so}(n)$-valued one-forms and two-forms, respectively, introduced using the Cartan structure equations. In the torsion-free case these equations are

$$d\mathrm{e}^a + \Gamma^a_b \wedge \mathrm{e}^b = 0 \quad \text{and} \quad R^a_b = d\Gamma^a_b + \Gamma^a_c \wedge \Gamma^c_b.$$

Thus Γ_{ab} and R_{ab} are both anti-symmetric when their indices are lowered. The curvature two-form can be expanded in the moving frame thus giving rise to a Riemann tensor R_{abcd} defined by

$$R^a_b = \frac{1}{2} R^a{}_{bcd} \mathrm{e}^c \wedge \mathrm{e}^d.$$

Assume the Riemannian manifold (M, g) is oriented by an invariant volume element

$$\mathrm{vol} = \mathrm{e}^1 \wedge \mathrm{e}^2 \wedge \cdots \wedge \mathrm{e}^n = \sqrt{|g|}dx^1 \wedge dx^2 \wedge \cdots \wedge dx^n.$$

The Hodge operator $* : \Lambda^p \to \Lambda^{n-p}$ is defined as

$$*(dx^{\mu_1} \wedge \cdots \wedge dx^{\mu_p}) = \frac{|g|^{1/2}}{(n-p)!}\varepsilon^{\mu_1\cdots\mu_p}{}_{\mu_{p+1}\cdots\mu_n}dx^{\mu_{p+1}} \wedge \cdots \wedge dx^{\mu_n}$$

or equivalently as

$$*(\mathrm{e}^{a_1} \wedge \cdots \wedge \mathrm{e}^{a_p}) = \frac{1}{(n-p)!}\varepsilon^{a_1\cdots a_p}{}_{a_{p+1}\cdots a_n}\mathrm{e}^{a_{p+1}} \wedge \cdots \wedge \mathrm{e}^{a_n}.$$

From now on we shall assume that $n = 4$. Given an oriented Riemannian four-manifold (M, g), the Hodge-$*$ operator is an involution on two-forms in the sense that $*^2 = \mathrm{Id}$. This induces a decomposition

$$\Lambda^2 = \Lambda^2_+ \oplus \Lambda^2_- \tag{9.2.8}$$

of two-forms into SD and ASD components corresponding to eigenvalues ± 1 of $*$. If F is a two-form then

$$F = F_+ + F_-, \tag{9.2.9}$$

where

$$F_+ = \frac{1}{2}(F + *F) \in \Lambda^2_+ \quad \text{and} \quad F_- = \frac{1}{2}(F - *F) \in \Lambda^2_-.$$

If the metric is given in terms of an orthonormal tetrad

$$g = (\mathrm{e}^1)^2 + (\mathrm{e}^2)^2 + (\mathrm{e}^3)^2 + (\mathrm{e}^4)^2 \tag{9.2.10}$$

then the bases of Λ^2_+ and Λ^2_- are given by

$$(\boldsymbol{\Omega}_1)_\pm = \mathrm{e}^4 \wedge \mathrm{e}^1 \pm \mathrm{e}^2 \wedge \mathrm{e}^3, \quad (\boldsymbol{\Omega}_2)_\pm = \mathrm{e}^4 \wedge \mathrm{e}^2 \pm \mathrm{e}^3 \wedge \mathrm{e}^1, \quad \text{and}$$
$$(\boldsymbol{\Omega}_3)_\pm = \mathrm{e}^4 \wedge \mathrm{e}^3 \pm \mathrm{e}^1 \wedge \mathrm{e}^2. \tag{9.2.11}$$

The Riemann tensor has the index symmetry $R_{abcd} = R_{[ab][cd]}$ so can be thought of as a map $\mathcal{R} : \Lambda^2 \to \Lambda^2$ given by

$$\mathcal{R}(F)_{ab} = R_{abcd}F^{cd}.$$

This map decomposes under (9.2.8) as follows:

$$\mathcal{R} = \left(\begin{array}{c|c} C_+ + \frac{R}{12} & \Phi \\ \hline \Phi & C_- + \frac{R}{12} \end{array} \right). \tag{9.2.12}$$

The $C_\pm$ terms are the SD and ASD parts of the Weyl tensor, the Φ terms are the trace-free Ricci curvature, and R is the scalar curvature which acts by scalar multiplication. The Weyl tensor is conformally invariant, so can be thought of as being defined by the conformal structure $[g] = \{c^2 g\}$ where c is a non-vanishing function on M.

Definition 9.2.1 *An ASD structure is a four-dimensional conformal structure such that the SD Weyl tensor C_+ vanishes.*

The Eguchi–Hanson and Taub-NUT gravitational instantons (9.1.3) and (9.1.4) are ASD in this sense. In fact more is true: These metrics are Ricci-flat and their scalar curvatures vanish. Thus the Weyl curvature is the only non-vanishing part of the Riemann tensor. In this case the ASD of the Weyl tensor implies that the Riemann tensor is ASD:

$$R_{abcd} = -\frac{1}{2}\varepsilon_{ab}{}^{pq} R_{pqcd} \tag{9.2.13}$$

or equivalently that the curvature two-form $R^a{}_b$ is ASD.

9.2.1 Two-component spinors in Riemannian signature

In the four-dimensional case it is convenient to use a modification of the tetrad formalism known as the null tetrad and write

$$g = 2(\mathrm{e}^{00'} \odot \mathrm{e}^{11'} - \mathrm{e}^{10'} \odot \mathrm{e}^{01'}),$$

where, in terms of the moving frame, the complex one-forms $\mathrm{e}^{AA'}$, $A = 0, 1$, $A' = 0, 1$, are defined by

$$\mathrm{e}^{00'} = \frac{1}{\sqrt{2}}(\mathrm{e}^1 + i\mathrm{e}^4), \quad \mathrm{e}^{11'} = \frac{1}{\sqrt{2}}(\mathrm{e}^1 - i\mathrm{e}^4),$$

$$\mathrm{e}^{01'} = -\frac{1}{\sqrt{2}}(\mathrm{e}^2 + i\mathrm{e}^3), \quad \text{and} \quad \mathrm{e}^{10'} = \frac{1}{\sqrt{2}}(\mathrm{e}^2 - i\mathrm{e}^3).$$

The null-tetrad representation is a convenient starting point for introducing the two-component spinor formalism [132] which is very well suited to study the ASD condition (9.2.13). Our discussion of the spinors will be a curved analogue of the formalism developed in Section 7.2.2. Consider the group isomorphism

$$SO(4, \mathbb{R}) \cong SU(2) \times SU(2)/\mathbb{Z}_2. \tag{9.2.14}$$

Locally there exist complex two-dimensional vector bundles $\mathbb{S}, \mathbb{S}'$ called spin bundles over M equipped with parallel symplectic structures $\varepsilon, \varepsilon'$ such that

$$\mathbb{C} \otimes TM \cong \mathbb{S} \otimes \mathbb{S}' \tag{9.2.15}$$

is a canonical bundle isomorphism, and

$$g(v_1 \otimes w_1, v_2 \otimes w_2) = \varepsilon(v_1, v_2)\varepsilon'(w_1, w_2) \tag{9.2.16}$$

for $v_1, v_2 \in \Gamma(\mathbb{S})$ and $w_1, w_2 \in \Gamma(\mathbb{S}')$. Here $\mathbb{C} \otimes TM$ is the complexifed tangent bundle, obtained by taking a union of complexifications of tangent spaces at all points of M. The spin bundles $\mathbb{S}$ and $\mathbb{S}'$ inherit connections from the Levi-Civita connection such that $\varepsilon, \varepsilon'$ are covariantly constant. We use the standard convention in which spinor indices are capital letters, unprimed for sections of $\mathbb{S}$ and primed for sections of $\mathbb{S}'$. For example, μ_A denotes a section of $\mathbb{S}^*$, the dual of $\mathbb{S}$, and $\nu^{A'}$ a section of $\mathbb{S}'$.

The symplectic structures on spin spaces ε_{AB} and $\varepsilon_{A'B'}$ (such that $\varepsilon_{01} = \varepsilon_{0'1'} = 1$) are used to raise and lower indices as in Section 7.2.2. For example given a section μ^A of $\mathbb{S}$ we define a section of $\mathbb{S}^*$ by $\mu_A := \mu^B \varepsilon_{BA}$. The complex conjugation maps $\mathbb{S}$ to itself by

$$\iota_A = (\alpha, \beta) \to \hat{\iota}_A = (-\overline{\beta}, \overline{\alpha}) \tag{9.2.17}$$

so that $\hat{\hat{\iota}}_A = -\iota_A$. This Hermitian conjugation induces a positive inner product

$$\iota_A \hat{\iota}^A = \varepsilon_{AB} \iota^A \hat{\iota}^B = |\alpha|^2 + |\beta|^2. \tag{9.2.18}$$

We define the inner product on the primed spinors in the same way.

Let the spin dyads (o^A, ι^A) and $(o^{A'}, \iota^{A'})$ span $\mathbb{S}$ and $\mathbb{S}'$, respectively. In the Euclidean signature a spinor and its complex conjugate form a basis of spin space. Thus we can always take $\iota^{A'} = \hat{o}^{A'}$, but we prefer not to do it as many of our formulae involving spin dyads are valid in other signatures where the properties of the complex conjugation are different.

We denote a normalized null-tetrad of vector fields on M by

$$\mathrm{e}_{AA'} = \begin{pmatrix} \mathrm{e}_{00'} & \mathrm{e}_{01'} \\ \mathrm{e}_{10'} & \mathrm{e}_{11'} \end{pmatrix}.$$

This tetrad is determined by the choice of spin dyads in the sense that

$$o^A o^{A'} e_{AA'} = e_{00'}, \quad -\iota^A o^{A'} e_{AA'} = e_{10'}, \quad -o^A \iota^{A'} e_{AA'} = e_{01'}, \quad \text{and}$$
$$\iota^A \iota^{A'} e_{AA'} = e_{11'}.$$

The dual tetrad of one-forms $e^{AA'}$ determine the metric by

$$g = \varepsilon_{AB}\varepsilon_{A'B'} e^{AA'} \otimes e^{BB'} = 2(e^{00'} \odot e^{11'} - e^{10'} \odot e^{01'}), \tag{9.2.19}$$

where $\odot$ is the symmetric tensor product. In terms of the spin bases we have

$$o_A o_{A'} e^{AA'} = e^{11'}, \quad \iota_A o_{A'} e^{AA'} = e^{01'}, \quad o_A \iota_{A'} e^{AA'} = e^{10'}, \quad \text{and} \quad \iota_A \iota_{A'} e^{AA'} = e^{00'}.$$

With indices, the above formula[1] for g becomes $g_{ab} = \varepsilon_{AB}\varepsilon_{A'B'}$.

A vector V can be decomposed as $V^{AA'} e_{AA'}$, where $V^{AA'}$ are the components of V in the basis. Its norm is given by $2\det(V^{AA'})$, which is unchanged under multiplication of the matrix $V^{AA'}$ by elements of $SU(2)$ on the left and right:

$$V^{AA'} \longrightarrow \Lambda^A_B V^{BB'} \tilde{\Lambda}^{A'}_{B'}, \quad \Lambda \in SU(2) \quad \text{and} \quad \tilde{\Lambda} \in \widetilde{SU}(2)$$

giving (9.2.14). The quotient by $\mathbb{Z}_2$ comes from the fact that multiplication on the left and right by -1 leaves $V^{AA'}$ unchanged.

A vector V is null when $\det(V^{AA'}) = 0$, so $V^{AA'} = \mu^A \nu^{A'}$ as the matrix V must be of rank one. The null vectors are necessarily complex in Riemannian signature.

The decomposition (9.2.9) of two-forms takes a simple form analogous to the flat situation (7.2.13) in the spinor notation. If

$$F = \frac{1}{2} F_{AA'BB'} e^{AA'} \wedge e^{BB'}$$

is a two-form then

$$F_{AA'BB'} = \phi_{AB}\varepsilon_{A'B'} + \tilde{\phi}_{A'B'}\varepsilon_{AB}, \tag{9.2.20}$$

where ϕ_{AB} and $\tilde{\phi}_{A'B'}$ are symmetric in their indices as in the flat case. This is precisely the decomposition of F into SD and ASD parts. Which is which depends on the choice of volume form; we choose $\tilde{\phi}_{A'B'}\varepsilon_{AB}$ to be the SD part. Thus

$$\Lambda^2_+ \cong \mathbb{S}'^* \odot \mathbb{S}'^* \quad \text{and} \quad \Lambda^2_- \cong \mathbb{S}^* \odot \mathbb{S}^*. \tag{9.2.21}$$

We shall often write $\Lambda^2_+ \cong \mathbb{S}' \odot \mathbb{S}'$, etc., because $\mathbb{S}'$ and its dual are naturally isomorphic vector bundles. The local bases Σ^{AB} and $\Sigma^{A'B'}$ of spaces of ASD and SD two-forms are defined by

$$e^{AA'} \wedge e^{BB'} = \varepsilon^{AB}\Sigma^{A'B'} + \varepsilon^{A'B'}\Sigma^{AB}. \tag{9.2.22}$$

[1] Note that we drop the prime on ε' when using indices, since it is already distinguished from ε by the primed indices.

Using (9.2.22) we can write $F = \phi_{AB}\Sigma^{AB} + \tilde{\phi}_{A'B'}\Sigma^{A'B'}$.

The first Cartan structure equations are

$$de^{AA'} = e^{BA'} \wedge \Gamma^{A}{}_{B} + e^{AB'} \wedge \Gamma^{A'}{}_{B'},$$

where Γ_{AB} and $\Gamma_{A'B'}$ are the $SU(2)$ and $\widetilde{SU}(2)$ spin connection one-forms. They are symmetric in their indices, and

$$\Gamma_{AB} = \Gamma_{CC'AB}e^{CC'}, \quad \Gamma_{A'B'} = \Gamma_{CC'A'B'}e^{CC'}, \quad \text{and} \qquad (9.2.23)$$
$$\Gamma_{CC'A'B'} = o_{A'}\nabla_{CC'}\iota_{B'} - \iota_{A'}\nabla_{CC'}o_{B'},$$

where $\nabla_{AA'} := \nabla_{e_{AA'}}$. The curvature of the spin connection

$$R^{A}{}_{B} = d\Gamma^{A}{}_{B} + \Gamma^{A}{}_{C} \wedge \Gamma^{C}{}_{B}$$

decomposes as

$$R^{A}{}_{B} = C^{A}{}_{BCD}\Sigma^{CD} + \frac{1}{12}R\Sigma^{A}{}_{B} + \Phi^{A}{}_{BC'D'}\Sigma^{C'D'},$$

and similarly for $R^{A'}{}_{B'}$. Here R is the Ricci scalar, $\Phi_{ABA'B'} = \Phi_{(AB)(A'B')}$ is the trace-free part of the Ricci tensor, and the symmetric spinor C_{ABCD} is the ASD part of the Weyl tensor:

$$C_{abcd} = \varepsilon_{A'B'}\varepsilon_{C'D'}C_{ABCD} + \varepsilon_{AB}\varepsilon_{CD}C_{A'B'C'D'}.$$

This leads to the following decomposition of the Riemann tensor:

$$\begin{aligned} R_{abcd} = {} & C_{ABCD}\varepsilon_{A'B'}\varepsilon_{C'D'} + C_{A'B'C'D'}\varepsilon_{AB}\varepsilon_{CD} \\ & + \Phi_{ABC'D'}\varepsilon_{A'B'}\varepsilon_{CD} + \Phi_{A'B'CD}\varepsilon_{AB}\varepsilon_{C'D'} \\ & + \frac{R}{12}(\varepsilon_{AC}\varepsilon_{BD}\varepsilon_{A'C'}\varepsilon_{B'D'} - \varepsilon_{AD}\varepsilon_{BC}\varepsilon_{A'D'}\varepsilon_{B'C'}). \end{aligned} \qquad (9.2.24)$$

A conformal structure is ASD iff $C_{A'B'C'D'} = 0$. The spinor form of the ASD condition (9.2.13) on the Riemann curvature is

$$R^{A'}{}_{B'} = 0.$$

Define the operators Δ_{AB} and $\Delta_{A'B'}$ by

$$[\nabla_a, \nabla_b] = \varepsilon_{AB}\Delta_{A'B'} + \varepsilon_{A'B'}\Delta_{AB}.$$

The spinor Ricci identities which follow from the definition of curvature are

$$\Delta_{AB}\iota_{A'} = \Phi_{ABA'B'}\iota^{B'} \quad \text{and} \qquad (9.2.25)$$

$$\Delta_{A'B'}\iota_{C'} = \left[C_{A'B'C'D'} - \frac{1}{12}R\varepsilon_{D'(A'}\varepsilon_{B')C'}\right]\iota^{D'} \qquad (9.2.26)$$

(and analogous equations for unprimed spinors). The Bianchi identities translate to

$$\nabla^{A'}{}_{A} C_{A'B'C'D'} = \nabla^{B}_{(B'} \Phi_{C'D')AB}, \quad \nabla^{AA'} \Phi_{ABA'B'} + \frac{1}{8} \nabla_{BB'} R = 0. \tag{9.2.27}$$

9.3 Hyper-Kähler metrics

The Ricci-flat ASD metrics in four dimensions have the property that they are Kähler with respect to three complex structures which satisfy the quaternionic algebra. This hyper-Kähler condition is a convenient way of imposing the ASD Ricci-flat equations on a given metric and we shall describe it in this section.

We shall start with discussing the Kähler structures. Our presentation follows the classical reference [94]. Let M be an even-dimensional manifold. An almost-complex-structure

$$I : TM \longrightarrow TM$$

is an endomorphism of the tangent bundle TM such that $I^2 = -\text{Id}$. Define the torsion of I by

$$N_I(X, Y) = [IX, IY] - [X, Y] - I[IX, Y] - I[X, IY], \quad \text{for} \quad X, Y \in TM.$$

Decompose the complexification of the tangent bundle

$$\mathbb{C} \otimes TM = T^{1,0}M \oplus T^{0,1}M,$$

where $T^{1,0}M$ and $T^{0,1}M$ are eigen-spaces of I corresponding to eigenvalues i and $-i$. If $X \in TM \otimes \mathbb{C}$ then the explicit decomposition is

$$X = \frac{1}{2}\,[X - iI(X)] + \frac{1}{2}\,[X + iI(X)]\,.$$

The Lie bracket of two vector fields in $T^{1,0}M$ does not have to be an element of $T^{1,0}M$. If it is, then one can introduce holomorphic coordinates on M. This is summarized in the following

Theorem 9.3.1 (Newlander–Nirenberg) *The following conditions are equivalent:*

1. *$T^{1,0}M$ spans an integrable distribution.*[2]
2. *$T^{0,1}M$ spans an integrable distribution.*
3. *$N_I(X, Y) = 0$ for any $X, Y \in TM$.*

[2] See Appendix C for a definition of what this means.

4. *M is a complex manifold and its complex structure induces an almost-complex-structure I such that in any holomorphic chart $(z^1, \ldots, z^n)$ the sub-bundle $T^{1,0}M$ is spanned by $\{\partial/\partial z^k\}$.*

The condition (4) in this theorem makes contact with the definition of a complex manifold given in Appendix B. If any of the conditions in the theorem is satisfied, I is called integrable. In view of this theorem an integrable almost-complex-structure is a complex structure.

Now assume that (M, I) is a complex manifold (thus $N_I = 0$) and g is a Riemannian metric on M which is Hermitian[3] with respect to I:

$$g(X, Y) = g(IX, IY).$$

Define a two-form $\boldsymbol{\Omega}$ by

$$\boldsymbol{\Omega}(X, Y) = g(X, IY).$$

Definition 9.3.2 *The triple (M, g, Ω) is a Kähler manifold if*

$$d\boldsymbol{\Omega} = 0.$$

In fact this condition together with the vanishing of N_I imply that the Kähler form is covariantly constant $\nabla_a \Omega_{bc} = 0$ with respect to the Levi-Civita connection of g [94].

A $4n$-dimensional Riemannian manifold is hyper-Kähler if it is Kähler with respect to three complex structures I_1, I_2, and I_3 such that

$$(I_1)^2 = (I_2)^2 = (I_3)^2 = -\text{Id}, \quad I_1 I_2 = I_3, \quad I_2 I_3 = I_1, \quad \text{and} \quad I_3 I_1 = I_2.$$

The next result shows that in four dimensions the hyper-Kähler condition is equivalent to ASD Ricci-flat equations on the metric.

Theorem 9.3.3 *Let (M, g) be a Riemannian four-manifold. Then g is ASD and Ricci-flat iff g is hyper-Kähler with respect to some triple of complex structures.*

Proof If the ASD Ricci-flat equations hold then the curvature $R^{A'}{}_{B'}$ of the connection on $\mathbb{S}'$ vanishes and there exists a covariantly constant basis $(o_{A'}, \iota_{A'})$. We can choose it to be normalized in the sense that $o_{A'}\iota^{A'} = 1$. Using the isomorphism $S^2(\mathbb{S}') \cong \Lambda^2_+$ we construct the covariantly constant two-forms

[3] Given any metric $\hat{g}$ on (M, I) we can construct a Hermitian metric g by

$$g(X, Y) = \hat{g}(X, Y) + \hat{g}(IX, IY).$$

spanning Λ^2_+:

$$\Sigma^{1'1'} = \frac{1}{2} o_{A'} o_{B'} \varepsilon_{AB} e^{AA'} \wedge e^{BB'} = e^{01'} \wedge e^{11'}$$

$$\Sigma^{1'0'} = -\frac{1}{2} o_{(A'} \iota_{B')} \varepsilon_{AB} e^{AA'} \wedge e^{BB'} = \frac{1}{2}(e^{01'} \wedge e^{10'} + e^{00'} \wedge e^{11'})$$

$$\Sigma^{0'0'} = \frac{1}{2} \iota_{A'} \iota_{B'} \varepsilon_{AB} e^{AA'} \wedge e^{BB'} = e^{00'} \wedge e^{10'}.$$

This basis of Λ^2_+ satisfies

$$\Sigma^{(A'B'} \wedge \Sigma^{C'D')} = 0 \quad \text{and} \quad d\Sigma^{A'B'} = 0. \tag{9.3.28}$$

The three Kähler forms which constitute the hyper-Kähler structure arise as linear combinations:

$$\mathbf{\Omega}_1 = -2i\, \Sigma^{1'0'}, \quad \mathbf{\Omega}_2 = i\,(\Sigma^{1'1'} - \Sigma^{0'0'}), \quad \text{and} \quad \mathbf{\Omega}_3 = \Sigma^{0'0'} + \Sigma^{1'1'}. \tag{9.3.29}$$

Conversely, given a hyper-Kähler structure (g, I_i) let E be a vector which is unit with respect to g. Thus, by the hermiticity of I_i, the vectors $I_i(E), i = 1, 2, 3$, are also unit and

$$g(E, I_i(E)) = \mathbf{\Omega}_i(E, E) = 0 \quad \text{and} \quad g(I_i(E), I_j(E)) = \delta_{ij}.$$

Let e^4 be a one-form dual to E. Extending the endomorphisms I_i to T^*M we conclude that $\{e^4, e^i = I_i(e^4)\}, i = 1, 2, 3$, forms an orthonormal tetrad so that g is given by (9.2.10). The relations

$$\mathbf{\Omega}_i(X, Y) = g(X, I_i Y), \quad i = 1, 2, 3,$$

where X, Y is any pair of vectors in the set $\{E, I_i(E)\}$, now imply that the Kähler forms are SD and given by (9.2.11) with the 'plus' sign. They are also covariantly constant so, again referring to the isomorphism $S^2(\mathbb{S}') \cong \Lambda^2_+$, there exists a covariantly constant frame for $\mathbb{S}'$. In this frame the primed connection vanishes and so $R^{A'}{}_{B'} = 0$. Thus the ASD Ricci-flat equations hold. □

As a by-product of this proof and the property (9.2.17) we deduce that a four-dimensional Riemannian manifold which admits a covariantly constant spinor has therefore to be hyper-Kähler, as the spinor and its complex conjugate give rise to a covariantly constant basis of Λ^2_+.

Theorem 9.3.3 admits an immediate and useful corollary.

Corollary 9.3.4 *Let* $\mathbf{\Omega}_j = (\mathbf{\Omega}_j)_+$, $j = 1, 2, 3$, *be a basis of SD two-forms* (9.2.11) *on a four-manifold* (M, g). *If*

$$d\mathbf{\Omega}_j = 0 \tag{9.3.30}$$

then $(M, g = (e^1)^2 + \cdots + (e^4)^2)$ *is hyper-Kähler (and thus ASD and Ricci-flat). Conversely, for any ASD Ricci-flat metric, there exists a covariantly constant basis of* $\mathbb{S}'$ *such that the SD two-forms are closed.*

The spinor form of the closure condition is of course $d\Sigma^{A'B'} = 0$. This formulation already assumes that $\mathbf{\Omega}_j$, or equivalently $\Sigma^{A'B'}$ are constructed from a tetrad. The algebraic condition $\Sigma^{(A'B'} \wedge \Sigma^{C'D')} = 0$ guarantees that the tetrad exists. This is a good way to impose the ASD Ricci-flat equations: choose a tetrad for a metric, construct the basis of SD two-forms, and impose the closure conditions. We shall use this formulation in the next two sections.

Let us finish this section with a historical remark. Given a hyper-Kähler (and therefore ASD Ricci-flat) metric we can choose one Kähler form $\mathbf{\Omega} = \mathbf{\Omega}_1$ and write the metric locally in terms of a Kähler potential $\Omega = \Omega(w, z, \bar{w}, \bar{z})$ where (w, z) are local holomorphic coordinates on an open ball in $\mathbb{C}^2$ and

$$g = \Omega_{w\bar{w}} dw\, d\bar{w} + \Omega_{w\bar{z}} dw\, d\bar{z} + \Omega_{z\bar{w}} dz\, d\bar{w} + \Omega_{z\bar{z}} dz\, d\bar{z}, \tag{9.3.31}$$

where $\Omega_{w\bar{w}} = \partial_w \partial_{\bar{w}} \Omega$, etc. The SD two-forms are given by

$$\mathbf{\Omega}_1 = \frac{i}{2}(\Omega_{w\bar{w}} dw \wedge d\bar{w} + \Omega_{w\bar{z}} dw \wedge d\bar{z} + \Omega_{z\bar{w}} dz \wedge d\bar{w} + \Omega_{z\bar{z}} dz \wedge d\bar{z}) \quad \text{and}$$

$$\mathbf{\Omega}_2 + i\mathbf{\Omega}_3 = dz \wedge dw. \tag{9.3.32}$$

The hyper-Kähler condition

$$\mathbf{\Omega}_1 \wedge \mathbf{\Omega}_1 = \mathbf{\Omega}_2 \wedge \mathbf{\Omega}_2 = \mathbf{\Omega}_3 \wedge \mathbf{\Omega}_3 \tag{9.3.33}$$

on g gives the non-linear Monge–Ampére equation on the function Ω

$$\Omega_{w\bar{w}} \Omega_{z\bar{z}} - \Omega_{w\bar{z}} \Omega_{z\bar{w}} = 1. \tag{9.3.34}$$

Plebański [135] demonstrated directly (without using the hyper-Kähler geometry) that any ASD Ricci-flat manifold is locally of the form (9.3.31) where Ω satisfies (9.3.34). In the context of ASD Ricci-flat metrics the Monge–Amperé equation (9.3.34) is known as the first heavenly equation.

In fact formula (9.3.31) and equation (9.3.34) first arose (in complexified setting) in the context of wave geometry [146] – a subject developed in Hiroshima during the 1930s. Wave geometry postulates the existence of a privileged spinor field which in the modern super-symmetric context would be called a Killing spinor. The integrability conditions come down to the ASD condition on the Riemannian curvature of the underlying complex space-time. This condition implies vacuum Einstein equations. The Institute at Hiroshima where wave geometry had been developed was completely destroyed by the atomic bomb in 1945. Two of the survivors wrote up the results of the theory in [120].

9.4 Multi-centred gravitational instantons

We shall discuss a large class of gravitational instantons which depend on a harmonic function on $\mathbb{R}^3$. Consider a metric of the form

$$g = V(dx_1{}^2 + dx_2{}^2 + dx_3{}^2) + V^{-1}(d\tau + A)^2, \tag{9.4.35}$$

where $V = V(x_i)$ and $A = A_i(x_j)dx_i$ are a function and a one-form, respectively, which do not depend on τ. This is known as the Gibbons–Hawking ansatz [71]. Choosing the tetrad

$$e^i = \sqrt{V}dx_i \quad \text{and} \quad e^4 = \frac{1}{\sqrt{V}}(d\tau + A), \quad i = 1, 2, 3$$

gives the basis of SD two-forms:

$$\mathbf{\Omega}_i = (d\tau + A) \wedge dx_i + \frac{1}{2}V\epsilon_{ijk}dx_j \wedge dx_k.$$

The ASD Ricci-flat equations are imposed as the closure condition (9.3.30). This gives the monopole equation

$$*_3 dV = dA, \tag{9.4.36}$$

where $*_3$ is the Hodge operator on $\mathbb{R}^3$ with its flat metric. Differentiating this equation implies that V has to be a solution to the Laplace equation on $\mathbb{R}^3$.

The vector field $K = \partial/\partial\tau$ is Killing, that is, $\text{Lie}_K g = 0$, which generates the S^1 action, and x_i are defined up to addition of a constant by $dx_i = K \lrcorner \mathbf{\Omega}_i$. In particular all SD two-forms (or equivalently the Kähler forms in the hyper-Kähler structure) are Lie derived along K. This means that K is tri-holomorphic. Conversely we have the following result:

Theorem 9.4.1 *Any four-dimensional hyper-Kähler metric which admits a tri-holomorphic Killing vector can be locally put in the form* (9.4.35) *where the function V and the one-form A on the space of orbits of the Killing vector satisfy* (9.4.36).

Proof Any Riemannian metric which admits a Killing vector takes the form

$$g = h + V^{-1}(d\tau + A)^2,$$

where τ parameterizes the orbits of $K = \partial/\partial\tau$ (see Appendix C for the proof that τ always exists locally) and (h, A, V) are a metric, a one-form, and a function on the space of orbits of K. The function V is defined by

$$V = \frac{1}{g(K, K)}. \tag{9.4.37}$$

The tri-holomorphic condition $\text{Lie}_K \mathbf{\Omega}_i = 0$ gives

$$K \lrcorner\, d\mathbf{\Omega}_i + d(K \lrcorner\, \mathbf{\Omega}_i) = 0.$$

The Kähler forms are closed, thus we can use the Poincaré lemma to deduce the local existence of functions x_i, $i = 1, 2, 3$, on M such that

$$dx_i = K \lrcorner\, \mathbf{\Omega}_i. \qquad (9.4.38)$$

These functions are Hamiltonians (also called moment maps) for the S^1 action generated by K. The metric is Hermitian with respect to all three complex structures, thus

$$g(I_i(K), I_i(K)) = g(K, K), \quad i = 1, 2, 3 \quad \text{(no summation)}.$$

But $g(I_i(K), I_i(K)) = |dx_i|^2$. Therefore

$$g_{ab}K^a K^b = g^{ab}\nabla_a x_1 \nabla_b x_1 = g^{ab}\nabla_a x_2 \nabla_b x_2 = g^{ab}\nabla_a x_3 \nabla_b x_3.$$

This gives $h = V(dx_1^2 + dx_2^2 + dx_3^2)$ and the metric is of the form (9.4.35). Imposing the hyper-Kähler conditions leads, as we have already demonstrated, to (9.4.36) which completes the proof. [4] □

- **Example.** Consider the flat metric on $\mathbb{R}^4 = \mathbb{C}^2$ with holomorphic coordinates (w, z)

$$g = |dz|^2 + |dw|^2$$

with the hyper-Kähler structure given by

$$\mathbf{\Omega}_1 = \frac{i}{2}(dz \wedge d\overline{z} + dw \wedge d\overline{w}) \quad \text{and} \quad \mathbf{\Omega}_2 + i\mathbf{\Omega}_3 = dz \wedge dw.$$

(This is of the form (9.3.31) with $\Omega = |z|^2 + |w|^2$.) Now consider an isometric and tri-holomorphic S^1 action

$$(z, w) \longrightarrow (e^{ic/2}z, e^{-ic/2}w),$$

where c is the constant group parameter. Using the formula (4.2.4) we find that this action is generated by the Killing vector

$$K = \frac{i}{2}\left(z\frac{\partial}{\partial z} - \overline{z}\frac{\partial}{\partial \overline{z}}\right) - \frac{i}{2}\left(w\frac{\partial}{\partial w} - \overline{w}\frac{\partial}{\partial \overline{w}}\right).$$

Formulae (9.4.38) give

$$x_1 = \frac{1}{4}(|z|^2 - |w|^2) \quad \text{and} \quad x_2 + ix_3 = \frac{1}{2}zw,$$

[4] A different proof of Theorem 9.4.1 based on exterior differential systems is given in Appendix C.

and (9.4.37) implies

$$V = g(K, K)^{-1} = \frac{4}{|z|^2 + |w|^2} = \frac{1}{r} \tag{9.4.39}$$

because $r^2 := x_1^2 + x_2^2 + x_3^2 = (|z|^2 + |w|^2)^2/16$. Thus the Gibbons–Hawking ansatz with the simple harmonic function $V = r^{-1}$ gives flat space. We also see that the apparent singularity $r = 0$ in the Gibbons–Hawking metric can be removed by a coordinate transformation. To find the one-form A in this case write the flat metric on $\mathbb{R}^3$ as

$$\mathbf{e}_1^2 + \mathbf{e}_2^2 + \mathbf{e}_3^2, \quad \text{where} \quad \mathbf{e}_1 = dr, \quad \mathbf{e}_2 = r d\theta, \quad \text{and} \quad \mathbf{e}_3 = r \sin\theta d\phi.$$

The relation $*_3\mathbf{e}_1 = \mathbf{e}_2 \wedge \mathbf{e}_3$ gives $*_3 dr = r^2 \sin\theta d\theta \wedge d\phi$ and so

$$*_3 dV = -\frac{1}{r^2} *_3 dr = d(\cos\theta d\phi).$$

Thus we can take

$$A = \cos\theta \, d\phi.$$

Allowing V to have point singularities modelled on r^{-1} and modifying V by adding a constant leads to non-flat metrics. We shall consider V of the form

$$V = V_0 + \sum_{m=1}^{N} \frac{1}{|\mathbf{x} - \mathbf{x}_m|},$$

where $\mathbf{x}_1, \ldots, \mathbf{x}_N$ are fixed points in $\mathbb{R}^3$ called the centres and $V_0 = \text{const}$. If (r_m, θ_m, ϕ_m) are spherical polar coordinates centred at the points $\mathbf{x}_m$ then, repeating the argument above, we find that there exists a trivialization of the S^1 bundle over $\mathbb{R}^3$ such that A is gauge equivalent to

$$A = \sum_{m=1}^{N} \cos\theta_m d\phi_m.$$

The behaviour of the corresponding metrics essentially depends on whether the constant V_0 vanishes or not.

- **Example. Taub-NUT.** Consider the Gibbons–Hawking metric with

$$V = V_0 + \frac{1}{r}, \qquad V_0 \neq 0.$$

This is equivalent to the Taub-NUT metric (9.1.4). To see it set $r = m + 2\hat{r}/m$ in (9.1.4). This gives

$$g = V\left[d\hat{r}^2 + \hat{r}^2(d\theta^2 + \sin^2\theta d\phi^2)\right] + V^{-1}(d\psi + \cos\theta\, d\phi)^2, \quad \text{where}$$

$$V = \frac{1}{m^2} + \frac{1}{\hat{r}},$$

which is in the Gibbons–Hawking form. In particular we have now verified that the Taub-NUT metric is indeed a solution to the ASD Ricci-flat equations.

Allowing more singularities leads to large classes of multi-instantons generalizing both the Taub-NUT and the Eguchi–Hanson solutions:

- The A_{N-1} ALE metrics [62, 71] are given by

$$V = \sum_{m=1}^{N} \frac{1}{|\mathbf{x} - \mathbf{x}_m|}.$$

We have seen that the case $N = 1$ is the flat metric on $\mathbb{R}^4$. Consider $N > 1$. The apparent singularities at $\mathbf{x} - \mathbf{x}_m$ are removable. To prove this first analyse the asymptotic behaviour $|\mathbf{x}| \to \infty$. This gives

$$V \sim \frac{N}{r}$$

and, by the previous calculation, $A \sim N\cos\theta d\phi$. The metric is

$$g \sim N\left\{\frac{1}{r}d\mathbf{x}^2 + r\,[d(\tau/N) + (A/N)]^2\right\}.$$

Setting $r = \rho^2/4$ yields

$$g \sim d\rho^2 + \frac{\rho^2}{4}\left\{\sigma_1^2 + \sigma_2^2 + [d(\tau/N) + \cos\theta\, d\phi]^2\right\}$$

which is regular if τ is periodic with a period $4\pi N$. Thus asymptotically we recover the flat metric on $\mathbb{R}^4/\mathbb{Z}_N$.

Now, focusing near any of the centres $\mathbf{x}_m$ and shifting the origin to $\mathbf{x}_m$ gives

$$V \sim \frac{1}{r}.$$

Therefore we can reverse our analysis of the flat metric (9.4.39) to deduce that $r = 0$ is a coordinate singularity and near the origin the metric looks like the flat metric on $\mathbb{R}^4$.

The Eguchi–Hanson metric (9.1.3) can be put in this form with $N = 2$ where the two centres lying on the x_3-axis are separated by $a^2/4$: Following

[126] define

$$r_+ = |\mathbf{x} - \mathbf{x}_+| \quad \text{and} \quad r_- = |\mathbf{x} - \mathbf{x}_-|,$$

and set

$$\tau = 2\phi, \quad r_\pm = \frac{1}{8}(r^2 \pm a^2 \cos\theta), \quad \phi_+ = \phi_- = \psi, \quad \text{and}$$

$$r_\pm \cos\theta_\pm = \frac{1}{8}(r^2 \cos\theta \pm a^2),$$

where $(r_\pm, \theta_\pm, \phi_\pm)$ are spherical polar coordinates around the two points $\mathbf{x}_\pm$. With this ansatz the two-centre ALE multi-instanton gives (9.1.3). This also shows that the Eguchi–Hanson metric is hyper-Kähler.

- The A_{N-1} ALF metrics [62, 71] are given by

$$V = 1 + \sum_{m=1}^{N} \frac{1}{|\mathbf{x} - \mathbf{x}_m|} \tag{9.4.40}$$

where we rescaled the potential to set the non-zero constant V_0 to one.

This time we first analyse the behaviour around the centres. Shifting the origin to $\mathbf{x}_m$ gives $V \sim r^{-1}$ when $r \to 0$. The analysis of this case is identical to the ALE case we have just considered – we find that the metric is flat near any of the centres. Now we consider the asymptotic region $r \to \infty$ where $V \sim 1$ and $A \sim N \cos\theta d\phi$. The metric is

$$g \sim dr^2 + r^2(d\theta^2 + \sin^2\theta d\phi^2) + N^2 \left[d(\tau/N) + \cos\theta d\phi\right]^2.$$

This is what we have called the ALF metric. For large r it approaches an S^1 bundle over S^2 where the radius of S^1 is fixed.

9.4.1 Belinskii–Gibbons–Page–Pope class

In this section we shall consider hyper-Kähler metrics admitting a tri-holomorphic action of $SU(2)$. This assumption will reduce the hyper-Kähler condition to the Euler equations which can be solved explicitly [13].

The hyper-Kähler metrics on $M = \mathbb{R} \times SU(2)$ with a transitive action of $SU(2)$ can be put in the form

$$g = w_1 w_2 w_3 d\rho^2 + \frac{w_2 w_3}{w_1}(\sigma_1)^2 + \frac{w_1 w_3}{w_2}(\sigma_2)^2 + \frac{w_1 w_2}{w_3}(\sigma_3)^2, \tag{9.4.41}$$

where w_1, w_2, and w_3 are functions of ρ, and σ_i are left-invariant one-forms on $SU(2)$ which satisfy (9.1.2). The ansatz (9.4.41) is a general one. Given a four-dimensional metric which is diagonal in the left-invariant basis one can always define a coordinate ρ such that g is of the form (9.4.41) for some functions w_i. Moreover the diagonalisability in the left-invariant basis can always be achieved if g is Ricci-flat: one first diagonalizes the metric on a surface $\rho =$ const and then shows that the Ricci-flat condition prohibits the off-diagonal terms. See the discussion in [160].

The SD two-forms are

$$\boldsymbol{\Omega}_1 = w_1\sigma_2 \wedge \sigma_3 + w_2 w_3 \sigma_1 \wedge d\rho,$$
$$\boldsymbol{\Omega}_2 = w_2\sigma_3 \wedge \sigma_1 + w_1 w_3 \sigma_2 \wedge d\rho, \quad \text{and}$$
$$\boldsymbol{\Omega}_3 = w_3\sigma_1 \wedge \sigma_2 + w_1 w_2 \sigma_3 \wedge d\rho. \tag{9.4.42}$$

We assume that the $SU(2)$ action fixes all complex structures, so the invariant frame is covariantly constant. Thus the closure condition (9.3.30) holds for ASD Ricci-flat metrics in this frame. It is equivalent to the Euler equations

$$\dot{w}_1 = w_2 w_3, \quad \dot{w}_2 = w_1 w_3, \quad \text{and} \quad \dot{w}_3 = w_1 w_2 \tag{9.4.43}$$

which are integrable and admit a Lax pair (8.1.7). These equations readily integrate to

$$(\dot{w}_3)^2 = ({w_3}^2 - C_1)({w_3}^2 - C_2),$$

where $C_1 = {w_3}^2 - {w_1}^2$ and $C_2 = {w_3}^2 - {w_2}^2$ are constants. The Belinskii–Gibbons–Page–Pope (BGPP) metric (9.4.41) is flat if $C_1 = C_2 = 0$, and is never complete if $C_1 C_2 (C_1 - C_2) \neq 0$. The remaining cases correspond to the Eguchi–Hanson solution (9.1.3).

Now choose a one-dimensional subgroup $U(1) \subset SU(2)$. Any hyper-Kähler metric with a tri-holomorphic $U(1)$ action can be put in the Gibbons–Hawking form (9.4.35). We shall now characterize the harmonic functions V for which (9.4.35) belongs to the BGPP class (9.4.41). Let g be such a metric and let K be the corresponding Killing vector which puts g in the Gibbons–Hawking form. We expand K in a left-invariant basis of $SU(2)$, eliminate the Euler angles and ρ in favour of (x_i, τ) and observe that $V^{-1} = g(K, K)$, where g is given by (9.4.41). The details can be found in [73] and [50], where it is shown that

$$V(x_1, x_2, x_3) = \left[\prod_{i=1}^{3}(\beta - \beta_i)\right]^{-1/2} \left[\sum_{i=1}^{3} \frac{{x_i}^2}{(\beta - \beta_i)^2}\right]^{-1},$$

where β is an algebraic root of

$$\sum_{i=1}^{3} \frac{{x_i}^2}{\beta - \beta_i} = C,$$

and $C; \beta_i$ are constants. Equivalently it can be shown [50] that the Gibbons–Hawking metric (9.4.35) belongs to the BGPP class (9.4.41) iff $V = \mathbf{r} \cdot \nabla \hat{V}$, and

$\hat{V}$ is a harmonic function constant on a central quadric, that is,

$$B_{ij}(\hat{V})x_i x_j = \text{const}$$

for some symmetric matrix $B_{ij} = B_{ij}(\hat{V})$.

- **Example.** Consider the Eguchi–Hanson metric which corresponds to the harmonic function

$$V = |\mathbf{r} + \mathbf{a}|^{-1} + |\mathbf{r} - \mathbf{a}|^{-1}, \quad \text{where} \quad \mathbf{a} = (0, 0, a).$$

We verify that $\mathbf{r} \cdot \nabla \hat{V} = V$, where the harmonic function

$$\hat{V} = -\frac{2}{a}\operatorname{arccoth}\left(\frac{|\mathbf{r} + \mathbf{a}| + |\mathbf{r} - \mathbf{a}|}{2a}\right)$$

is constant on the ellipsoid

$$\frac{x_1^2 + x_2^2}{a^2\{[\coth(a\hat{V}/2)]^2 - 1\}} + \frac{x_3^2}{a^2[\coth(a\hat{V}/2)]^2} = 1.$$

9.5 Other gravitational instantons

The positive action theorem [183] states that Ricci-flat manifolds (M, g) which have the topology of $\mathbb{R}^4$ at infinity and approach the flat Euclidean metric

$$\eta_{\mu\nu}dx^\mu dx^\nu, \quad \text{where} \quad \eta = \text{diag}(1, 1, 1, 1)$$

sufficiently fast, in the sense that

$$g_{\mu\nu} = \eta_{\mu\nu} + O(r^{-4}), \quad (\partial_\mu)^p(g_{\nu\lambda}) = O(r^{-4-p}), \quad \text{and} \quad r^2 = \eta_{\mu\nu}x^\mu x^\nu, \tag{9.5.44}$$

have to be flat. A weaker asymptotic condition one can impose on g is asymptotically locally Euclidean as in Definition 9.1.1. Globally the neighbourhood of infinity must look like $S^3/\Gamma \times \mathbb{R}$, where $\Gamma \subset SO(4)$ is a finite group of isometries acting freely on S^3 (a Kleinian group).

In the following we shall use the isomorphism (9.2.14) and consider Γ as a finite subgroup of $SU(2)$. Finite subgroups of $\Gamma \subset SU(2)$ correspond to Platonic solids in $\mathbb{R}^3$. They are the cyclic groups, and the binary dihedral, tetrahedral, octahedral, and icosahedral groups (one can think about the last three as Möbius transformations of $S^2 = \mathbb{CP}^1$ which leave the points corresponding to vertices of a given Platonic solid fixed). All Kleinian groups act on $\mathbb{C}^2$, and the 'infinity' $S^3 \subset \mathbb{C}^2$. Let $(z_1, z_2) \in \mathbb{C}^2$. For each Γ there exist three invariants x, y, and z which are polynomials in (z_1, z_2) invariant under Γ. These invariants satisfy some algebraic relations which we list below:

Group	Relation $F_\Gamma(x, y, z) = 0$
Cyclic	$xy - z^k = 0$
Dihedral	$x^2 + y^2 z + z^k = 0$
Tetrahedral	$x^2 + y^3 + z^4 = 0$
Octahedral	$x^2 + y^3 + yz^3 = 0$
Icosahedral	$x^2 + y^3 + z^5 = 0$

In each case

$$\mathbb{C}^2/\Gamma \subset \mathbb{C}^3 = \{(x, y, z) \in \mathbb{C}^3,\ F_\Gamma(x, y, z) = 0\}.$$

The manifold M on which an ALE metric is defined is obtained by minimally resolving the singularity at the origin of $\mathbb{C}^2/\Gamma$. This desingularization is achieved by taking M to be the zero set of

$$\widetilde{F}_\Gamma(x, y, z) = F_\Gamma(x, y, z) + \sum_{i=1}^{r} a_i\, f_i(x, y, z),$$

where f_i span the ring of polynomials in (x, y, z) which do not vanish when $\partial_x F_\Gamma = \partial_y F_\Gamma = \partial_z F_\Gamma = 0$. The dimension r of this ring is equal to the number of non-trivial conjugacy classes of Γ which is $k-1$, $k+1$, 6, 7, and 8, respectively [22]. Kronheimer [99, 100] proved that for each Γ a unique hyper-Kähler metric exists on a minimal resolution M, and that this metric is precisely the ALE metric with $\mathbb{R}^4/\Gamma$ as its infinity.

- **Example.** Consider the cyclic group Γ of matrices of order two generated by

$$\begin{pmatrix} 1 & 0 \\ 0 & 1 \end{pmatrix} \text{ and } \begin{pmatrix} -1 & 0 \\ 0 & -1 \end{pmatrix}.$$

This subgroup of $SU(2)$ acts linearly on $\mathbb{C}^2$. If $(z_1, z_2) \in \mathbb{C}^2$ then the monomials

$$x = (z_1)^2, \quad y = (z_2)^2, \quad \text{and} \quad z = z_1 z_2$$

are invariant under Γ. These monomials satisfy the algebraic relation

$$xy = z^2$$

which gives the function $F_\Gamma(x, y, z)$ in this case. The quotient $\mathbb{C}^2/\Gamma$ is singular, but this singularity can be resolved by adding lower order terms

$$xy = (z - p_1)(z - p_2)$$

which gives $\widetilde{F}_\Gamma(x, y, z)$. The non-singular zero set of $\widetilde{F}_\Gamma$ in $\mathbb{C}^3$ is the Eguchi–Hanson manifold.

An example of an ALF gravitational instanton (see Definition 9.1.2) is the Atiyah–Hitchin metric [11]. Consider the metric of the form (9.4.41). The action of $SU(2)$ lifts to the bundle of SD two-forms Λ^2_+. If the action is trivial, so that all SD two-forms are Lie-derived along the Killing vector fields generating the action, the ASD Ricci-flat equations reduce to the Euler system (9.4.43). Here we assume that the action is non-trivial and instead rotates the SD two-forms. This means that the two-forms $\mathbf{\Omega}_i$ given by (9.4.42) are not in the covariantly constant spin-frame, and the connection one-form $\Gamma^{A'}_{B'}$ does not vanish. Thus instead of the closure condition one has $d\Sigma^{A'B'} + 2\Gamma^{(A'}_{C'} \wedge \Sigma^{B')C'} = 0$ or, equivalently, $d\mathbf{\Omega}_i = \alpha_{ij} \wedge \mathbf{\Omega}_j$ where $\alpha_{ij} = -\alpha_{ji}$ are one-forms constructed out of $\Gamma^{A'}_{B'}$. The equations resulting from the ASD vacuum condition $R^{B'}_{A'} = 0$ are

$$\dot{w}_1 = w_2 w_3 - w_1(w_2 + w_3),$$
$$\dot{w}_2 = w_1 w_3 - w_2(w_1 + w_3), \quad \text{and} \tag{9.5.45}$$
$$\dot{w}_3 = w_1 w_2 - w_3(w_1 + w_2).$$

Following [11] this system can be solved in fairly closed-form using elliptic functions. Redefining the coordinate ρ one has

$$g = \frac{w_1 w_2 w_3}{W^4} d\rho^2 + \frac{w_2 w_3}{w_1}(\sigma_1)^2 + \frac{w_1 w_3}{w_2}(\sigma_2)^2 + \frac{w_1 w_2}{w_3}(\sigma_3)^2,$$

where

$$w_1 = -W\frac{dW}{d\rho} - \frac{1}{2}W^2 \operatorname{cosec}(\rho),$$
$$w_2 = -W\frac{dW}{d\rho} + \frac{1}{2}W^2 \cot(\rho), \quad \text{and}$$
$$w_3 = -W\frac{dW}{d\rho} + \frac{1}{2}W^2 \operatorname{cosec}(\rho),$$

and W satisfies the ODE

$$\frac{d^2 W}{d\rho^2} + \frac{1}{4} W \operatorname{cosec}^2(\rho) = 0.$$

The solution corresponding to the complete metric is

$$W = \frac{1}{\pi}\sqrt{\sin\rho\, K\left[\sin^2(\rho/2)\right]}, \quad \text{where} \quad K(k) = \int_0^{\pi/2} \frac{d\phi}{\sqrt{1 - k\sin^2\phi}}.$$

The importance of the Atiyah–Hitchin metric is that it arises as a natural metric on the moduli space of charge-two non-abelian magnetic $SU(2)$ monopole and its geodesics describe low-energy monopole scattering [11].

9.5.1 Compact gravitational instantons and $K3$

Compact solutions to Einstein equations with non-zero cosmological constant are relatively easy to find: the conformally flat metric on S^4 and the Fubini–Study metric on $\mathbb{CP}^2$ are two examples. The first one arises as the conformal rescaling of the flat metric on $\mathbb{R}^4$. The metric on S^4 with radius r_0

$$g = \frac{1}{\left[1 + (\rho/r_0)^2\right]^2} \left[d\rho^2 + \frac{\rho^2}{4}\left(\sigma_1^2 + \sigma_2^2 + \sigma_3^2\right)\right]$$

is conformally flat, and so obviously ASD. The second one is the standard Kähler metric on the complex projective space. It is given by the expression (9.3.31) with the Kähler potential

$$\Omega = \log(1 + |z|^2 + |w|^2),$$

where (z, w) are holomorphic coordinates on $\mathbb{CP}^2$ defined by

$$w = \frac{Z_1}{Z_3} \quad \text{and} \quad z = \frac{Z_2}{Z_3}$$

in the neighbourhood of $Z_3 \neq 0$ in terms of the homogeneous coordinates (Z_1, Z_2, Z_3). Notice that Ω does not satisfy the Monge–Ampere equation (9.3.34) as the metric is not Ricci-flat. This is the one example when one needs to be careful about the choice of orientation. The Kähler metric on $\mathbb{CP}^2$ we have just constructed is SD (rather that ASD) with respect to the orientation picked by the SD Kähler two-form. Setting

$$z = r\cos\left(\frac{\theta}{2}\right)\exp\left[\frac{i}{2}(\psi + \phi)\right] \quad \text{and} \quad w = r\sin\left(\frac{\theta}{2}\right)\exp\left[\frac{i}{2}(\psi - \phi)\right]$$

yields

$$g = \frac{dr^2}{(1 + r^2)^2} + \frac{1}{4}\frac{r^2\sigma_3^2}{(1 + r^2)^2} + \frac{1}{4}\frac{r^2}{1 + r^2}(\sigma_1^2 + \sigma_2^2).$$

The apparent singularity at $r = 0$ is a NUT singularity resulting from using spherical polars. To analyse the behaviour of the metric when $r \to \infty$ set $r = u^{-1}$. Fixing (θ, ϕ) gives

$$g \sim du^2 + \frac{1}{4}u^2 d\psi^2$$

near $u = 0$. Thus $u = 0$ is a bolt singularity which can be removed if $0 \leq \psi \leq 4\pi$.

In view of these two simple compact gravitational instantons it may therefore seem natural to look for compact Ricci-flat gravitational instantons. The flat four-torus $T^4 = S^1 \times S^1 \times S^1 \times S^1$ is a trivial example, but no other examples are explicitly known. The point is that Ricci-flat metrics on compact spaces cannot admit Killing vectors, except the trivial cases with S^1 factors. To see this consider a Killing vector K^a. The Killing equations $\nabla_{(a}K_{b)} = 0$ imply the identity

$$\nabla_a \nabla_b K_c = R_{bca}{}^d K_d.$$

Contracting this identity on the (ab) indices and contracting the resulting formula with K^c yields

$$R_{ab}K^a K^b - K^b \Box K_b = 0.$$

Integrating this by parts over a compact manifold M gives

$$\int_M (R_{ab}K^a K^b + |\nabla_a K_b|^2)\sqrt{g} d^4 x = 0.$$

If the metric on g is Ricci-flat $R_{ab} = 0$ then we necessarily have $\nabla_a K_b = 0$, as in the Riemannian case there are no non-zero tensors with zero norm. This condition can only hold if the components of the Riemann tensor vanish along K_a. If that is the case then g is of the form $d\tau^2 + g_3$, where $K = \partial/\partial\tau$ and g_3 is a Ricci-flat metric on $\mathbb{R}^3$. Thus g_3 is necessarily flat, and so is the four-dimensional metric.

A Ricci-flat metric on a simply connected compact manifold is known to exist on the so-called $K3$ surface.[5] The Riemann tensor of this metric is ASD and the metric is Kähler and does not admit any Killing vectors. The existence of this metric follows from Yau's proof [189] of the Calabi conjecture, but finding the explicit expression for the metric is one of the biggest outstanding open problems in Riemannian geometry and the theory of gravitational instantons.

9.6 Einstein–Maxwell gravitational instantons

In this section we shall discuss the Einstein–Maxwell theory which admits regular multi-centred instantons where the geometry is not determined by fixing the asymptotics – there exists non-flat, but asymptotically flat solutions. These solutions have been studied in [53, 190, 182]. Our presentation follows [53].

The Einstein–Maxwell gravitational instantons on a four-dimensional manifold M are solutions to the Einstein–Maxwell equations with Riemannian

[5] Named after three geometers: Kummer, Kähler, and Kodaira. The metric on $K3$ has so far remained even more elusive than the notoriously difficult mountain $K2$ in the Karakorum range of the Himalayas.

signature

$$R_{ab} - \frac{1}{2} R g_{ab} = T_{ab}, \tag{9.6.46}$$

where the Einstein–Maxwell energy–momentum tensor is given by

$$T_{ab} = 2 F_a{}^c F_{bc} - \frac{1}{2} g_{ab} F^{cd} F_{cd},$$

and the Maxwell equations

$$\nabla_a F^{ab} = 0$$

hold for the closed two-form F_{ab}. The metric is given by

$$g = \frac{1}{U\widetilde{U}} (d\tau + \boldsymbol{\omega})^2 + U\widetilde{U} d\mathbf{x}^2, \tag{9.6.47}$$

where the functions $U, \widetilde{U}$ and the one-form ω depend on $\mathbf{x} = (x_1, x_2, x_3)$ and satisfy

$$\begin{aligned} \nabla^2 U = \nabla^2 \widetilde{U} = 0 \quad \text{and} \\ \nabla \times \omega = \widetilde{U} \nabla U - U \nabla \widetilde{U}. \end{aligned} \tag{9.6.48}$$

We will work with the orthonormal tetrad

$$\mathbf{e}^4 = \frac{1}{(U\widetilde{U})^{1/2}} (d\tau + \boldsymbol{\omega}) \quad \text{and} \quad \mathbf{e}^i = (U\widetilde{U})^{1/2} dx_i\,. \tag{9.6.49}$$

With respect to this tetrad the electromagnetic field strength may now be written as

$$\begin{aligned} F_{4i} &= \frac{1}{2} \partial_i \left(U^{-1} - \widetilde{U}^{-1}\right) \quad \text{and} \\ F_{ij} &= \frac{1}{2} \varepsilon_{ijk} \partial_k \left(U^{-1} + \widetilde{U}^{-1}\right), \end{aligned} \tag{9.6.50}$$

where the derivatives are partial derivatives with respect to the corresponding space-time indices. One can check that this field strength satisfies the Bianchi identities, and thus locally at least we can write $F = dA$. The metrics with $U = \widetilde{U}$ have purely magnetic field strength $F = -2 *_3 dU$ and are the Riemannian analogues of the Majumdar–Papapetrou solutions [152] of the Einstein–Maxwell equations.

The solutions (9.6.47) were first found in the Lorentzian regime by Israel and Wilson [89] and by Perjés [133] as a stationary generalization of the static Majumdar–Papapetrou multi-black-hole solutions. However, it was shown by Hartle and Hawking that all the non-static solutions suffered from naked singularities [33, 76]. With Riemannian signature however, regular solutions exist [190, 182]. We can take

$$U = \frac{4\pi}{\beta} + \sum_{m=1}^{N} \frac{a_m}{| \mathbf{x} - \mathbf{x}_m |} \quad \text{and} \quad \widetilde{U} = \frac{4\pi}{\tilde{\beta}} + \sum_{n=1}^{\tilde{N}} \frac{\tilde{a}_n}{| \mathbf{x} - \tilde{\mathbf{x}}_n |}, \tag{9.6.51}$$

in these expressions $\beta, \tilde{\beta}, a_m, \mathbf{x}_m, \tilde{a}_n, \tilde{\mathbf{x}}_n$, N, and $\tilde{N}$ are constants. For the signature to remain positive throughout we can require $U, \widetilde{U} > 0$ which in turn requires $a_m, \tilde{a}_n > 0$.

If there is at least one non-coincident centre, $\mathbf{x}_m \neq \tilde{\mathbf{x}}_n$, regularity requires that τ is identified with period 4π and that the constants satisfy the following constraints at all the non-coincident centres

$$U(\tilde{\mathbf{x}}_n)\tilde{a}_n = 1 \quad \text{and} \quad \widetilde{U}(\mathbf{x}_m)a_m = 1, \qquad \forall m, n. \tag{9.6.52}$$

Given the locations of the centres $\{\mathbf{x}_n, \tilde{\mathbf{x}}_m\}$, these constraints may be solved uniquely for the $\{a_n, \tilde{a}_m\}$ [190]. When $\frac{4\pi}{\beta} = \frac{4\pi}{\tilde{\beta}} = 0$ the solution is only unique up to the overall scaling

$$U \to e^s U \quad \text{and} \quad \widetilde{U} \to e^{-s}\widetilde{U}. \tag{9.6.53}$$

In general this scaling leaves the metric invariant and induces a linear duality transformation on the Maxwell field mapping solutions to solutions:

$$\mathbf{E} \to \cosh s\, \mathbf{E} + \sinh s\, \mathbf{B} \quad \text{and} \quad \mathbf{B} \to \sinh s\, \mathbf{E} + \cosh s\, \mathbf{B}. \tag{9.6.54}$$

The rescaling does not leave the action and other properties of the solutions invariant.

The constants β and $\tilde{\beta}$ determine the asymptotics of the solution. There are three possibilities:

- The case $\frac{4\pi}{\beta} = \frac{4\pi}{\tilde{\beta}} \neq 0$ gives an ALF metric, tending to an S^1 bundle over S^2 at infinity, with first Chern number $N - \tilde{N}$. Without loss of generality we have rescaled the harmonic functions using (9.6.53) so that $\beta = \tilde{\beta}$. Equations (9.6.52) now imply that $\sum a_m - N = \sum \tilde{a}_n - \tilde{N}$. If $N = \tilde{N}$ the asymptotic bundle is trivial and we obtain asymptotically flat ($\sim \mathbb{R}^3 \times S^1$) solutions.
- The case $\frac{4\pi}{\beta} = 0$, $\frac{4\pi}{\tilde{\beta}} = 1$ gives an ALE metric, tending to $\mathbb{R}^4/\mathbb{Z}_{|N-\tilde{N}|}$. We have used the rescaling (9.6.53) to set $\frac{4\pi}{\tilde{\beta}} = 1$ without loss of generality. In this case the constraints (9.6.52) require that $\sum a_m = N - \tilde{N}$. Of course we can reverse the roles of β and $\tilde{\beta}$. If $N = \tilde{N} + 1$ the solution is asymptotically Euclidean ($\sim \mathbb{R}^4$).
- The case $\frac{4\pi}{\beta} = \frac{4\pi}{\tilde{\beta}} = 0$ leads to an asymptotically locally Robinson–Bertotti metric, tending to $AdS_2 \times S^2$ or $AdS_2/\mathbb{Z} \times S^2$.

  The former case only arises if all of the centres are coincident, so that $U = \widetilde{U}$, and τ need not be made periodic. For both these asymptotics, the constraints (9.6.52) require that $N = \widetilde{N}$. We may further use the rescaling (9.6.53) to set $\sum a_m = \sum \tilde{a}_n$.

As Riemannian solutions, the metrics are naturally thought of as generalizations of the Gibbons–Hawking multi-centre metrics (9.4.35) which in fact they include as the special case $\widetilde{U} = 1$, albeit with an additional ASD Maxwell field. A crucial new aspect of the ALE and ALF Israel–Wilson–Perjés solutions is that when

$$N = \widetilde{N} \pm 1 \text{ (for ALE)} \qquad \text{or} \qquad N = \widetilde{N} \text{ (for ALF)},$$

the fibration of the τ circle over S^2 at infinity is trivial and the metrics do not require the $\mathbb{Z}_N$ identifications at infinity that are needed in the Gibbons–Hawking case. The space-times are therefore strictly asymptotically Euclidean and asymptotically flat, respectively, in these cases.

We shall now use the arguments of Tod given in the Lorentzian setting [157] and show that the solution (9.6.47) and (9.6.50) with the harmonic functions described by (9.6.51) and satisfying the constraints (9.6.52) is the most general regular Einstein–Maxwell instanton with a 'charged' covariantly constant spinor.

Theorem 9.6.1 [53] *Let (g, A) be a regular solution to the Riemannian Einstein–Maxwell equations such that there exist spinors $(\alpha_A, \beta_{A'})$ (which do not vanish identically) satisfying*

$$\nabla_{AA'}\alpha_B - i\sqrt{2}\phi_{AB}\beta_{A'} = 0 \quad \text{and} \quad \nabla_{AA'}\beta_{B'} + i\sqrt{2}\tilde{\phi}_{A'B'}\alpha_A = 0, \tag{9.6.55}$$

where the spinors ϕ and $\tilde{\phi}$ are symmetric in their respective indices and give the ASD and SD parts of the electromagnetic field

$$F_{ab} = \phi_{AB}\varepsilon_{A'B'} + \tilde{\phi}_{A'B'}\varepsilon_{AB}. \tag{9.6.56}$$

Then the metric g admits a Killing vector and (g, A) can be put in the form (9.6.47) where U and $\widetilde{U}$ are harmonic functions of the form (9.6.51) and ω is a one-form which satisfies (9.6.48).

Proof We shall use the conjugation properties of Euclidean spinors (9.2.17) to define

$$U = (\alpha_A\hat{\alpha}^A)^{-1} \quad \text{and} \quad \widetilde{U} = (\beta_{A'}\hat{\beta}^{A'})^{-1}. \tag{9.6.57}$$

In the positive-definite case U and $\widetilde{U}$ are bounded unless α or β have zeros. In the Lorentzian case their possible vanishing leads to plane wave space-times [157]. Now define a (complex) null tetrad

$$X_a = \alpha_A\beta_{A'}, \quad \overline{X}_a = \hat{\alpha}_A\hat{\beta}_{A'}, \quad Y_a = \alpha_A\hat{\beta}_{A'}, \quad \text{and} \quad \overline{Y}_a = -\hat{\alpha}_A\beta_{A'}. \tag{9.6.58}$$

We can check that $(\hat{\alpha}_A, \hat{\beta}_{A'})$ is also a solution to the Killing spinor equation (9.6.55) (note that $\hat{\phi}_{AB} = -\phi_{AB}$ so that F is real when expanded in a complex null tetrad). It therefore follows from (9.6.55) that X_a, $\overline{X}_a$, and $Y_a - \overline{Y}_a$ are gradients and that $K_a = Y_a + \overline{Y}_a$ is a Killing vector. Define local coordinates $(x_i, \tau) = (x, y, z, \tau)$ by

$$X = \frac{1}{\sqrt{2}}(dx + idy), \quad (Y - \overline{Y}) = i\sqrt{2}dz, \quad \text{and} \quad K^a\nabla_a = \sqrt{2}\frac{\partial}{\partial\tau}, \tag{9.6.59}$$

where $X = X_a e^a = X_{AA'}e^{AA'}$ and similarly for $Y, \overline{Y}$. The vector K Lie derives the spinors $(\alpha_A, \beta_{A'})$, implying that U and $\tilde{U}$ are independent of τ.

The metric is now given by $g = \varepsilon_{AB}\varepsilon_{A'B'}e^{AA'}e^{BB'}$. This expression may be evaluated by noting that from (9.6.57) we have $\varepsilon_{AB} = U(\alpha_A\hat{\alpha}_B - \alpha_B\hat{\alpha}_A)$ and similarly for $\varepsilon_{A'B'}$. Using the fact that from the above definitions $K_aK^a = 2(U\tilde{U})^{-1}$, we find that the metric takes the form (9.6.47) for some one form $\boldsymbol{\omega}$. The next step is to find $\boldsymbol{\omega}$.

The definitions of U, $\tilde{U}$, and K together with (9.6.55) imply

$$\nabla_a K_b = i\sqrt{2}\left(\tilde{U}^{-1}\phi_{AB}\varepsilon_{A'B'} + U^{-1}\tilde{\phi}_{A'B'}\varepsilon_{AB}\right), \tag{9.6.60}$$

and

$$\nabla_a U^{-1} = i\sqrt{2}\phi_{AB}K^B_{A'} \quad \text{and} \quad \nabla_a\tilde{U}^{-1} = -i\sqrt{2}\tilde{\phi}_{A'B'}K^{B'}_A. \tag{9.6.61}$$

The formulae in (9.6.61) may be inverted to find expressions for ϕ_{AB} and $\tilde{\phi}_{A'B'}$, using

$$K^{A'}_B K^{BC'} = \frac{1}{2}\varepsilon^{A'C'}K_{DE'}K^{DE'}.$$

Substituting the result into (9.6.60) yields the expression (9.6.48) for $\nabla \times \boldsymbol{\omega}$.

Finally, differentiating the relations (9.6.55) shows that the energy–momentum tensor is that of Einstein–Maxwell theory: $T_{ab} = 2\phi_{AB}\tilde{\phi}_{A'B'}$. The Maxwell equations

$$\nabla^{AA'}\phi_{AB} = 0 \quad \text{and} \quad \nabla^{AA'}\tilde{\phi}_{A'B'} = 0 \tag{9.6.62}$$

now imply that U and $\tilde{U}$ are harmonic on $\mathbb{R}^3$. This completes the local reconstruction of the solution from the Killing spinors.

So far everything has proceeded as in [157] with minor differences in signs and the reality conditions. The main difference arises in global regularity considerations which lead us to consider the invariant

$$\begin{aligned}|F_{ab}F^{ab}| &= |2(\phi_{AB}\phi^{AB} + \tilde{\phi}_{A'B'}\tilde{\phi}^{A'B'})| \\ &= |\nabla U^{-1}|^2 + |\nabla\tilde{U}^{-1}|^2,\end{aligned} \tag{9.6.63}$$

where the norm of the gradients is taken with respect to the flat metric on $\mathbb{R}^3$, and we have used (9.6.61). Regularity requires this invariant be bounded. Therefore both $|\nabla U^{-1}|$ and $|\nabla \widetilde{U}^{-1}|$ must be bounded. The various boundary conditions we have described imply that U and $\widetilde{U}$ are regular as $|\mathbf{x}| \to \infty$. In particular, they are both regular outside a ball B_R of sufficiently large radius R in $\mathbb{R}^3$.

The coordinates $\{\mathbf{x}, \tau\}$ cover $\mathbb{R} \times (\mathbb{R}^3 \setminus \mathcal{S})$, where $\mathcal{S}$ is the compact subset of B_R on which U or $\widetilde{U}$ blow up. A theorem from [32] can now be applied separately to both harmonic functions to prove that $\mathcal{S}$ consists of a finite number of points. In fact

$$\#\mathcal{S} < \max\{|\nabla U^{-1}|, |\nabla \widetilde{U}^{-1}|\}|U(p) + \widetilde{U}(p)|\ R + 1, \qquad (9.6.64)$$

where p is any point in B_R which does not belong to $\mathcal{S}$. This combined with the maximum principle shows that (9.6.51) are the most general harmonic functions leading to regular metrics. It also follows from (9.6.57) and the positivity of the spinor inner product that a_m and $\tilde{a}_n$ in (9.6.51) are all non-negative. □

9.7 Kaluza–Klein monopoles

In the gauge-theoretic context, instantons in D dimensions can be interpreted as solitons in $(D+1)$ dimensions. This remains true in the context of gravity. The point is that if g is a Riemannian Ricci-flat metric in D dimensions then the product metric $-dt^2 + g$ is a static Lorentizian Ricci-flat metric in $(D+1)$ dimensions. To construct gravitational solitons – static, non-singular solutions – to Lorentzian Einstein equations one of course has to worry about the asymptotic behaviour. The simplest example of such soliton is the five-dimensional Kaluza–Klein monopole[6] of Gross–Perry [74] and Sorkin [148].

In its original version the Kaluza–Klein theory is five-dimensional general relativity such that the fifth, space-like, dimension is compactified on a circle. The Kaluza–Klein monopole is given by

$$g^{(5)} = -dt^2 + V({dx_1}^2 + {dx_2}^2 + {dx_3}^2) + V^{-1}(d\tau + A)^2 \qquad (9.7.65)$$

where we have taken the gravitational instanton in four dimensions to be the multi-Taub-NUT gravitational solution (9.4.40).

From the (3+1)-dimensional perspective the solutions (9.7.65) give rise to solutions of Einstein–Maxwell theory with a dilaton. This is the standard

[6] The existence of solitons in (3+1)-dimensions is ruled out by the Birkhoff theorem which states that every non-flat static solutions to vacuum Einstein equations is diffeomorphic to the Schwarzchild black-hole.

Kaluza–Klein reduction where the fifth dimension compactifies to a circle of small radius $\mathcal{R}$. This corresponds to τ in (9.4.35) being periodic. If the radius is sufficiently small then low-energy experiments will average over the fifth dimension thus leading to an effective four-dimensional theory with the Maxwell potential and a scalar field.

We perform the Kaluza–Klein reduction with respect to the space-like Killing vector $\partial/\partial\tau$. The four-dimensional theory is invariant under the general coordinate transformations independent of τ. The translation of the fibre coordinate $\tau \rightarrow \tau + \Lambda(x^\mu)$ where $x^\mu = (\mathbf{x}, t)$ induces the $U(1)$ gauge transformations of the Maxwell one-form. The scaling symmetry

$$\tau \longrightarrow c\tau, \quad [g^{(5)}]_{\mu\tau} \longrightarrow c^{-2}[g^{(5)}]_{\mu\tau},$$

is spontaneously broken by the Kaluza–Klein vacuum, since τ is a coordinate on a circle with a fixed radius. The scalar field corresponding to this symmetry breaking is called the dilaton.

It is the usual practice to conformally rescale the resulting $(3+1)$-dimensional metric, and the dilaton so that the multiple of the Ricci scalar of the Lorentzian metric $G_{\mu\nu}$ in the reduced Lagrangian is equal to $\sqrt{|\det(G_{\mu\nu})|}$. The corresponding Maxwell field is $F = dA$, and the physical metric $G_{\mu\nu}$ in $(3+1)$ signature is given by

$$g^{(5)} = \exp(-2\phi/\sqrt{3})G_{\mu\nu}dx^\mu dx^\nu + \exp(4\phi/\sqrt{3})(d\tau + A)^2, \qquad (9.7.66)$$

where the triple

$$G = \sqrt{V}d\mathbf{x}^2 - \frac{1}{\sqrt{V}}dt^2, \qquad \phi = -\frac{\sqrt{3}}{4}\log V, \quad \text{and} \quad F = *_3 dV$$

satisfies the Einstein–Maxwell dilaton equations. These equations arise from the Einstein–Hilbert Lagrangian in five dimensions:

$$\int \sqrt{|\det g^{(5)}|}\, R^{(5)} d\tau d^4x,$$

where $R^{(5)}$ is the Ricci scalar of $g^{(5)}$. Substituting the anzatz (9.7.66) into this Lagrangian yields the four-dimensional Lagrangian

$$2\pi\mathcal{R}\int \sqrt{|\det G|}\left(R - 2G^{\mu\nu}\nabla_\mu\phi\,\nabla_\nu\phi - \frac{1}{4}e^{2\sqrt{3}\phi}F_{\mu\nu}F^{\mu\nu}\right)d^4x, \qquad (9.7.67)$$

where $\mathcal{R}$ is the radius of the Kaluza–Klein circle, $F = dA$, and R is the Ricci scalar of G.

9.7.1 Kaluza–Klein solitons from Einstein–Maxwell instantons

The details of the Kaluza–Klein lift are more complicated in the case of Einstein–Maxwell gravitational instantons (9.6.47). The Einstein–Maxwell theory without a dilaton cannot be consistently lifted to pure gravity in five dimensions. However, Einstein–Maxwell configurations may be lifted to solutions of five-dimensional Einstein–Maxwell theory with a Chern–Simons term. This lift is the bosonic sector of the lift from $\mathcal{N} = 2$ supergravity in four dimensions to $\mathcal{N} = 2$ supergravity in five dimensions [31].

We are interested in lifting the four-dimensional Riemannian theory to a Lorentzian theory on a five-dimensional manifold M_5. The four-dimensional action is

$$S_4 = \int d^4x\sqrt{g}\left(R - F_{ab}F^{ab}\right), \tag{9.7.68}$$

with equations of motion given by (9.6.46). The five-dimensional action is

$$S_5 = \int d^5x\sqrt{-g^{(5)}}\left[R^{(5)} - H_{\alpha\beta}H^{\alpha\beta}\right] - \frac{8}{3\sqrt{3}}\int H \wedge H \wedge W, \tag{9.7.69}$$

where $H = dW$ is the five-dimensional Maxwell field. In this section we use Greek indices ranging from 0 to 4 in five dimensions. The equations of motion in five dimensions are

$$G_{\alpha\beta} = 2H_\alpha{}^\gamma H_{\beta\gamma} - \frac{1}{2}g^{(5)}_{\alpha\beta}H^{\gamma\delta}H_{\gamma\delta} \quad \text{and} \quad d*_5 H = -\frac{2}{\sqrt{3}}H \wedge H. \tag{9.7.70}$$

Given a solution, g and $F = dA$, to the four-dimensional equations (9.6.46), we may lift the solution to five dimensions as follows:

$$g^{(5)} = g - (dt + \Phi)^2 \quad \text{and} \quad W = \frac{\sqrt{3}}{2}A, \tag{9.7.71}$$

where Φ is a one-form determined by g and F through

$$d\Phi = *_4 F. \tag{9.7.72}$$

One may then check that the five-dimensional configuration (9.7.71) solves the equations of motion (9.7.70). Note that solutions to (9.7.72) exist because $d{*_4}F = 0$ on shell. In the cases (9.6.47) and (9.6.50) we may solve for Φ explicitly to find

$$\Phi = -\frac{1}{2}\left(U^{-1} + \widetilde{U}^{-1}\right)(d\tau + \omega) + \chi, \tag{9.7.73}$$

where χ satisfies

$$\nabla \times \chi = \frac{1}{2}\nabla\left(U - \widetilde{U}\right). \tag{9.7.74}$$

To investigate regularity and causality of the five-dimensional metrics we shall analyse the behaviour near the centres where $U \to \infty$ or $\widetilde{U} \to \infty$. In the four-dimensional Riemannian Israel–Wilson–Perjés solutions these can always be made to be regular points [190, 182] as we discussed in Section 9.6. We shall follow [53] and re-examine the regularity of the metric around these points and check for the possible occurrence of closed time-like curves.

Before analysing the behaviour near the centres note the following. Firstly, that

$$g^{(5)}{}_{\tau\tau} \equiv g^{(5)}\left(\frac{\partial}{\partial\tau}, \frac{\partial}{\partial\tau}\right) = -\frac{(U-\widetilde{U})^2}{(2U\widetilde{U})^2} < 0, \tag{9.7.75}$$

if $U \neq \widetilde{U}$. Therefore, to avoid closed time-like curves throughout the five-dimensional space-time we must not identify τ. Secondly, possible candidates for the location of horizons are where the metric becomes degenerate:

$$0 = g^{(5)}{}_{tt} g^{(5)}{}_{\tau\tau} - [g^{(5)}{}_{t\tau}]^2 = -\frac{1}{U\widetilde{U}}. \tag{9.7.76}$$

This occurs at the centres where U or $\widetilde{U}$ diverge.

In order to understand the geometry near the centres, there are three different cases we need to consider separately. The first is that $U \to \infty$ while $\widetilde{U}$ remains finite. Using polar coordinates $(r = \rho^2/4, \theta, \phi)$ centred on the point $\mathbf{x}_m$ and requiring that $a_m \widetilde{U}(\mathbf{x}_m) = 1$, the metric becomes

$$g^{(5)} \sim d\rho^2 + \frac{\rho^2}{4}\left[(d\tau + \cos\theta d\phi)^2 + d\Omega^2_{S^2}\right] - (dt - a_m d\tau/2)^2 \tag{9.7.77}$$

as $\rho \to 0$, with $d\Omega^2_{S^2} = d\theta^2 + \sin^2\theta d\phi^2$. The metric may be made regular about this point if we identify τ with period 4π. Unfortunately this introduces closed time-like curves as we discussed. If we choose not to identify τ we are left with time-like naked singularities at the centres. We see that there is no horizon at these points, but rather a (singular) origin of polar coordinates. Therefore, metrics with this behaviour at the centres cannot lift to causal, regular solitons in five dimensions.

The remaining two possibilities involve coincident centres where both U and $\widetilde{U}$ go to infinity, so that $\mathbf{x}_m = \tilde{\mathbf{x}}_m$. One needs to treat separately the cases where $a_m = \tilde{a}_m$ and where $a_m \neq \tilde{a}_m$. In the latter case we again find regularity at the expense of closed time-like curves going out to infinity, or alternatively naked singularities. This leaves only the former case with $a_m = \tilde{a}_m$ for all m. That is, $\widetilde{U} = U + k$, with k being some constant.

By considering the asymptotic regime, one can see that in order to obtain a regular asymptotic geometry without closed time-like curves, one requires that either both U and $\widetilde{U}$ go to a constant at infinity or they both go to zero. Rescaling the harmonic functions and performing a duality rotation on the Maxwell field, as we discussed in four dimensions above, implies that without loss of generality $U = \widetilde{U}$. Therefore the only lift that leads to a globally regular and causal five-dimensional space-time is the case $U = \widetilde{U}$, which corresponds to the Euclidean Majumdar–Papapetrou metric in four dimensions. The five-dimensional metric can be written as

$$g^{(5)} = -(dt - d\tau/U)^2 + \frac{d\tau^2}{U^2} + U^2 d\mathbf{x}^2 . \tag{9.7.78}$$

Away from the centres, the space-times approach either $\mathbb{R}^{1,4}$ or $AdS_3 \times S^2$, with U going to a constant or zero at infinity, respectively.

With a rescaling of coordinates, the geometry near the centres where $U \to \infty$ may be written as

$$g^{(5)} \sim a_m^2 \left(\frac{dr^2}{r^2} + 2r\,dt\,d\tau - dt^2 + d\Omega^2_{S^2} \right) . \tag{9.7.79}$$

Calculating the curvature shows that this metric locally describes $AdS_3 \times S^2$. The Killing vector $\partial/\partial t$ is everywhere regular and time-like. This remains true in the full space-time (9.7.78). There is no horizon and the degeneration of the metric at the centres is analogous to the origin of polar coordinates.

The coordinates in (9.7.79) may be mapped to Poincaré coordinates as follows:

$$\begin{aligned} Y &= \frac{1}{r^{1/2}\cos\frac{t}{2}}, \\ X &= \frac{\tau}{2} - \frac{1}{2}\left[\frac{1}{r} - 1\right]\tan\frac{t}{2}, \quad \text{and} \\ T &= \frac{\tau}{2} - \frac{1}{2}\left[\frac{1}{r} + 1\right]\tan\frac{t}{2}, \end{aligned} \tag{9.7.80}$$

so that the metric becomes

$$g^{(5)} = \frac{4a_m^2}{Y^2}\left(-dT^2 + dX^2 + dY^2\right) + a_m^2 d\Omega^2_{S^2} . \tag{9.7.81}$$

There is no singularity at $t = \pm\pi$ as may be checked by writing down the embedding of AdS_3 as a quadric in $\mathbb{R}^{2,2}$ in terms of these coordinates. The map (9.7.80) is periodic in t. Taking t with infinite range corresponds to passing to the (causal) universal cover of AdS_3. There is no need to identify τ and therefore the space-time is causal.

The metrics (9.7.78) give causal, regular solutions to the five-dimensional theory with an everywhere defined time-like Killing vector. Writing the metric in the form (9.7.78) suggests that the space-times should be thought of as containing N parallel 'solitonic strings'. The strings have world-volumes in the $t - \tau$ plane. There is a plane-fronted wave [68] carrying momentum along the $\partial/\partial\tau$ direction of the string. We call these plane-fronted waves solitonic-strings to emphasize that the fields are localized along strings and there are no horizons. The strings are magnetic sources for the two-form field strength

$$H = -\sqrt{3} *_3 dU . \tag{9.7.82}$$

This is possible because of the topologically non-trivial S^2 at each centre (9.7.79).

9.7.2 Solitons in higher dimensions

The higher dimensional theories which are far-reaching generalizations of the Kaluza–Klein attempt to unify gravity and electromagnetism dominate modern theoretical physics. The currently fashionable M theory postulates that the dimension of space-time is 11.[7] At the classical and low energy level this theory is equivalent to 11-dimensional supergravity. The bosonic sector of this theory consists of a metric g of signature (10, 1) and a three-form C. With the definition $G = dC$ the action is

$$S = \int d^{11}x\sqrt{|g|}R - \int \left(\frac{1}{2}G \wedge *G + \frac{1}{6}C \wedge G \wedge G\right).$$

Varying this action with respect to (g, C) gives the equations of motion:

$$\begin{aligned} R_{\mu\nu} &= \frac{1}{12}(G_{\mu\alpha\beta\gamma}G_{\nu}{}^{\alpha\beta\gamma} - \frac{1}{12}g_{\mu\nu}G_{\alpha\beta\gamma\delta}G^{\alpha\beta\gamma\delta}), \\ dG &= 0, \quad \text{and} \\ d * G &= -\frac{1}{2}G \wedge G, \end{aligned}$$

where $*$ is taken with respect to the 11-dimensional metric g.

These equations resemble the Einstein–Maxwell–Chern–Simons theory (9.7.70) and should be thought of as higher dimensional analogues of Einstein–Maxwell theory where the Maxwell equations are replaced by non-linear equations for the four-form field G. If the four-form vanishes then the 11-dimensional metric satisfies the Einstein vacuum equation and the Kaluza–Klein monopole

$$g = -dt^2 + dy^2 + g_{\text{Taub-NUT}}$$

provides an example of a soliton in this theory. Here $x^\mu = (t, y_1, \ldots, y_6, x_1, \ldots, x_3, \tau)$ where $(x^1, \ldots x^3, \tau)$ are local coordinates on a Taub-NUT gravitational instanton (9.4.40).

The more interesting solutions have non-vanishing four-form. The celebrated example is the five-brane soliton [75]:

$$g = V^{-1/3}(-dt^2 + dy_1^2 + \cdots + dy_5^2) + V^{2/3}(dx_1^2 + \cdots + dx_5^2) \quad \text{and } G = *_5 dV,$$

[7] Those conservative readers who object to the higher dimensionality of space-time and would prefer to settle on four dimensions will find themselves in a minority. They may be comforted by the quote of Anatole France, a French novelist (1844–1924), who said 'If fifty million people say a foolish thing, it is still a foolish thing'.

where $V = V(x_1, \ldots, x_5)$ is a harmonic function on $\mathbb{R}^5$ given by

$$V = 1 + \sum_{m=1}^{N} \frac{1}{|\mathbf{x} - \mathbf{x}_m|^3}$$

and $*_5$ is the Hodge operator on $\mathbb{R}^5$ taken with respect to the flat metric. The five-brane is a flat six-dimensional surface[8] in the 11-dimensional space-time M obtained by fixing the values of the $\mathbf{x}$ coordinates. The coordinates $(t, \mathbf{y})$ are adapted to the isometries of the world-volume of the brane. Borrowing terminology from Maxwell theory we say that the five-brane is an example of a purely magnetic solution as there is no dt in the field G.

It is interesting to examine the asymptotic geometry of the five-brane solution in the simplest case when

$$V = 1 + \frac{1}{r^3}.$$

At large r the metric approaches the flat metric on $\mathbb{R}^{10,1}$. Let us consider the 'near horizon geometry' when r is small. Expanding the metric near $r = 0$ gives

$$g \sim r(-dt^2 + dy_1^2 + \cdots + dy_5^2) + \frac{1}{r^2}(dr^2 + r^2 d\Omega_{S^4}^2),$$

where $d\Omega_{S^4}^2$ is the constant curvature metric on the round four-sphere. Defining a new coordinate $r = \rho^{-2}/4$ and rescaling $(t, \mathbf{y})$ gives

$$g \sim 4\frac{-dt^2 + dy_1^2 + \cdots + dy_5^2 + d\rho^2}{\rho^2} + d\Omega_{S^4}^2,$$

This is a metric on $AdS_7 \times S^4$ – a Cartesian product of a four-dimensional sphere and a seven-dimensional Lorentzian anti-deSitter space. Thus the five-brane solution is indeed regular and gives an example of a soliton in 11-dimensional supergravity.

Most other known solutions to 11-dimensional supergravity (and its 10-dimensional string-theory reductions) also involve harmonic functions or lifts of gravitational instantons studied in this chapter. See [69] for a clear exposition of these ideas.

Exercises

1. Verify that the Eguchi–Hanson metric (9.1.3) and the Taub-NUT metrics (9.1.4) are regular at $r = a$ and $r = m$, respectively.

[8] In general a p-brane is a surface with with $(p + 1)$-dimensional world-volume. A particle is a zero-brane with one-dimensional world-line, a string is a one-brane, etc.

2. Consider a Kähler metric (9.3.31) with the null tetrad given by

$$e^{A1'} = dw^A \quad \text{and} \quad e^{A0'} = \frac{\partial^2 \Omega}{\partial w_A \partial \overline{w}_B} d\overline{w}_B,$$

where $w^A = (w, z)$.

Verify that the traceless Ricci tensor vanishes if Ω satisfies (9.3.32).

Calculate the ASD Weyl spinor C_{ABCD} in terms of Ω and its derivatives.

3. Show that the BGPP metric (9.4.41) is hyper-Kähler if the Euler equations (9.4.43) hold.

Derive the Eguchi–Hanson metric (9.1.3) as a special case of the BGPP class.

4. Verify that the Einstein–Maxwell equations for (9.6.47) and (9.6.50) reduce to (9.6.48).
5. Let K_a satisfy the Killing equations $\nabla_{(a} K_{b)} = 0$. Show that

$$\nabla_a \nabla_b K_c = R_{bca}{}^d K_d$$

and deduce the spinor identity

$$\nabla^A{}_{A'} \psi_{B'C'} = \alpha C^{D'}_{A'B'C'} K^A_{D'} + \beta K^B{}_{(A'} \Phi^A_{B'C')B} + \gamma R \varepsilon_{A'(B'} K^A{}_{C')}$$
$$+ \delta \varepsilon_{A'(B'} \Phi_{C')}{}^{D'DA} K_{DD'}, \tag{9.7.83}$$

where $\psi_{A'B'}$ is the SD part of dK and $\alpha, \beta, \gamma, \delta$ are constants which should be determined.

6. Derive the Einstein–Maxwell dilaton equations arising from the Lagrangian (9.7.67).

Find the solution to these equations from a Kaluza–Klein soliton $-dt^2 + g$ where g is the Taub-NUT metric, and verify that the equations are satisfied.

10 Anti-self-dual conformal structures

All gravitational instantons discussed in Chapter 9 (with the exception of the analytically continued Schwarzschild solution (9.1.1)) had ASD Riemannian tensor. This ASD property underlies the existence of many explicit examples as well as 'implicit' solution generation techniques. In this chapter we shall consider a more general situation where only the conformal Weyl curvature in four dimensions is ASD. This contains the ASD of the Riemann tensor as a special case. The motivation for studying ASD conformal structures is two-fold:

- The hyper-Kähler metrics studied in Chapter 9 are examples of ASD conformal structures where the Ricci tensor vanishes. The only non-trivial compact example is the $K3$ surface. There are interesting differential geometric generalizations of hyper-Kähler conditions which are non-Ricci flat but still ASD: scalar-flat Kähler or hyper-Hermitian metrics in four dimensions.
- In Section 8.1 we pointed out that many integrable systems admitting soliton solutions arise as symmetry reductions of the ASDYM equations in four dimensions. The Riemann–Hilbert factorization problem introduced in Section 3.3.1 underlies this approach to integrability.

  There is a large class of integrable systems (the dispersionless integrable systems in $2+1$ and 3 dimensions) which do not fit into this framework: they do not admit soliton solutions and there is no associated Riemann–Hilbert problem where the corresponding Lie group is finite dimensional. These systems can nevertheless be described as symmetry reductions of ASD conditions on a four-dimensional conformal structure.

What happens to the ASD condition in other space-time signatures? In Lorentzian signature $(+++-)$ the Hodge $*$ is not an involution (it squares to -1 instead of 1) and there is no decomposition of two-forms into real SD and ASD parts analogous to (9.2.8). In neutral $(++--)$ signature the Hodge $*$ is an involution, and there is a decomposition exactly as in the Riemannian case, depending on $[g]$. Thus ASD conformal structures exist in the neutral signature. These are relevant in the theory of integrable systems.

We shall be interested in finding local solutions to the conformal ASD equations in both Riemannian and neutral signatures. In the real analytic case both signatures can be treated on an equal footing by going to the complexification – a device already used in Chapter 7 for ASDYM. Thus in most of this chapter $(M, [g])$ denotes a complex four-dimensional manifold with a holomorphic conformal structure – the components of a metric in a conformal class only depend on the coordinates on M and not on their complex conjugates.

10.1 α-surfaces and anti-self-duality

Recall from Section 7.2.3 that an α-plane in complexified Minkowski space is a null two-plane spanned by vectors of the form $V^{AA'} = \kappa^A \pi^{A'}$ with fixed $\pi^{A'}$. Fixing κ^A and varying $\pi^{A'}$ gives rise to a β-plane. These definitions make sense in flat space-time $\mathbb{M}$. If M carries a curved metric there will be integrability conditions (coming from the Frobenius theorem (Theorem C.2.4) proved in Appendix C) for an α-plane to be tangent to a two-dimensional surface.

Definition 10.1.1 *An α-surface is a two-dimensional surface in M such that its tangent plane at every point is an α-plane.*

Let $g \in [g]$ and let ∇ denote the Levi-Civita connection of g on M. Let $e^{AA'}$ be a null tetrad of one-forms such that

$$g = \varepsilon_{AB}\varepsilon_{A'B'} e^{AA'} e^{BB'}$$

and let $e_{AA'}$ be a dual tetrad of vector fields (see Section 9.2.1 for a summary of two-component spinor formalism).

Consider the connection induced by ∇ on the primed spin bundle $\mathbb{S}' \to M$. The connection coefficients $\Gamma_{AA'B'}{}^{C'}$ of ∇ can be read off from Cartan's structure equations and are given by (9.2.23). Equivalently

$$\nabla_{AA'}\mu^{C'} = e_{AA'}(\mu^{C'}) + \Gamma_{AA'B'}{}^{C'} \mu^{B'},$$

where $\mu^{A'}$ is a section of $\mathbb{S}'$ in coordinates determined by the basis $e_{AA'}$. Given a connection on a vector bundle, one can lift a vector field on the base to a horizontal vector field on the total space. Let $\pi_{A'}$ denote the local coordinates on the fibres of $\mathbb{S}'$. Then the horizontal lifts $\tilde{e}_{AA'}$ of $e_{AA'}$ are given explicitly by

$$\tilde{e}_{AA'} := e_{AA'} - \Gamma_{AA'B'}{}^{C'} \pi^{B'} \frac{\partial}{\partial \pi^{C'}}.$$

In the flat case there is a two-sphere of α-planes through each point spanned by $\pi^{A'}\partial_{AA'}$. This gives a three-parameter family of α-planes in $M_{\mathbb{C}}$ and leads to the Definition 7.2.1 of the twistor space.

Given a curved metric g, define a two-dimensional distribution on $\mathbb{S}'$ by $\mathcal{D} = \text{span}\{L_0, L_1\}$, where

$$L_A := \pi^{A'}\tilde{\mathbf{e}}_{AA'}. \tag{10.1.1}$$

The vector fields $\pi^{A'}\mathbf{e}_{AA'}$ span a sphere of α-planes (one for each $\pi^{A'}$) at each point of M and (L_0, L_1) are horizontal lifts of these vectors to the spin bundle.

Theorem 10.1.2 (Penrose [131]) *There exists a three-parameter family of α-surfaces in M iff the conformal structure $[g]$ is ASD.*

Proof There will exist a maximal (i.e. three-parameter) family of α-surfaces if each α-plane is tangent to some α-surface. Therefore the distribution $\mathcal{D}$ must be integrable in the sense of Frobenius theorem (Theorem C.2.4):

$$[L_0, L_1] \in \text{span}\{L_0, L_1\}.$$

Using the formula for horizontal lifts of $\mathbf{e}_{AA'}$ and the spinor decomposition of the curvature (9.2.24) we find

$$[\pi^{A'}\tilde{\mathbf{e}}_{AA'}, \pi^{B'}\tilde{\mathbf{e}}_{BB'}] = (\Gamma_{AA'B}{}^{D} - \Gamma_{BA'A}{}^{D})\pi^{A'}\pi^{B'}\tilde{\mathbf{e}}_{DB'} + \pi^{A'}\pi^{B'}\varepsilon_{AB}\varepsilon^{F'Q'}C_{A'B'E'Q'}\pi^{E'}\frac{\partial}{\partial\pi^{F'}}.$$

One can see from this that if the SD Weyl spinor $C_{A'B'C'D'} = 0$ then $\pi^{A'}\tilde{\mathbf{e}}_{AA'}$, $A = 0, 1$, form an integrable distribution. The projection of a leaf of this distribution to M gives an α-surface. □

We deduce that the existence of α-surfaces depends on the conformally invariant Weyl spinor. It is therefore a property of the conformal structure $[g]$, rather than a chosen metric $g \in [g]$.

10.2 Curvature restrictions and their Lax pairs

Theorem 10.1.2 can be used in the context of integrable systems by interpreting (L_0, L_1) as a Lax pair for the ASD conformal structure. This Lax pair consists of vector fields, and thus is fundamentally different than the matrix Lax pair (7.1.6) for the ASDYM equations.[1]

[1] In Proposition 10.3.2 we shall however see that there is a connection if one allows infinite-dimensional gauge groups.

We shall work with the projective spin bundle $P\mathbb{S}'$, with inhomogeneous fibre coordinate $\lambda = \pi_{0'}/\pi_{1'}$. The Frobenius integrability conditions for $\mathcal{D}$ give compatibility conditions for the pair of linear equations

$$L_0 f = \left(\mathbf{e}_{00'} - \lambda \mathbf{e}_{01'} + l_0 \frac{\partial}{\partial \lambda} \right) f = 0$$

$$L_1 f = \left(\mathbf{e}_{10'} - \lambda \mathbf{e}_{11'} + l_1 \frac{\partial}{\partial \lambda} \right) f = 0$$

to have a solution f for all $\lambda \in \mathbb{CP}^1$, where f is a function on $P\mathbb{S}'$ and $l_A = \Gamma_{AA'B'C'} \pi^{A'} \pi^{B'} \pi^{C'}$ are two cubic polynomials in λ with coefficients given by components of the connection. In the integrable systems language λ is the spectral parameter.

We shall now describe various conditions that one can place on a metric $g \in [g]$ on top of ASD of the Weyl tensor. This provides a more direct link with integrable systems, as in each case described below one can choose a spin frame and local coordinates to reduce the special ASD condition to an integrable scalar PDE with corresponding Lax pair.

10.2.1 Hyper-Hermitian structures

Consider a structure $(M, I_j, j = 1, 2, 3)$, where M is a four-dimensional manifold and $I_i : TM \to TM$ are anti-commuting endomorphisms of the tangent bundle satisfying the algebra of quaternions:

$$(I_1)^2 = (I_2)^2 = (I_3)^2 = -\mathrm{Id}, \quad I_1 I_2 = I_3, \quad I_2 I_3 = I_1, \quad \text{and} \quad I_3 I_1 = I_2. \tag{10.2.2}$$

Consider the sphere of almost-complex-structures on M given by $\sum_j u_j I_j$, for $\mathbf{u} = (u_1, u_2, u_3)$ such that $|\mathbf{u}| = 1$. If each of these almost-complex-structures is integrable, we call (M, I_j) a hyper-complex manifold.

So far we have not introduced a metric. A natural restriction on a metric given a hyper-complex structure is to require it to be Hermitian with respect to each of the complex structures. This is equivalent to the requirement

$$g(X, Y) = g(I_j X, I_j Y), \quad j = 1, 2, 3 \tag{10.2.3}$$

for all vectors X, Y. Given a hyper-complex manifold, we call a metric satisfying (10.2.3) a hyper-Hermitian metric. There are two reality conditions one can impose:

- In Riemannian signature the complex structures I_j define a unique conformal structure obtained by picking a vector V and letting $(V, I_1(V), I_2(V), I_3(V))$ be an orthonormal basis of a metric in the conformal class.

- In neutral signature not all endomorphisms I_j compatible with the metric are real and $I_1 = I, I_2 = iS, I_3 = iT$ where the real anti-commuting endomorphisms (I, S, T) define a pseudo-hyper-complex structure

$$S^2 = T^2 = -I^2 = \mathrm{Id} \quad \text{and} \quad ST = -TS = \mathrm{Id}. \tag{10.2.4}$$

The neutral metric is Hermitian with respect to I and anti-Hermitian with respect to (S, T).

Hyper-Hermitian metrics are necessarily ASD. One way to formulate this is via the Lax pair formalism as follows:

Theorem 10.2.1 [44] *Let $e_{AA'}$ be four independent holomorphic vector fields on a four dimensional complex manifold M. Put*

$$L_0 = e_{00'} - \lambda e_{01'} \quad \text{and} \quad L_1 = e_{10'} - \lambda e_{11'}.$$

If

$$[L_0, L_1] = 0 \tag{10.2.5}$$

for every value of the parameter λ, then g given by (9.2.19) *is a hyper-Hermitian metric on M. Given any four-dimensional hyper-Hermitian metric there exists a null tetrad such that* (10.2.5) *holds.*

Interpreting λ as the projective primed spin bundle coordinate, we see that hyper-Hermitian metric must be ASD from Theorem 10.1.2. Theorem 10.2.1 characterizes hyper-Hermitian metrics as those which possess a Lax pair containing no ∂_λ terms.

We shall now discuss the local formulation of the hyper-Hermitian condition as a PDE. Expanding (10.2.5) in powers of λ gives

$$[\mathrm{e}_{00'}, \mathrm{e}_{10'}] = 0, \quad [\mathrm{e}_{00'}, \mathrm{e}_{11'}] + [\mathrm{e}_{01'}, \mathrm{e}_{10'}] = 0, \quad \text{and} \quad [\mathrm{e}_{01'}, \mathrm{e}_{11'}] = 0. \tag{10.2.6}$$

It follows from (10.2.6), using the Frobenius theorem and the Poincaré lemma, that one can choose coordinates (x^A, w^A), $(A = 0, 1)$, in which $\mathrm{e}_{AA'}$ take the form

$$\mathrm{e}_{A0'} = \frac{\partial}{\partial x^A} \quad \text{and} \quad \mathrm{e}_{A1'} = \frac{\partial}{\partial w^A} - \frac{\partial \Theta^B}{\partial x^A}\frac{\partial}{\partial x^B},$$

where $\Theta^B = \Theta^B(x^0, x^1, w^0, w^1)$ is a pair of functions satisfying a system of coupled non-linear PDEs:

$$\frac{\partial^2 \Theta_C}{\partial x_A \partial w^A} + \frac{\partial \Theta_B}{\partial x^A}\frac{\partial^2 \Theta_C}{\partial x_A \partial x_B} = 0. \tag{10.2.7}$$

Note the indices here are not spinor indices, they are simply a convenient way of labelling coordinates and the functions Θ^A. We raise and lower them in the usual way using the standard anti-symmetric matrix ε_{AB}. Given a solution to (10.2.7) the conformal class is represented by the metric

$$g = dx_A dw^A + \frac{\partial \Theta_A}{\partial x^B} dw^A dw^B.$$

- **Example.** Put $w^A = (w, z)$ and $x^A = (y, -x)$. A simple class of solutions to (10.2.7) is provided by

$$\Theta_0 = ax^l \quad \text{and} \quad \Theta_1 = by^k, \qquad k, l \in \mathbb{Z}, \quad a, b \in \mathbb{C},$$

as both the linear and non-linear part of (10.2.7) vanish separately. The corresponding metric is

$$g = dwdx + dzdy + (alx^{l-1} + bky^{k-1})dwdz.$$

10.2.2 ASD Kähler structures

Let (M, g) be an ASD four-manifold and let J be a complex structure such that the corresponding fundamental two-form is closed, so that the metric is Kähler (see Definition 9.3.2 in Section 9.3). There exist local coordinates $(w^A, \tilde{w}^A)$ and a complex-valued Kähler potential $\Omega = \Omega(w^A, \tilde{w}^A)$ such that g is given by

$$g = \frac{\partial^2 \Omega}{\partial w^A \partial \tilde{w}^B} dw^A d\tilde{w}^B. \qquad (10.2.8)$$

Choose a spin frame $(o_{A'}, \iota_{A'})$ such that the null tetrad of vector fields $\mathbf{e}_{AA'}$ is

$$\mathbf{e}_{A0'} = o^{A'} \mathbf{e}_{AA'} = \frac{\partial}{\partial w^A} \quad \text{and} \quad \mathbf{e}_{A1'} = \iota^{A'} \mathbf{e}_{AA'} = \frac{\partial^2 \Omega}{\partial w^A \partial \tilde{w}^B} \frac{\partial}{\partial \tilde{w}_B}.$$

The Lax pair (10.1.1) becomes

$$L_A = \frac{\partial}{\partial w^A} - \lambda \frac{\partial^2 \Omega}{\partial w^A \partial \tilde{w}^B} \frac{\partial}{\partial \tilde{w}_B} + l_A \frac{\partial}{\partial \lambda}$$

for some functions l_0, l_1 which depend on $(w^A, \tilde{w}^A, \lambda)$. Consider the Lie bracket

$$[L_0, L_1] = \lambda^2 \frac{\partial^2 \Omega}{\partial w^A \partial \tilde{w}^B} \frac{\partial^3 \Omega}{\partial w_A \partial \tilde{w}_B \partial \tilde{w}^C} \frac{\partial}{\partial \tilde{w}_C} + l^A \frac{\partial^2 \Omega}{\partial w^A \partial \tilde{w}^B} \frac{\partial}{\partial \tilde{w}_B}$$
$$+ \left(\frac{\partial l^A}{\partial w^A} - \lambda \frac{\partial^2 \Omega}{\partial w^A \partial \tilde{w}^B} \frac{\partial l^A}{\partial \tilde{w}_B} + l_A \frac{\partial l^A}{\partial \lambda} \right) \frac{\partial}{\partial \lambda}.$$

The ASD condition is equivalent to integrability of the distribution L_A, therefore

$$[L_A, L_B] = \varepsilon_{AB}\alpha^C L_C$$

for some α^C. The lack of any $\partial/\partial w^A$ term in the Lie bracket above implies $\alpha^C = 0$. Analysing other terms we deduce the existence of $k = k(w^A, \tilde{w}^A) \in \ker \Box$ such that $l_A = \lambda^2 \partial k/\partial w^A$, and

$$\frac{\partial^2\Omega}{\partial w^A \partial \tilde{w}^B}\frac{\partial}{\partial \tilde{w}^C}\left(\frac{\partial^2\Omega}{\partial w_A \partial \tilde{w}_B}\right) = \frac{\partial k}{\partial w_A}\frac{\partial^2\Omega}{\partial w^A \partial \tilde{w}^C}, \tag{10.2.9}$$

where

$$\Box = \frac{\partial^2\Omega}{\partial w^A \partial \tilde{w}^B}\frac{\partial^2}{\partial w_A \partial \tilde{w}_B}.$$

There are two possible reality conditions

- Real-analytic $(++--)$ slices are obtained if $\mathbf{e}_{AA'}$, ν, k are all real. In this case we alter our definition of the complex structure J by

  $$J(\mathbf{e}_{A1'}) = -\mathbf{e}_{A1'} \quad \text{and} \quad J(\mathbf{e}_{A0'}) = -\mathbf{e}_{A0'}.$$

  Therefore $J^2 = \mathrm{Id}$, and g is pseudo-Kähler.
- In the Euclidean case the quadratic-form g and the complex structure

  $$J = i(\mathbf{e}^{A0'} \otimes \mathbf{e}_{A0'} - \mathbf{e}^{A1'} \otimes \mathbf{e}_{A1'})$$

  are real but the vector fields $\mathbf{e}_{AA'}$ are complex and $J^2 = -\mathrm{Id}$.

Solving the algebraic system (10.2.9) for $\partial k/\partial w^A$ we can deduce a formulation of the ASD Kähler condition as a fourth-order PDE. ASD Kähler metrics are locally given by (10.2.8) where $\Omega(w^A, \tilde{w}^A)$ is a solution to a fourth-order PDE (which we write as a system of two second-order PDEs):

$$\frac{\partial k}{\partial w^A} = \frac{\partial^2\Omega}{\partial w^A \partial \tilde{w}^B}\frac{\partial \ln c}{\partial \tilde{w}_B} \quad \text{and} \tag{10.2.10}$$

$$\Box k = \frac{\partial^2\Omega}{\partial w^A \partial \tilde{w}^B}\frac{\partial^2 k}{\partial w_A \partial \tilde{w}_B} = 0, \tag{10.2.11}$$

where

$$c = \det(g) = \frac{1}{2}\frac{\partial^2\Omega}{\partial w_A \partial \tilde{w}_B}\frac{\partial^2\Omega}{\partial w^A \partial \tilde{w}^B}.$$

Moreover (10.2.10) and (10.2.11) arise as the integrability condition for the linear system $L_0 f = L_1 f = 0$, where $f = f(w^A, \tilde{w}^A, \lambda)$ and

$$L_A = \frac{\partial}{\partial w^A} - \lambda \frac{\partial^2 \Omega}{\partial w^A \partial \tilde{w}^B} \frac{\partial}{\partial \tilde{w}_B} + \lambda^2 \frac{\partial k}{\partial w^A} \frac{\partial}{\partial \lambda}. \tag{10.2.12}$$

As a spin-off we have

Proposition 10.2.2 *A four-dimensional Kähler metric is ASD iff its scalar curvature vanishes.*

Proof Any scalar-flat Kähler metric is locally of the form (10.2.8) (as it is Kähler). Calculating the Ricci scalar shows that it vanishes iff (10.2.10) and (10.2.11) hold. Thus the metric is ASD. Conversely, if g is ASD and Kähler then (10.2.10) and (10.2.11) hold by the integrability of the Lax pair. Therefore g is scalar-flat. □

In view of this result ASD Kähler metrics are often called scalar-flat Kähler.

10.2.3 Null-Kähler structures

The structure we are about to describe does not exist in Riemannian signature and neutral signature is the only allowed reality condition which admits real solutions. We shall impose this condition from the start.

A null-Kähler structure [47] on a real four-manifold M consists of an inner product g of signature $(++--)$ and a real rank-two endomorphism $N : TM \to TM$ parallel with respect to this inner product such that

$$N^2 = 0 \quad \text{and} \quad g(NX, Y) + g(X, NY) = 0$$

for all $X, Y \in TM$.

The parallel two-form $\mathbf{\Omega} = g(N, \ldots)$ is simple (i.e. $\mathbf{\Omega} \wedge \mathbf{\Omega} = 0$) therefore it is SD or ASD by the argument given in Section 7.1.1. We chose the orientation such that $\mathbf{\Omega}$ is SD. The isomorphism $\Lambda^2{}_+(M) \cong \mathrm{Sym}^2(\mathbb{S}')$ between the bundle of SD two-forms and the symmetric tensor product of two spin bundles implies that the existence of a null-Kähler structure is in four dimensions equivalent to the existence of a parallel real spinor. If the spinor is $\iota_{A'}$ then $\mathbf{\Omega}_{ab} = \iota_{A'}\iota_{B'}\varepsilon_{AB}$. The Ricci identity (9.2.26) implies the vanishing of the curvature scalar.

In [24] and [47] it was shown that null-Kähler structures are locally given by one arbitrary function $\Theta : M \to \mathbb{R}$ of four variables, and admit a canonical form

$$g = dwdx + dzdy - \Theta_{xx}dz^2 - \Theta_{yy}dw^2 + 2\Theta_{xy}dwdz, \qquad (10.2.13)$$

with $N = dw \otimes \partial/\partial y - dz \otimes \partial/\partial x$.

Further conditions can be imposed on the curvature of g to obtain non-linear PDEs for the potential function Θ. Define

$$k := \Theta_{wx} + \Theta_{zy} + \Theta_{xx}\Theta_{yy} - \Theta_{xy}^2. \qquad (10.2.14)$$

- The Ricci-flat condition implies that

  $$k = xP(w, z) + yQ(w, z) + R(w, z),$$

  where P, Q, and R are arbitrary functions of (w, z). In fact the number of the arbitrary functions can be reduced down to one by redefinition of Θ and the coordinates. This is the hyper-heavenly equation of Plebański and Robinson [136] for non-expanding metrics of type $[N]\times[\text{Any}]$. (A manifold (M, g) is called hyper-heavenly if the SD Weyl spinor is algebraically special and the Einstein equations hold.)
- The conformal ASD condition implies a fourth-order PDE for Θ

  $$\Box k = 0, \qquad (10.2.15)$$

  where

  $$\Box = \partial_x\partial_w + \partial_y\partial_z + \Theta_{yy}\partial_x^2 + \Theta_{xx}\partial_y^2 - 2\Theta_{xy}\partial_x\partial_y.$$

  This equation is integrable: It admits a Lax formulation $[L_0, L_1] = 0$ with

  $$L_0 = (\partial_w - \Theta_{xy}\partial_y + \Theta_{yy}\partial_x) - \lambda\partial_y + k_y\partial_\lambda \quad \text{and}$$
  $$L_1 = (\partial_z + \Theta_{xx}\partial_y - \Theta_{xy}\partial_x) + \lambda\partial_x - k_x\partial_\lambda,$$

  and its solutions can in principle be constructed by twistor methods [47] or the dressing method [16].

10.2.4 ASD Einstein structures

If there exists a metric g in the ASD conformal class $[g]$ which is Einstein with non-zero cosmological constant, that is $R_{ab} = 6\Lambda g_{ab}$ then [139] the coordinates can be chosen so that

$$g = \frac{1}{\Lambda}\left[\mathcal{K}_{w\tilde{w}}dwd\tilde{w} + \mathcal{K}_{w\tilde{z}}dwd\tilde{z} + \mathcal{K}_{z\tilde{w}}dzd\tilde{w} + (\mathcal{K}_{z\tilde{z}} + 2e^{\mathcal{K}})dzd\tilde{z}\right],$$

where $\mathcal{K} = \mathcal{K}(w, z, \tilde{w}, \tilde{z})$ satisfies the Przanowski equation

$$\mathcal{K}_{w\tilde{w}}\mathcal{K}_{z\tilde{z}} - \mathcal{K}_{w\tilde{z}}\mathcal{K}_{z\tilde{w}} + (2\mathcal{K}_{w\tilde{w}} - \mathcal{K}_w\mathcal{K}_{\tilde{w}})e^{\mathcal{K}} = 0.$$

10.2.5 Hyper-Kähler structures and heavenly equations

In Theorem 9.3.3 it was shown that the hyper-Kähler condition on a four-metric is equivalent to the ASD Ricci-flat condition. The resulting Lax pair arises as a special case of the hyper-Hermitian Lax pair.

Theorem 10.2.3 [6, 116] *Let $e_{AA'}$ be four independent holomorphic vector fields on a four-dimensional complex manifold M, and let ν be a holomorphic four-form. Put*

$$L_0 = e_{00'} - \lambda e_{01'} \quad \text{and} \quad L_1 = e_{10'} - \lambda e_{11'}.$$

If

$$[L_0, L_1] = 0 \tag{10.2.16}$$

for every $\lambda \in \mathbb{CP}^1$, and

$$Lie_{L_A}\nu = 0, \tag{10.2.17}$$

then $c^{-1}e_{AA'}$ is a null tetrad for a hyper-Kähler metric on M, where

$$c^2 = \nu(e_{00'}, e_{01'}, e_{10'}, e_{11'}).$$

Given any four-dimensional hyper-Kähler metric such a null tetrad and four-form exists.

Proof Given a Lax pair of vector fields as in Theorem 10.2.3 define SD two-forms $\Sigma^{A'B'}$ by

$$\Sigma(\lambda) = \nu(L_0, L_1, .., .) = \Sigma^{A'B'}\pi_{A'}\pi_{B'}.$$

We shall show that $\Sigma(\lambda)$ is closed for any fixed λ. This will imply that there exist a closed basis $\Sigma^{A'B'}$ of SD two-forms and the Corollary 9.3.4 can be used to deduce that g is hyper-Kähler. Let d_h be a total derivative on $M \times \mathbb{CP}^1$ which holds λ = const:

$$\begin{aligned} d_h\Sigma(\lambda) &= d_h\,[\nu(L_0, L_1, \ldots)] = d_h\left\{L_0 \lrcorner\, [\nu(L_1, \ldots, \ldots)]\right\} \\ &= \mathrm{Lie}_{L_0}\,[\nu(L_1, \ldots, \ldots)] - L_0 \lrcorner\, [d_h\nu(L_1, \ldots, \ldots)] \\ &= [L_0, L_1] \lrcorner\, \nu + L_1 \lrcorner\, \mathrm{Lie}_{L_0}(\nu) - L_0 \lrcorner\, (L_1 \lrcorner\, d\nu) = 0. \end{aligned}$$

Conversely, given a hyper-Kähler metric we can choose coordinates $(w^A, \tilde{w}^A) = (w, z, \tilde{w}, \tilde{z})$ such that the metric is given in terms of a Kähler potential by (10.2.8). The SD two-forms are given by the complexification of (9.3.32) and the hyper-Kähler condition (9.3.33) gives the first heavenly equation of Plebański [135]:

$$\Omega_{w\tilde{z}}\Omega_{z\tilde{w}} - \Omega_{w\tilde{w}}\Omega_{z\tilde{z}} = 1 \quad \text{or} \quad \frac{1}{2}\frac{\partial^2\Omega}{\partial w_A \partial \tilde{w}_B}\frac{\partial^2\Omega}{\partial w^A \partial \tilde{w}^B} = 1. \tag{10.2.18}$$

We can take the tetrad of vector fields to be

$$\mathrm{e}_{AA'} = \begin{pmatrix} \Omega_{w\tilde{w}}\partial_{\tilde{z}} - \Omega_{w\tilde{z}}\partial_{\tilde{w}} & \partial_w \\ \Omega_{z\tilde{w}}\partial_{\tilde{z}} - \Omega_{z\tilde{z}}\partial_{\tilde{w}} & \partial_z \end{pmatrix} = \left(\frac{\partial^2 \Omega}{\partial w^A \partial \tilde{w}^B} \frac{\partial}{\partial \tilde{w}_B} \quad \frac{\partial}{\partial w^A} \right). \tag{10.2.19}$$

These vector fields preserve the volume form

$$\nu = dw \wedge dz \wedge d\tilde{w} \wedge d\tilde{z}.$$

The dual tetrad is

$$\mathrm{e}^{A1'} = dw^A \quad \text{and} \quad \mathrm{e}^{A0'} = \frac{\partial^2 \Omega}{\partial w_A \partial \tilde{w}_B} d\tilde{w}_B$$

with the flat solution $\Omega = w^A \tilde{w}_A$. The Lax pair for the first heavenly equation

$$\begin{aligned} L_0 &:= \Omega_{w\tilde{w}}\partial_{\tilde{z}} - \Omega_{w\tilde{z}}\partial_{\tilde{w}} - \lambda \partial_w \quad \text{and} \\ L_1 &:= \Omega_{z\tilde{w}}\partial_{\tilde{z}} - \Omega_{z\tilde{z}}\partial_{\tilde{w}} - \lambda \partial_z \end{aligned} \tag{10.2.20}$$

satisfies (10.2.16) and (10.2.17) where the volume form is $\nu = dw \wedge dz \wedge d\tilde{w} \wedge d\tilde{z}$. This ends the proof. □

As a spin-off from the proof we have deduced Plebański's first heavenly equation and its Lax pair. Plebański also gave the alternative local form of the ASD Ricci-flat condition [135] called the second heavenly equation. This equation is the special case of the ASD null-Kähler equations (10.2.14 and 10.2.15) corresponding to $k = 0$.

Below we shall derive the second heavenly equation by fixing the residual gauge freedom in the Lax pair (10.2.3). Consider the hyper-Kähler Lax equations (10.2.16 and 10.2.17). The Frobenius theorem (Theorem C.2.4) applied to the equations

$$[\mathrm{e}_{00'}, \mathrm{e}_{10'}] = 0 \quad \text{and} \quad [\mathrm{e}_{00'}, \mathrm{e}_{11'}] + [\mathrm{e}_{01'}, \mathrm{e}_{10'}] = 0$$

implies the existence of a complex-valued function Θ and coordinate system $(w^A, x_A) := (w, z, x, y)$, such that

$$\mathrm{e}_{AA'} = \begin{pmatrix} \partial_y & \partial_w + \Theta_{yy}\partial_x - \Theta_{xy}\partial_y \\ -\partial_x & \partial_z - \Theta_{xy}\partial_x + \Theta_{xx}\partial_y \end{pmatrix} = \left(\frac{\partial}{\partial x^A} \quad \frac{\partial}{\partial w^A} + \frac{\partial^2 \Theta}{\partial x^A \partial x^B} \frac{\partial}{\partial x_B} \right). \tag{10.2.21}$$

Finally equation $[\mathrm{e}_{01'}, \mathrm{e}_{11'}] = 0$ implies that Θ satisfies second heavenly equation

$$\Theta_{xw} + \Theta_{yz} + \Theta_{xx}\Theta_{yy} - {\Theta_{xy}}^2 = 0 \quad \text{or} \quad \frac{\partial^2 \Theta}{\partial w^A \partial x_A} + \frac{1}{2} \frac{\partial^2 \Theta}{\partial x^B \partial x^A} \frac{\partial^2 \Theta}{\partial x_B \partial x_A} = 0. \tag{10.2.22}$$

The dual frame is given by

$$e^{A0'} = dx^A + \frac{\partial^2 \Theta}{\partial x^B \partial x_A} dw^B \quad \text{and} \quad e^{A1'} = dw^A,$$

and g is of the form (10.2.13) with $\Theta = 0$ defining the flat metric. The Lax pair corresponding to (10.2.22) is

$$\begin{aligned} L_0 &= \partial_y - \lambda(\partial_w - \Theta_{xy}\partial_y + \Theta_{yy}\partial_x) \quad \text{and} \\ L_1 &= \partial_x + \lambda(\partial_z + \Theta_{xx}\partial_y - \Theta_{xy}\partial_x). \end{aligned} \tag{10.2.23}$$

This is a special case of the Lax pair for equation (10.2.15) with $k = 0$ (to reach an agreement one needs to make a transformation $\lambda \to 1/\lambda$ in the ASD null-Kähler Lax pair).

- **Example.** One solution to the second heavenly equation is

$$\Theta = \frac{t}{wx + zy}, \qquad t = \text{const.}$$

The corresponding metric is given by

$$g = dwdx + dzdy - \frac{2t}{(wx + zy)^3}(wdz - zdw)^2.$$

This metric was first constructed by Sparling and Tod [149] using the $\mathcal{H}$-space formalism. The metric appears to be singular on the light cone of the origin, but this singularity may be removed by a coordinate transformation [149].

10.2.5.1 *Recursion operator*

The recursion operator R is a map from the space of linearized solutions of the ASD Ricci-flat equations to itself. Linearized solutions can be regarded as vector fields on the solutions space. Thus R is a natural generalization of the recursion operator we introduced in Section 3.2.1 in the context of bi-Hamiltonian systems. We shall present a theory of recursion operator in the following [45].

We shall first identify the space of linearized solutions to the ASD Ricci-flat equations with the space of solutions to the background coupled wave equation in two ways as follows.

Lemma 10.2.4 *Let $\Box_\Omega$ and $\Box_\Theta$ denote wave operators on the ASD Ricci-flat background determined by Ω and Θ, respectively. Linearized solutions to*

(10.2.18) *and* (10.2.22) *satisfy*

$$\Box_{\Omega}\delta\Omega = 0 \quad \textit{and} \quad \Box_{\Theta}\delta\Theta = 0. \tag{10.2.24}$$

Proof In both cases $\Box_g = \varepsilon^{AB} \mathbf{e}_{A1'} \mathbf{e}_{B0'}$ since

$$\Box_g = \frac{1}{\sqrt{g}} \partial_a (g^{ab} \sqrt{g} \partial_b) = g^{ab} \partial_a \partial_b + (\partial_a g^{ab}) \partial_b$$

but $\partial_a g^{ab} = 0$ for both heavenly coordinate systems. The linearized first heavenly equation takes the form $\left[\partial\tilde{\partial}(\Omega + \delta\Omega)\right]^2 = \nu$, where $\partial = \mathbf{e}^{A0'} \otimes \mathbf{e}_{A0'}$ and $\tilde{\partial} = \mathbf{e}^{A1'} \otimes \mathbf{e}_{A1'}$ so that $d = \partial + \tilde{\partial}$. This implies

$$0 = \left(\partial\tilde{\partial}\Omega \wedge \partial\tilde{\partial}\right) \delta\Omega = d\left[\partial\tilde{\partial}\Omega \wedge (\partial - \tilde{\partial})\delta\Omega\right] = d * d\delta\Omega.$$

Here $*$ is the Hodge star operator corresponding to g. For the second equation we make use of the tetrad (10.2.21) and perform coordinate calculations. □

From now on we identify tangent spaces to the spaces of solutions to (10.2.18) and (10.2.22) with the space of solutions to the curved background wave equation, $\mathcal{W}_g$. We will define the recursion operator on the space $\mathcal{W}_g$.

Lemma (10.2.4) shows that we can consider a linearized perturbation as an element of $\mathcal{W}_g$ in two ways. These two will be related by the square of the recursion operator. The linearized ASD Ricci-flat metrics corresponding to $\delta\Omega$ and $\delta\Theta$ are

$$h^I{}_{AA'BB'} = \iota_{(A'} o_{B')} \nabla_{(A1'} \nabla_{B)0'} \delta\Omega \quad \text{and} \quad h^{II}{}_{AA'BB'} = o_{A'} o_{B'} \nabla_{A0'} \nabla_{B0'} \delta\Theta,$$

where $o^{A'} = (1, 0)$ and $\iota^{A'} = (0, 1)$ are the constant spin frame associated to the null tetrads (10.2.21) and (10.2.19).

Given $\phi \in \mathcal{W}_g$ we use the first of these equations to find h^I. If we put the perturbation obtained in this way on the LHS of the second equation and add an appropriate gauge term we obtain ϕ' – the new element of $\mathcal{W}_g$ that provides the $\delta\Theta$ which gives rise to

$$h^{II}_{ab} = h^I_{ab} + \nabla_{(a} V_{b)}. \tag{10.2.25}$$

To extract the recursion relations we must find V such that $h^I{}_{AA'BB'} - \nabla_{(AA'} V_{BB')} = o_{A'} o_{B'} \chi_{AB}$. Take $V_{BB'} = o_{B'} \nabla_{B1'} \delta\Omega$, which gives

$$\nabla_{(AA'} V_{BB')} = -\iota_{(A'} o_{B')} \nabla_{(A0'} \nabla_{B)1'} \delta\Omega + o_{A'} o_{B'} \nabla_{A1'} \nabla_{B1'} \delta\Omega.$$

This reduces (10.2.25) to

$$\nabla_{A1'} \nabla_{B1'} \phi = \nabla_{A0'} \nabla_{B0'} \phi'. \tag{10.2.26}$$

Both heavenly formulations use the covariantly constant spin frame (see Theorem 9.3.3) so $\nabla_{AA'} = \mathbf{e}_{AA'}$.

Definition 10.2.5 *Define the recursion operator* $R : \mathcal{W}_g \longrightarrow \mathcal{W}_g$ *by*

$$e_{A1'}\phi = e_{A0'} R\phi, \tag{10.2.27}$$

so formally $R = (e_{A0'})^{-1} \circ e_{A1'}$ *(no summation over the index* A*).*

Remarks

- From (10.2.27) and from the field equations it follows that if ϕ belongs to $\mathcal{W}_g$ then so does $R\phi$.
- If $R^2\delta\Omega = \delta\Theta$ then $\delta\Omega$ and $\delta\Theta$ correspond to the same variation in the metric up to gauge.
- The operator $\phi \mapsto e_{A0'}\phi$ is overdetermined, and its consistency follows from the wave equation on ϕ.
- This definition is formal in that in order to invert the operator $\phi \mapsto e_{A0'}\phi$ we need to specify boundary conditions.

To summarize

Proposition 10.2.6 [45] *Let* $\mathcal{W}_g$ *be the space of solutions of the wave equation on the curved ASD background given by g:*

1. *Elements of* $\mathcal{W}_g$ *can be identified with linearized perturbations of the heavenly equations.*
2. *There exists a (formal) map* $R : \mathcal{W}_g \longrightarrow \mathcal{W}_g$ *given by* (10.2.27).

10.2.5.2 *Heavenly hierarchies*

The generators of higher flows are first obtained by applying powers of the recursion operator to the linearized perturbations corresponding to the evolution along coordinate vector fields. This embeds the second heavenly equation into an infinite system of overdetermined, but consistent, PDEs (which we will truncate at some arbitrary but finite level). These equations in turn can be naturally embedded into a system of equations that are the consistency conditions for an associated linear system that extends (10.2.16) and (10.2.17). This yields a hierarchy of flows of the ASD Ricci-flat equations [17, 45, 49, 155].

We shall discuss the hierarchy for the second Plebański equation [45]; that for the first arises from a different coordinate and gauge choice. The first few iterations of the recursion operator (10.2.27) with the tetrad (10.2.21) can be explicitly integrated to give

$$w \longrightarrow y \longrightarrow -\Theta_x \longrightarrow \Theta_z \longrightarrow \dots \quad \text{and}$$

$$z \longrightarrow -x \longrightarrow -\Theta_y \longrightarrow -\Theta_w \longrightarrow \cdots.$$

Introduce the coordinates x^{Ai}, where for $i = 0, 1, x^{Ai} = x^{AA'}$ are the original coordinates on M, and for $1 < i \leq n$, x^{Ai} are the parameters for the new flows (with $(2n-2)$-dimensional parameter space $\mathbb{X}$). The propagation of Θ along these parameters is determined by the recursion relations (10.2.27):

$$\partial_y(\partial_{Bi+1}\Theta) = (\partial_w - \Theta_{xy}\partial_y + \Theta_{yy}\partial_x)\partial_{Bi}\Theta \quad \text{and}$$
$$-\partial_x(\partial_{Bi+1}\Theta) = (\partial_z + \Theta_{xx}\partial_y - \Theta_{xy}\partial_x)\partial_{Bi}\Theta. \tag{10.2.28}$$

The hyper-Kähler hierarchy is the system arising as the consistency conditions for (10.2.28):

$$\partial_{Ai}\partial_{Bj-1}\Theta - \partial_{Bj}\partial_{Ai-1}\Theta + \{\partial_{Ai-1}\Theta, \partial_{Bj-1}\Theta\}_{yx} = 0, \quad i, j = 1, \ldots, n. \tag{10.2.29}$$

Here $\{\ldots,\ldots\}_{yx}$ is the Poisson bracket with respect to the Poisson structure $\partial/\partial x^A \wedge \partial/\partial x_A = 2\partial_x \wedge \partial_y$.

Lemma 10.2.7 *The linear system for equations* (10.2.29) *is*

$$L_{Ai} f = (-\lambda D_{Ai+1} + \delta_{Ai}) f = 0, \quad i = 0, \ldots, n-1, \tag{10.2.30}$$

where

1. $f := f(x^{Ai}, \lambda)$ *is a function on* $\mathbb{CP}^1 \times \mathcal{N}$, *where* $\mathcal{N} = M \times \mathbb{X}$.
2. $D_{Ai+1} := \partial_{Ai+1} + [\partial_{Ai}, V]$, $(V = \varepsilon^{AB}\partial_{A0}\Theta\partial_{B0})$, *and* $\delta_{Ai} := \partial_{Ai}$ *are* $4n$ *vector fields on* $\mathcal{N}$.

Proof This follows by direct calculation. The compatibility conditions for (10.2.30) are

$$[D_{Ai+1}, D_{Bj+1}] = 0, \tag{10.2.31}$$
$$[\delta_{Ai}, \delta_{Bj}] = 0, \quad \text{and} \tag{10.2.32}$$
$$[D_{Ai+1}, \delta_{Bj}] - [D_{Bj+1}, \delta_{Ai}] = 0. \tag{10.2.33}$$

It is straightforward to see that equations (10.2.32) and (10.2.33) hold identically with the above definitions and (10.2.31) is equivalent to (10.2.29). □

The concept of hierarchy is useful in finding solutions to the heavenly equations. This is analogous to the finite-gap integration of KdV described in Section 3.4. We say that an ASD Ricci-flat metric admits a hidden symmetry if the associated heavenly potential is stationary with respect to some direction in the extended parameter space $\mathbb{X}$.

- **Example.** Let us demonstrate how to use the recursion procedure to find metrics with hidden symmetries. Let $\partial_{t_n}\Omega := \phi_n$ be a linearization of the first heavenly equation. We have $R : z \longrightarrow \Omega_w = \partial_{t_1}\Omega$. Look for solutions to (10.2.18) with an additional constraint $\partial_{t_2}\Omega = 0$. The recursion relations (10.2.27) imply $\Omega_{wz} = \Omega_{ww} = 0$, therefore

$$\Omega(w, z, \tilde{w}, \tilde{z}) = wq(\tilde{w}, \tilde{z}) + P(z, \tilde{w}, \tilde{z}).$$

The heavenly equation yields $dq \wedge dP \wedge dz = d\tilde{z} \wedge d\tilde{w} \wedge dz$. With the definition $\partial_z P = p$ the metric is

$$g = 2dwdq + 2dzdp + Qdz^2,$$

where $Q = -2P_{zz}$. We adopt (w, z, q, p) as a new coordinate system. The heavenly equations imply that $Q = Q(q, z)$ is an arbitrary function of two variables. These are the null ASD plane wave solutions [135]. They have no Riemannian real sections but restricting to real coordinates and real Q yields a metric in neutral signature.

10.2.5.3 *Hamiltonian and Lagrangian formalisms*

In this section we shall investigate the Lagrangian and Hamiltonian formulations of the hyper-Kähler equations in their 'heavenly' forms. Rather than considering the equations as a real system of elliptic or ultra-hyperbolic equations, we complexify and consider the equations locally as evolving initial data from a three-dimensional hypersurface δM and it is this space of initial data that leads to local solutions on a neighbourhood of such a hypersurface that is denoted by $\mathcal{S}$ and is endowed with a (conserved) symplectic form.

For the first equation we have the Lagrangian density

$$\mathcal{L}_\Omega = \Omega\left[\nu - \frac{1}{3}(\partial\tilde{\partial}\Omega)^2\right] = \left(\Omega - \frac{1}{3}\Omega\{\Omega_{\tilde{z}}, \Omega_{\tilde{w}}\}_{wz}\right)\nu \qquad (10.2.34)$$

and for the second equation

$$\mathcal{L}_\Theta = \left[\frac{1}{3}\Theta\{\Theta_x, \Theta_y\}_{xy} - \frac{1}{2}(\Theta_x\Theta_w + \Theta_y\Theta_z)\right]\nu. \qquad (10.2.35)$$

If the field equations are assumed, the variation of these Lagrangians will yield only a boundary term. Starting with the first equation, this defines a potential one-form P (compare (5.1.10)) on the solution space $\mathcal{S}$ and hence a symplectic structure $\mathbf{\Omega} = dP$ on $\mathcal{S}$. Starting with the second we find a symplectic structure with the same expression on perturbations $\delta\Theta$ as we had for $\delta\Omega$. However, since their relation to perturbations of the hyper-Kähler structure are different, they define different symplectic structures on $\mathcal{S}$. These are related by the recursion operator since we have $R^2\delta\Omega = \delta\Theta$ from the construction of the recursion operator. In order to see that these structures yield the bi-Hamiltonian framework, these symplectic structures need to be compatible with the recursion operator in the sense that $\mathbf{\Omega}(R\phi, \phi') = \mathbf{\Omega}(\phi, R\phi')$.

We shall discuss this using the first heavenly formulation (10.2.18) which is easier as one can use identities from Kähler geometry.

Proposition 10.2.8 *The symplectic form on the space of solutions $\mathcal{S}$ derived from the boundary term in the variational principle for the first Lagrangian is*

$$\mathbf{\Omega}(\delta_1\Omega, \delta_2\Omega) = \frac{2}{3}\int_{\delta M} \delta_1\Omega * d(\delta_2\Omega) - \delta_2\Omega * d(\delta_1\Omega). \qquad (10.2.36)$$

Proof Varying (10.2.34) we obtain

$$\delta L = \delta\Omega\left[\nu - \frac{1}{3}(\partial\tilde{\partial}\Omega)^2\right] - \frac{2}{3}\Omega\partial\tilde{\partial}\Omega \wedge \partial\tilde{\partial}\delta\Omega = \frac{2}{3}\partial\tilde{\partial}\Omega \wedge (\delta\Omega\partial\tilde{\partial}\Omega - \Omega\partial\tilde{\partial}\delta\Omega).$$

We use the relation between the complex and Kähler structures and the field equation to obtain

$$\begin{aligned}\delta L &= -\frac{1}{3}\partial\tilde{\partial}\Omega \wedge \left[\delta\Omega d(\partial - \tilde{\partial})\Omega - \Omega d(\partial - \tilde{\partial})\delta\Omega\right]\\ &= \frac{1}{3}dA(\delta\Omega) - \frac{1}{3}\partial\tilde{\partial}\Omega\left[- * \partial\tilde{\partial}\Omega(\partial - \tilde{\partial})\delta\Omega(\partial - \tilde{\partial})\Omega + *\partial\tilde{\partial}\Omega(\partial - \tilde{\partial})\Omega(\partial - \tilde{\partial})\delta\Omega\right]\\ &= \frac{1}{3}dA(\delta\Omega), \quad \text{where } A(\delta\Omega) = \Omega * d\delta\Omega - \delta\Omega * d\Omega.\end{aligned}$$

Define the one-form on $\mathcal{M}$:

$$P(\delta\Omega) = \int_{\delta M} A(\delta\Omega).$$

The symplectic structure $\mathbf{\Omega}$ is the (functional) exterior derivative of P:

$$\begin{aligned}\mathbf{\Omega}(\delta_1\Omega, \delta_2\Omega) &= \delta_1\left[P(\delta_2\Omega)\right] - \delta_2\left[P(\delta_1\Omega)\right] - P([\delta_1\Omega, \delta_2\Omega])\\ &= \frac{2}{3}\int_{\delta M} \delta_1\Omega * d(\delta_2\Omega) - \delta_2\Omega * d(\delta_1\Omega). \qquad \square\end{aligned}$$

Thus $\mathbf{\Omega}$ coincides with the symplectic form on the solution space to the wave equation on the ASD vacuum background.

The existence of the recursion operator allows the construction of an infinite sequence of symplectic structures. The key property which can be estabished [45] by an application of Stokes theorem is the following: Let ϕ, $\phi' \in W_g$ and let $\mathbf{\Omega}$ be given by (10.2.36). Then

$$\mathbf{\Omega}(R\phi, \phi') = \mathbf{\Omega}(\phi, R\phi') \qquad (10.2.37)$$

where the recursion operator is defined by (10.2.27). This property guarantees that the bilinear forms

$$\mathbf{\Omega}^k(\phi, \phi') \equiv \mathbf{\Omega}(R^k\phi, \phi') \qquad (10.2.38)$$

are skew. Furthermore they are symplectic and lead to the bi-Hamiltonian formulation. In this context formula (10.2.37) and the closure condition for $\mathbf{\Omega}^k$ are an algebraic consequence of the fact that R comes from two Poisson structures.

10.3 Symmetries

By a symmetry of a metric, we mean a conformal Killing vector, that is, a vector field K satisfying

$$\mathrm{Lie}_K g = cg, \tag{10.3.39}$$

where c is a function. If c vanishes, K is called a Killing vector. If c is a non-zero constant K is called a homothety. If we are dealing with a conformal structure $[g]$, a symmetry is a vector field K satisfying (10.3.39) for some $g \in [g]$. Then (10.3.39) will be satisfied for any $g \in [g]$, where the function c will depend on the choice of $g \in [g]$. Such a K is referred to as a conformal Killing vector for the conformal structure.

In the complex category (and in the neutral signature) there are two types of Killing vectors: non-null where $g(K, K) \neq 0$ and null where $g(K, K) = 0$. Note that a null vector for $g \in [g]$ is null for all $g \in [g]$, so nullness of a vector with respect to a conformal structure makes sense.

10.3.1 Einstein–Weyl geometry

Given a four-dimensional ASD conformal structure $(M, [g])$ with a non-null conformal Killing vector K, the three-dimensional space $\mathcal{W}$ of trajectories of K inherits a conformal structure $[h]$, due to (10.3.39). The ASD condition on $[g]$ results in extra geometrical structure on $(\mathcal{W}, [h])$; it becomes an Einstein–Weyl (EW) space [93].

Let $\mathcal{W}$ be a complex three-dimensional manifold. Given a conformal structure $[h]$, a torsion-free connection D is said to preserve $[h]$ if

$$D_i h_{jk} = \omega_i h_{jk}, \tag{10.3.40}$$

for some $h \in [h]$ and a one-form ω. If (10.3.40) holds for a single $h \in [h]$ it holds for all, where ω will depend on the particular $h \in [h]$. Under conformal rescaling

$$h \longrightarrow \phi^2 h \quad \text{and} \quad \omega \longrightarrow \omega + 2d(\log \phi).$$

The condition (10.3.40) is equivalent to the requirement that null geodesics of any $h \in [h]$ be geodesics of D. Given D we can define its Riemann $W^i{}_{jkl}$ and Ricci W_{ij} curvature tensors in the usual way (9.2.6). The notion of a curvature scalar must be modified, because there is no distinguished metric in the conformal class to contract W_{ij} with. Given some $h \in [h]$ we can form $W = h^{ij} W_{ij}$. Under a conformal transformation $h \rightarrow \phi^2 h$, W transforms as $W \rightarrow \phi^{-2} W$. This is because W_{ij} is unaffected by any conformal rescaling, being formed entirely out of the connection D. W is an example of a conformally weighted function, with weight -2.

One can now define a conformally invariant analogue of the Einstein equation as follows:

$$W_{(ij)} - \frac{1}{3} W h_{ij} = 0. \tag{10.3.41}$$

This is the EW equation. Notice that the LHS is a well-defined tensor (i.e. weight 0), since the weights of W and h_{ij} cancel. Equation (10.3.41) is the EW equation for $(D, [h])$. It says that given any $h \in [h]$, the trace-free part of the Ricci tensor of D is zero if one defines the trace using h. Notice also that W_{ij} is not necessarily symmetric, unlike the Ricci tensor for a Levi-Civita connection.

In the special case that D is the Levi-Civita connection of some metric $h \in [h]$, (10.3.41) reduces to the Einstein equation. This happens when ω is exact and can be set to zero by a conformal transformation. All Einstein metrics in three dimensions have constant curvature. On the other hand the EW condition allows non-trivial degrees of freedom: The general solution to (10.3.41) depends on four arbitrary functions of two variables [30]. In what follows, we refer to an EW structure by (h, ω). The connection D is fully determined by this data using (10.3.40).

The EW equations are equivalent [30, 78, 128] to the existence of a two-dimensional family of surfaces $Z \subset \mathcal{W}$ which are null with respect to $[h]$, and totally geodesic with respect to D (this means that any geodesic which passes through $p \in Z$ and is tangent to Z at p lies in Z). This condition has been used in [46] to construct a Lax representation for the EW equation. The details are as follows: Let V_1, V_2, and V_3 be three independent vector fields on $\mathcal{W}$, and let e^1, e^2, and e^3 be the dual one-forms. Assume that

$$h = e^2 \otimes e^2 - 2(e^1 \otimes e^3 + e^3 \otimes e^1)$$

and some one-form ω give an EW structure. Let $V(\lambda) = V_1 - 2\lambda V_2 + \lambda^2 V_3$ where $\lambda \in \mathbb{CP}^1$. Then $h(V(\lambda), V(\lambda)) = 0$ for all $\lambda \in \mathbb{CP}^1$ so $V(\lambda)$ determines a sphere of null vectors.

Consider a null totally geodesic surface in $\mathcal{W}$ with a normal vector given by $V(\lambda)$ for some λ. The vectors $V_1 - \lambda V_2$ and $V_2 - \lambda V_3$ form a basis of the orthogonal complement of $V(\lambda)$. For each $\lambda \in \mathbb{CP}^1$ they span a null two-surface. Therefore the Frobenius theorem implies that the horizontal lifts

$$L = V_1 - \lambda V_2 + l\partial_\lambda \quad \text{and} \quad M = V_2 - \lambda V_3 + m\partial_\lambda \tag{10.3.42}$$

of these vectors to $T(\mathcal{W} \times \mathbb{CP}^1)$ span an integrable distribution, and (10.3.41) is equivalent to

$$[L, M] = \alpha L + \beta M$$

for some α and β. The functions l and m are third order in λ, because the Möbius transformations of $\mathbb{CP}^1$ are generated by vector fields quadratic in λ. Any EW space arises from such a Lax pair.

The following result, due to Jones and Tod [93], relates ASD conformal structures in four dimensions to EW structures in three dimensions.

Theorem 10.3.1 (Jones–Tod [93]) *Let $(M, [g])$ be an ASD four-manifold with a non-null conformal Killing vector K. An EW structure on the space $\mathcal{W}$ of trajectories of K is defined by*

$$h := |K|^{-2}g - |K|^{-4}\mathbb{K} \odot \mathbb{K} \quad \textit{and} \quad \omega = 2|K|^{-2} *_g (\mathbb{K} \wedge d\mathbb{K}), \tag{10.3.43}$$

where $|K|^2 := g(K, K)$, $\mathbb{K} := g(K, .)$, and $*_g$ is the Hodge-$*$ of g. All EW structures arise in this way. Conversely, let (h, ω) be a three-dimensional EW structure on $\mathcal{W}$, and let (V, η) be a function of weight -1 and a one-form on $\mathcal{W}$ satisfying the generalized monopole equation

$$*_h \left(dV + \frac{1}{2}\omega V \right) = d\eta, \tag{10.3.44}$$

where $*_h$ is the Hodge-$*$ of h. Then

$$g = V^2 h + (d\tau + \eta)^2$$

is an ASD metric with non-null Killing vector ∂_τ.

Applying the Jones–Tod correspondence to the special ASD conditions discussed in Section 10.2 will yield special integrable systems in three dimensions. Choosing the neutral reality conditions will give hyperbolic equations in (2+1) dimensions. In each case of interest we shall assume that the symmetry preserves the special geometric structure in four dimensions. This will give rise to special EW backgrounds, together with general solutions of the generalized monopole equation (10.3.44) on these backgrounds.

10.3.1.1 *Scalar-flat Kähler with symmetry: $SU(\infty)$-Toda equation*

Choose the Riemannian reality conditions. Let (M, g) be a scalar-flat Kähler metric, with a symmetry K Lie deriving the Kähler form $\mathbf{\Omega}$. One can follow the steps of LeBrun [104] to reduce the problem to a pair of coupled PDEs: the $SU(\infty)$-Toda equation and its linearization. The key step in the construction is to use the moment map for K as one of the coordinates, that is, define a function $z : M \longrightarrow \mathbb{R}$ by $dz = K \lrcorner\, \mathbf{\Omega}$. Then x, y arise as isothermal coordinates on two-dimensional surfaces orthogonal to K and dz. The metric takes the form

$$g = V\left[e^u(dx^2 + dy^2) + dz^2\right] + \frac{1}{V}(d\tau + \eta)^2, \tag{10.3.45}$$

where the function u satisfies the $SU(\infty)$-Toda equation

$$(e^u)_{zz} + u_{xx} + u_{yy} = 0, \tag{10.3.46}$$

and V is a solution to its linearization – the generalized monopole equation (10.3.44). The corresponding EW space from the Jones–Tod construction is

$$h = e^u(dx^2 + dy^2) + dz^2 \quad \text{and} \quad \omega = 2u_z dz. \tag{10.3.47}$$

Given the $SU(\infty)$-Toda EW space, any solution to the monopole equation will yield a scalar-flat Kähler metric. The special solution $V = cu_z$, where c is a constant, will lead to a hyper-Kähler metric with symmetry. The analytic continuation $z = it$ in (10.3.46) gives the Lorentzian $SU(\infty)$-Toda equation and an EW structure in $(++-)$ signature.

10.3.1.2 *ASD Einstein with symmetry*

This reduction also leads to the $SU(\infty)$-Toda equation [140, 159]. The following argument has been given in [159]. Let K^a be a Killing vector for an ASD Einstein metric with non-zero Λ, where $\Lambda = R/24$. The identity (9.7.83) implies that

$$\nabla_{AA'}\psi_{B'C'} = 2\Lambda\varepsilon_{A'(B'}K_{C')A}, \tag{10.3.48}$$

where $\psi_{A'B'} = (1/2)\nabla_{AA'}K^A_{B'}$ is the SD derivative of K. Let

$$J^a{}_b = \psi^{-1}\delta^A_B\psi^{A'}_{B'}, \quad \text{where} \quad \psi^2 = \frac{1}{2}\psi^{A'B'}\psi_{A'B'}$$

be an endomorphism of TM which, from its definition, squares to $-\text{Id}$. The formula (10.3.48) implies that this almost-complex-structure is integrable and one can introduce a complex coordinate $\xi = x + iy$ on the plane orthogonal to K and $J(K)$ such that the metric takes the form

$$g = \frac{P}{z^2}\left[e^u(dx^2 + dy^2) + dz^2\right] + \frac{1}{Pz^2}(d\tau + \eta)^2,$$

where the coordinate z is defined by $z = \Lambda\psi^{-1}$, the function $u = u(x, y, z)$ satisfies the $SU(\infty)$-Toda equation (10.3.46) and

$$P = \frac{1}{4\Lambda}zu_z - \frac{1}{2\Lambda}.$$

The one-form η can now be found by solving the generalized monopole equation (10.3.44) with $V = P$.

10.3.1.3 *ASD null-Kähler with symmetry: dKP equation*

Choose the neutral reality conditions. Let (M, g, N) be an ASD null-Kähler structure with a Killing vector K such that $\mathrm{Lie}_K N = 0$. In [47] it was demonstrated that there exist smooth real-valued functions $H = H(x, y, t)$ and $W = W(x, y, t)$ such that

$$g = W_x(dy^2 - 4dxdt - 4H_x dt^2) - W_x^{-1}(d\tau - W_x dy - 2W_y dt)^2 \qquad (10.3.49)$$

is an ASD null-Kähler metric on a circle bundle $M \to \mathcal{W}$ if

$$H_{yy} - H_{xt} + H_x H_{xx} = 0 \quad \text{and} \qquad (10.3.50)$$

$$W_{yy} - W_{xt} + (H_x W_x)_x = 0. \qquad (10.3.51)$$

All real analytic ASD null-Kähler metrics with symmetry arise from this construction. With the definition $u = H_x$ the x-derivative of equation (10.3.50) becomes

$$(u_t - uu_x)_x = u_{yy}, \qquad (10.3.52)$$

which is the dispersionless Kadomtsev–Petviashvili (dKP) equation [46]. The corresponding EW structure is

$$h = dy^2 - 4dxdt - 4udt^2 \quad \text{and} \quad \omega = -4u_x dt$$

(this metric has the property that the linearized dKP equation for $u + \delta u$ can be written as $h^{ij}\partial_i\partial_j\delta u + \cdots$, where $(\cdots)$ denotes lower order terms).

This EW structure possesses a covariantly constant null vector with weight $-\frac{1}{2}$, and every such EW structure with this property can be put into the above form [46]. The covariant-constancy is with respect to a derivative on weighted vectors that preserves their weight. Details can be found in [46].

The linear equation (10.3.51) is the (derivative of the) generalized monopole equation with $V = W_x$ from the Jones–Tod construction. Given a dKP EW structure, any solution to this monopole equation will yield an ASD null-Kähler structure in four dimensions. The special monopole $V = H_{xx}/2$ will yield a pseudo-hyper-Kähler structure with symmetry whose SD derivative is null.

10.3.1.4 *Hyper-Hermitian metrics with symmetry: Diff(S^1) equation*

Let us assume that a hyper-Hermitian four-manifold admits a symmetry which Lie derives all complex structures. This implies [51] that the EW structure is locally given by

$$h = (dy + udt)^2 - 4(dx + wdt)dt \quad \text{and} \quad \omega = u_x dy + (uu_x + 2u_y)dt,$$

where $u(x, y, t)$ and $w(x, y, t)$ satisfy a system of quasi-linear PDEs [127, 111, 112, 63]:

$$u_t + w_y + uw_x - wu_x = 0 \quad \text{and} \quad u_y + w_x = 0. \tag{10.3.53}$$

This system is equivalent to $[L, M] = 0$ where the Lax pair of vector fields is given by

$$L = \partial_t - w\partial_x - \lambda\partial_y \quad \text{and} \quad M = \partial_y + u\partial_x - \lambda\partial_x. \tag{10.3.54}$$

The corresponding hyper-Hermitian metric will arise from any solution to this coupled system, and its linearization (the generalized monopole (10.3.44)). The special monopole $V = u_x/2$ leads to hyper-Kähler metric with triholomorphic homothety. Both neutral and Riemannian reality conditions can be imposed. In the former case the EW space has $(++-)$ signature, and (u, w) are real-valued functions of real coordinates (x, y, t).

- **Example.** Let us choose the Lorentzian reality conditions on the underlying EW structure, and assume that u and w in (10.3.53) do not depend on y. One needs to consider the two cases $w = 0$ and $w = w(t) \neq 0$ separately. The corresponding equations can now be easily integrated to give (in the $w \neq 0$ case one needs to change variables)

$$h = (dy + Adt)^2 - 4dxdt \quad \text{and} \quad \omega = A'dy + AA'dt, \tag{10.3.55}$$

  where $A = A(x)$ is an arbitrary function. Some interesting complete solutions belong to this class. For example, $A = x$ leads to the EW structure on Thurston's nil manifold $S^1 \times \mathbb{R}^2$ [128]:

$$h = (dy + xdt)^2 - 4dxdt \quad \text{and} \quad \omega = dy + xdt.$$

The system (10.3.53) is called the Diff(S^1) system, where Diff(S^1) is the group of diffeomorphisms of a circle. This terminology is justified by the following result which we formulate and prove using the Lorentzian reality conditions.

Proposition 10.3.2 [52] *The system* (10.3.53) *arises as a symmetry reduction of the ASDYM equations in neutral signature with the infinite-dimensional gauge group Diff(S^1) and two commuting translational symmetries exactly one of which is null. Any such symmetry reduction is gauge equivalent to* (10.3.53).

Proof Consider the flat metric of neutral signature on $\mathbb{R}^4$ which in real double null coordinates $(w, z, \tilde{w}, \tilde{z})$ takes the form (7.1.1). Let $\mathfrak{g}$ be a Lie algebra of some (possibly infinite dimensional) gauge group. The ASDYM equations (7.1.3) on a connection $A \in T^*\mathbb{R}^4 \otimes \mathfrak{g}$ are equivalent to the commutativity of the Lax pair (7.1.6).

We shall require that the connection possesses two commuting translational symmetries, one null and one non-null which in our coordinates are in $\partial_{\tilde{w}}$ and

$\partial_{\tilde{y}}$ directions, where $z = y + \tilde{y}$ and $\tilde{z} = y - \tilde{y}$. Choose a gauge such that $A_{\tilde{z}} = 0$ and one of the Higgs fields $\Phi = A_{\tilde{w}}$ is constant and rename the coordinate $w = t$. The Lax pair (7.1.6) has so far been reduced to

$$L = \partial_t - W - \lambda\partial_y \quad \text{and} \quad M = \partial_y - U - \lambda\Phi, \tag{10.3.56}$$

where $W = -A_t$ and $U = -A_z$ are functions of (y, t) with values in the Lie algebra $\mathfrak{g}$, and Φ is an element of $\mathfrak{g}$ which does not depend on (y, t). The reduced ASDYM equations are

$$\partial_y W - \partial_t U + [W, U] = 0 \quad \text{and} \quad \partial_y U + [W, \Phi] = 0.$$

Now choose $G = \text{Diff}(S^1)$, so that (U, W, Φ) become vector fields on S^1. We can choose a local coordinate x on S^1 such that

$$\Phi = \partial_x, \quad W = w(x, y, t)\partial_x, \quad \text{and} \quad U = -u(x, y, t)\partial_x, \tag{10.3.57}$$

where u and w are smooth functions on $\mathbb{R}^3$. The reduced Lax pair (10.3.56) is identical to (10.3.54) and the ASDYM equations reduce to the pair of PDEs (10.3.53). □

As discussed in Section 8.1, reductions of the ASDYM equations with $G = SU(1, 1)$ or $G = SL(2, \mathbb{R})$ by two translations (one of which is null) lead to well-known integrable systems KdV and NLS. The group $SU(1, 1)$ is a subgroup of $\text{Diff}(S^1)$ which can be seen by considering the Mobius action of $SU(1, 1)$:

$$\zeta \longrightarrow M(\zeta) = \frac{\alpha\zeta + \beta}{\overline{\beta}\zeta + \overline{\alpha}}, \quad |\alpha|^2 - |\beta|^2 = 1$$

on the unit disc. This restricts to the action on the circle as $|M(\zeta)| = 1$ if $|\zeta| = 1$. We should therefore expect that equation (10.3.53) contains KdV and NLS as its special cases (but not necessarily symmetry reduction). To find explicit classes of solutions to (10.3.53) out of solutions of NLS (we leave KdV as an exercise) we proceed as follows. Consider the matrices

$$\tau_+ = \begin{pmatrix} 0 & 1 \\ 0 & 0 \end{pmatrix}, \quad \tau_- = \begin{pmatrix} 0 & 0 \\ 1 & 0 \end{pmatrix}, \quad \text{and} \quad \tau_0 = \begin{pmatrix} 1 & 0 \\ 0 & -1 \end{pmatrix}$$

with the commutation relations

$$[\tau_+, \tau_-] = \tau_0, \quad [\tau_0, \tau_+] = 2\tau_+, \quad \text{and} \quad [\tau_0, \tau_-] = -2\tau_-.$$

The NLS equation

$$i\phi_t = -\frac{1}{2}\phi_{yy} + \phi|\phi|^2, \quad \text{where} \quad \phi = \phi(y, t), \tag{10.3.58}$$

arises from the reduced Lax pair (10.3.56) with

$$W = \frac{1}{2i}(-|\phi|^2\tau_0 + \phi_y\tau_- - \overline{\phi}_y\tau_+), \quad U = -\phi\tau_- - \overline{\phi}\tau_+, \quad \text{and} \quad \Phi = i\tau_0.$$

Now we replace the matrices by vector fields on S^1 corresponding to the embedding of $\mathfrak{su}(1,1)$ in Diff(S^1)

$$\tau_+ \longrightarrow \frac{1}{2i}e^{2ix}\frac{\partial}{\partial x}, \quad \tau_- \longrightarrow -\frac{1}{2i}e^{-2ix}\frac{\partial}{\partial x}, \quad \text{and} \quad \tau_0 \longrightarrow \frac{1}{i}\frac{\partial}{\partial x},$$

and read off the solution to (10.3.53) from (10.3.57):

$$u = \frac{1}{2i}(\overline{\phi}e^{2ix} - \phi e^{-2ix}) \quad \text{and} \quad w = \frac{1}{2}|\phi|^2 + \frac{1}{4}(e^{2ix}\overline{\phi}_y + e^{-2ix}\phi_y). \qquad (10.3.59)$$

The second equation in (10.3.53) is satisfied identically, and the first is satisfied if $\phi(y,t)$ is a solution to the NLS equation (10.3.58).

10.3.2 Null symmetries and projective structures

If the conformal symmetry K is null, the three-dimensional space of trajectories of K inherits a degenerate conformal structure, and the connection with EW geometry is lost. The situation was investigated in detail in [29] and [57]. The conformal symmetry K defines a pair of totally null foliations of M, one by α-surfaces and one by β-surfaces; these foliations intersect along integral curves of K which are null geodesics. The main result from [57] is that there is a canonically defined *projective structure* (see the formula (C27)) on the two-dimensional space of β-surfaces U. One can explicitly write down all ASD conformal structures with null-conformal Killing vectors in terms of their underlying projective structures. The details are as follows.

Consider a four-dimensional conformal structure $(M,[g])$ with a null-conformal Killing vector K, that is,

$$\text{Lie}_K g = c\, g, \quad \text{where} \quad g(K,K) = 0.$$

The second condition implies that $K^a = \iota^A o^{A'}$ for some spinors ι^A and $o^{A'}$.

Lemma 10.3.3 *Let*

$$K = \iota^A o^{A'} e_{AA'}$$

be a null-conformal Killing vector. Then the two-dimensional distributions spanned by $\iota^A e_{AA'}$ and $o^{A'} e_{AA'}$ are Frobenius integrable.

Proof Using $K_{AA'} = \iota_A o_{A'}$, the conformal Killing equations become

$$o_{A'}\nabla_{BB'}\iota_A + \iota_A\nabla_{BB'}o_{A'} = \psi_{A'B'}\varepsilon_{AB} + \phi_{AB}\varepsilon_{A'B'} + \frac{1}{2}c\varepsilon_{AB}\varepsilon_{A'B'}, \qquad (10.3.60)$$

where $\psi_{A'B'}$ and ϕ_{AB} are the SD and ASD parts of dK. Contracting both sides with $\iota^A o^{A'}$ gives

$$0 = o^{A'}\iota_B\psi_{A'B'} + \iota^A o_{B'}\phi_{AB} + \frac{1}{2}c\iota_B o_{B'}.$$

Multiplying by ι^B and $o^{B'}$, respectively, leads to the algebraic identities

$$\iota^A\iota^B\phi_{AB} = 0 \quad \text{and} \quad o^{A'}o^{B'}\psi_{A'B'} = 0.$$

Multiplying (10.3.60) by $\iota^A\iota^B$ and $o^{A'}o^{B'}$ gives the so-called geodesic shear-free conditions:

$$\iota^A\iota^B\nabla_{BB'}\iota_A = 0 \quad \text{and} \quad o^{A'}o^{B'}\nabla_{BB'}o_{A'} = 0. \tag{10.3.61}$$

The relations (10.3.61) are equivalent to the integrability of distributions defined by $o^{A'}$ and ι^A. We will show it for the $o^{A'}$ case; the ι^A case is identical. Let $X = \alpha^A o^{A'}e_{AA'}$ and $Y = \beta^A o^{A'}e_{AA'}$ be vector fields, which by definition are in the α-planes determined by $o^{A'}$. The Frobenius condition (C9) is

$$[X, Y]_{AA'} = (f\alpha_A + g\beta_A)o_{A'},$$

for some functions f and g. Multiplying by $o^{A'}$ gives

$$o^{A'}[X, Y]_{AA'} = o^{A'}(X^{BB'}\nabla_{BB'}Y_{AA'} - Y^{BB'}\nabla_{BB'}X_{AA'}) = 0.$$

Substituting the spinor expressions for $X^{AA'}$ and $Y^{AA'}$ results in

$$o^{A'}o^{B'}\nabla_{BB'}o_{A'} = 0,$$

which is (10.3.61), and it is easy to show this is sufficient as well as necessary. □

Let $\mathcal{D}_\iota$ be two-dimensional distribution spanned by $\iota^A e_{AA'}$. The leaves of this distribution are β-surfaces in M. Consider the two-dimensional space of β-surfaces $U = M/\mathcal{D}_\iota$. Its projectivized tangent bundle admits a one-dimensional distribution $\Theta = \{L_0, L_1, K\}/\mathcal{D}_\iota$ defined on

$$P(TU) = [P(\mathbb{S}') = M \times \mathbb{CP}^1]/\mathcal{D}_\iota.$$

This leads to the following sequence of projections:

$$\begin{array}{ccc} M & \longleftarrow & M \times \mathbb{CP}^1 \\ \mathcal{D}_\iota \downarrow & & \downarrow \\ U & \longleftarrow & P(TU) \end{array} \tag{10.3.62}$$

and equips U with a projective structure – an equivalence class of torsion-free connections on TU such that two connections are equivalent if they share the same unparametrized geodesics (see Appendix C). Two-dimensional projective structures are equivalent to second-order ODEs of the form

$$\frac{d^2y}{dx^2} = A_3(x, y)\left(\frac{dy}{dx}\right)^3 + A_2(x, y)\left(\frac{dy}{dx}\right)^2 + A_1(x, y)\left(\frac{dy}{dx}\right) + A_0(x, y), \tag{10.3.63}$$

obtained by choosing local coordinates (x, y) and eliminating the affine parameter from the geodesic equation; the integral curves of this ODE are geodesics of the projective structure. The projective spray corresponding to this ODE is

$$\boldsymbol{\Theta} = \partial_x + \lambda\partial_y + (A_0 + \lambda A_1 + \lambda^2 A_2 + \lambda^3 A_3)\partial_\lambda. \tag{10.3.64}$$

Conversely, a spray $\boldsymbol{\Theta}$ which arises from (10.3.62) is homogeneous of degree one in $\pi^{A'}$, thus it is given by

$$\boldsymbol{\Theta} = \pi^{A'}\frac{\partial}{\partial x^{A'}} - \Gamma^{C'}_{A'B'}\pi^{A'}\pi^{B'}\frac{\partial}{\partial \pi^{C'}}$$

which is a homogeneous form of (10.3.64).

Reconstruction of the ASD conformal structure with null symmetry comes down to extending the spray on $\mathbb{P}(TU)$ to a Lax pair on $P(\mathbb{S}')$. This can be done and the result is summarized in the following.

Theorem 10.3.4 [57] *Let $(M, [g], K)$ be a smooth neutral-signature ASD-conformal structure with null-conformal Killing vector. Then there exist local coordinates (ϕ, x, y, z) and $g \in [g]$ such that $K = \partial_\phi$ and g has one of the following two forms, according to whether the twist $\mathbb{K} \wedge d\mathbb{K}$ vanishes or not ($\mathbb{K} := g(K, .)$):*

1. $\mathbb{K} \wedge d\mathbb{K} = 0$:

$$g = [d\phi + (zA_3 - Q)dy](dy - \beta dx) - \{dz - [z(-\beta_y + A_1 + \beta A_2 + \beta^2 A_3)]dx - [z(A_2 + 2\beta A_3) + P]dy\}dx, \tag{10.3.65}$$

where A_1, A_2, A_3, β, Q, and P are arbitrary functions of (x, y).

2. $\mathbb{K} \wedge d\mathbb{K} \neq 0$:

$$g = \{d\phi + A_3\partial_z G dy + [A_2\partial_z G + 2A_3(z\partial_z G - G) - \partial_z\partial_y G]dx\}(dy - zdx) - \partial_z^2 G dx[dz - (A_0 + zA_1 + z^2 A_2 + z^3 A_3)dx], \tag{10.3.66}$$

where A_0, A_1, A_2, and A_3 are arbitrary functions of (x, y), and G is a function of (x, y, z) satisfying the following PDE:

$$[\partial_x + z\partial_y + (A_0 + zA_1 + z^2 A_2 + z^3 A_3)\partial_z]\partial_z^2 G = 0. \tag{10.3.67}$$

In (10.3.66) all the $A_i, i = 0, 1, 2, 3$, functions occur explicitly in the metric. In (10.3.65) the function A_0 does not explicitly occur. It is determined by the following equation:

$$A_0 = \beta_x + \beta\beta_y - \beta A_1 - \beta^2 A_2 - \beta^3 A_3. \qquad (10.3.68)$$

If the projective structure is flat, that is, $A_i = 0$ and $\beta = P = 0$ then (10.3.65) is Ricci-flat [135], and in fact this is the most general ASD Ricci-flat metric with a null Killing vector which preserves the pseudo-hyper-Kähler structure. The twisting metrics (10.3.66) generalize those found in [123]. More generally, if the projective structure comes from a Riemannian metric on U then there will always exist a (pseudo-)Kähler structure in the conformal class (10.3.66) if $G = z^2/2 + \gamma(x, y)z + \delta(x, y)$ for certain γ and δ [26].

It is interesting that integrable systems are not involved in the null case, given their ubiquity in the non-null case.

10.3.3 Dispersionless integrable systems

Dispersionless integrable systems can arise from solitonic systems in a following way: Let

$$A\left(\frac{\partial}{\partial X}\right) = \frac{\partial^n}{\partial X^n} + a_1(X^i)\frac{\partial^{n-1}}{\partial X^{n-1}} + \cdots + a_n(X^i) \quad \text{and}$$

$$B\left(\frac{\partial}{\partial X}\right) = \frac{\partial^m}{\partial X^m} + b_1(X^i)\frac{\partial^{m-1}}{\partial X^{m-1}} + \cdots + b_m(X^i)$$

be differential operators on $\mathbb{R}$ with coefficients depending on local coordinates $X^i = (X, Y, T)$ on $\mathbb{R}^3$. The overdetermined linear system

$$\Psi_Y = A\left(\frac{\partial}{\partial X}\right)\Psi \quad \text{and} \quad \Psi_T = B\left(\frac{\partial}{\partial X}\right)\Psi$$

admits a solution $\Psi(X, Y, T)$ on a neighbourhood of an initial point (X, Y_0, T_0) for arbitrary initial data $\Psi(X, Y_0, T_0) = f(X)$ iff the integrability conditions $\Psi_{YT} = \Psi_{TY}$, or

$$A_T - B_Y + [A, B] = 0 \qquad (10.3.69)$$

are satisfied. The non-linear system (10.3.69) for $a_1, \ldots, a_n$ and $b_1, \ldots, b_m$ can be solved by a (2+1)-dimensional version of the IST discussed in Chapter 2.

The dispersionless limit [192] is obtained by substituting

$$\frac{\partial}{\partial X^i} = \varepsilon\frac{\partial}{\partial x^i} \quad \text{and} \quad \Psi(X^i) = \exp\left[\psi(x^i/\varepsilon)\right],$$

and taking the limit $\varepsilon \longrightarrow 0$. In the limit the commutators of differential operators are replaced by the Poisson brackets of their symbols according to the relation

$$\frac{\partial^k}{\partial X^k}\Psi \longrightarrow (\psi_x)^k\Psi, \quad [A, B] \longrightarrow \frac{\partial A}{\partial \lambda}\frac{\partial B}{\partial x} - \frac{\partial A}{\partial x}\frac{\partial B}{\partial \lambda} = \{A, B\}, \quad \text{and} \quad \lambda = \psi_x,$$

where A and B are polynomials in λ, with coefficients depending on $x^i = (x, y, t)$. The dispersionless limit of the system (10.3.69) is

$$A_t - B_y + \{A, B\} = 0. \tag{10.3.70}$$

Non-linear DEs of the form (10.3.70) are called dispersionless integrable systems.

There are several methods of integrating (10.3.70) [15, 46, 63, 64, 96, 97, 98, 111, 112, 155]. We shall focus on the one which makes contact with EW geometry and twistor theory. The system (10.3.70) is equivalent to the integrability $[L, M] = 0$ of a two-dimensional distribution of vector fields

$$L = \partial_t - B_\lambda\partial_x + B_x\partial_\lambda \quad \text{and} \quad M = \partial_y - A_\lambda\partial_x + A_x\partial_\lambda \tag{10.3.71}$$

on $\mathbb{R}^3 \times \mathbb{RP}^1$. This is similar to the EW Lax pair (10.3.42) in Lorentzian signature.

We shall therefore generalize the notion of the dispersionless integrable systems by allowing distributions of vector fields more general than (10.3.71). The derivatives A_λ, A_x, B_λ, and B_x of the symbols (A, B) of operators can be replaced by independent polynomials A_1, A_2, B_1, and B_2 in λ with coefficients depending on (x, y, t):

$$L = \partial_t - B_1\partial_x + B_2\partial_\lambda \quad \text{and} \quad M = \partial_y - A_1\partial_x + A_2\partial_\lambda, \tag{10.3.72}$$

where A_1, B_1 are linear in λ and A_2, B_2 are at most cubic in λ . We take the integrability of this generalized distribution (10.3.72) as our definition of a dispersionless integrable system. The definition is intrinsic in the sense that it does not refer to an underlying soliton equation.

10.3.3.1 *Interpolating integrable system*

In Section 10.3 we have seen how certain dispersionless systems (dKP, $SU(\infty)$-Toda, and Diff(S^1) equation) arise from ASD conformal equations in four dimensions and lead to EW structures. In each case it was possible to choose a solution of the generalized monopole equation (10.3.44) so that the resulting ASD conformal structure is Ricci-flat.

It should therefore be possible to obtain these dispersionless equations directly as reductions of the ASD Ricci-flat condition, for example, in its second heavenly form (10.2.22). This is indeed the case and the details can be found in [46, 51, 65]. Here we shall present a dispersionless integrable system which arises as a symmetry reduction of (10.2.22) and contains the dKP equation (10.3.52) and the Diff(S^1) equation (10.3.53) as special cases. This *interpolating integrable system* is given by [59]

$$u_y + w_x = 0 \quad \text{and} \quad u_t + w_y - c(uw_x - wu_x) + buu_x = 0, \tag{10.3.73}$$

where b and c are constants and u, w are smooth functions of (x, y, t). It admits a Lax pair

$$\begin{aligned} L &= \frac{\partial}{\partial t} + (cw + bu - \lambda cu - \lambda^2)\frac{\partial}{\partial x} + b(w_x - \lambda u_x)\frac{\partial}{\partial \lambda} \quad \text{and} \\ M &= \frac{\partial}{\partial y} - (cu + \lambda)\frac{\partial}{\partial x} - bu_x\frac{\partial}{\partial \lambda} \end{aligned} \tag{10.3.74}$$

with a spectral parameter $\lambda \in \mathbb{CP}^1$. A linear combination of L and M is of the form (10.3.72).

Setting $b = 0$ and $c = -1$ gives the Diff(S^1) equation (10.3.53) and setting $c = 0$ and $b = 1$ gives the dKP equation (10.3.52). In fact one constant can always be eliminated from (10.3.73) by redefining the coordinates and it is only the ratio of b/c which remains. We prefer to keep both constants as it makes the limits more transparent.

Proposition 10.3.5 [59] *The most general* $(++--)$ *ASD Ricci-flat metric with a conformal Killing vector whose SD derivative is null can be put in the form*

$$g = e^{c\tau}[Vh - V^{-1}(d\tau + \eta)^2], \tag{10.3.75}$$

where (u, w) *solve the system* (10.3.73), τ *parameterizes the orbits of the conformal Killing vector* $K = \partial/\partial\tau$, *and*

$h = (dy + cudt)^2 - 4[dx + cudy - (cw + bu)dt]dt$, $\eta = -\frac{1}{2}u_x dy + \left(\frac{c}{2}uu_x - u_y\right)dt$, *and* $V = \frac{1}{2}u_x$.

The proof uses the second heavenly formalism. The coordinates can be chosen so that the conformal Killing vector is given by $K = (cz + b)\partial_z + (cx - 2bz)\partial_x$ and the ASD Ricci-flat metric is given by (10.2.13) where Θ satisfies the second heavenly equation (10.2.22). One then imposes the conformal Killing equations on Θ, and (after a series of Legendre transforms and coordinate transformations) arrives at the statement of Proposition 10.3.5. In particular the proof explains the origin of the two parameters (b, c) in (10.3.73) as the conformal symmetry is

$$K = c \times (\text{dilatation}) + b \times \text{(rotation with null SD derivative)}.$$

The details of the proof are given in [59].

10.3.3.2 *Manakov–Santini system*

The system (10.3.73) is a special case of the Manakov–Santini system [111, 112]:

$$U_{xt} - U_{yy} + (UU_x)_x + V_x U_{xy} - V_y U_{xx} = 0 \quad \text{and} \tag{10.3.76}$$
$$V_{xt} - V_{yy} + U V_{xx} + V_x V_{xy} - V_y V_{xx} = 0,$$

where $U = U(x, y, t)$ and $V = V(x, y, t)$. To see this, notice that the first equation in (10.3.73) implies the existence of $v(x, y)$ such that $u = v_x$, $w = -v_y$, and v satisfies

$$v_{xt} - v_{yy} + c(v_x v_{xy} - v_y v_{xx}) + buv_{xx} = 0. \tag{10.3.77}$$

Differentiating the second equation in (10.3.73) and eliminating w yields

$$u_{xt} - u_{yy} - c(v_y u_{xx} - v_x u_{xy}) + b(uu_x)_x = 0. \tag{10.3.78}$$

Now assume the generic case when the constants c and b are non-zero, and set $U = bu$ and $V = cv$. Then the systems (10.3.77) and (10.3.78) are equivalent to (10.3.76) with an additional constraint

$$cU - bV_x = 0. \tag{10.3.79}$$

The Manakov–Santini system corresponds to an EW structure:

$$h = (dy - V_x dt)^2 - 4[dx - (U - V_y)dt]dt \quad \text{and} \tag{10.3.80}$$
$$\omega = -V_{xx} dy + (4U_x - 2V_{xy} + V_x V_{xx})dt.$$

To verify this set $x^i = (y, x, t)$. The (11), (12), (22), and (23) components of the EW equations hold identically. The (13) component vanishes if the second equation in (10.3.76) holds, and finally the (33) component vanishes if both equations in (10.3.76) are satisfied.

In [111, 112] Manakov and Santini have solved the initial value problem for the system (10.3.76) using their version of IST applicable to Lax pairs containing vector fields.

10.3.3.3 *Quadric ansatz*

The usual way of reducing PDEs to ODEs is to determine a symmetry group of transformations acting on dependent and independent variables, and reduce the number of independent variables down to one. This was discussed in Section 4.3.

Tod [158] has proposed a non-symmetric way of reducing PDEs to ODEs which can be used to construct solutions to some dispersionless integrable systems. Our presentation of this follows [48, 50]. Let $u = u(x^1, \dots, x^n) \in \mathbb{R}$

be a solution to a PDE

$$F(u, u_i, u_{ij}, \ldots, u_{ij\cdots k}, x^i) = 0, \tag{10.3.81}$$

where $u_i = \partial u/\partial x^i$ and $(x^1, \ldots, x^n) \in \mathbb{R}^n$. The ansatz is to seek solutions constant on a polynomial hypersurface $\Sigma \subset \mathbb{R}^n$, or equivalently to seek symmetric objects

$$M(u),\ M_i(u),\ M_{ij}(u), \ldots, M_{ij\cdots k}(u),$$

so that a solution of equation (10.3.81) is determined implicitly by

$$Q(x^i, u) := M(u) + M_i(u)x^i + M_{ij}(u)x^i x^j + \cdots + M_{ij\cdots k}(u)x^i x^j \cdots x^k = C, \tag{10.3.82}$$

where C is a constant. Here Σ should be regarded as the zero locus of a polynomial $Q(x^i, u) - C$ in $\mathbb{R}^n$. If u satisfies (10.3.81) and the algebraic constraint (10.3.82), then so does $g^t(u)$, where g^t is a flow generated by any section of $T\Sigma$. Note however that vectors tangent to Σ do not generate symmetries of (10.3.81), as the choice of Σ depends on u.

We shall concentrate on the quadric ansatz

$$Q(x^i, u) := M_{ij}(u)x^i x^j = C. \tag{10.3.83}$$

This ansatz can be made whenever we have a non-linear PDE of the form

$$\frac{\partial}{\partial x^j}\left[b^{ij}(u)\frac{\partial u}{\partial x^i}\right] = 0, \tag{10.3.84}$$

where u is a function of coordinates $x^i, i = 1, \ldots, n$. We shall seek solutions constant on central quadrics or equivalently seek a matrix $\mathbf{M}(u) = [M_{ij}(u)]$ so that a solution of equation (10.3.84) is determined implicitly as in (10.3.83). We differentiate (10.3.83) implicitly to find

$$\frac{\partial u}{\partial x^i} = -\frac{2}{\dot{Q}}M_{ij}x^j, \quad \text{where} \quad \dot{Q} = \frac{\partial Q}{\partial u}. \tag{10.3.85}$$

Now we substitute this into (10.3.84) and integrate once with respect to u. Introducing $g(u)$ by

$$\dot{g} = \frac{1}{2}b^{ij}M_{ij} = \frac{1}{2}\mathrm{Tr}\,(\mathbf{bM}) \tag{10.3.86}$$

we obtain

$$(g\dot{M}_{ij} - M_{ik}b^{km}M_{mj})x^i x^j = 0,$$

so that as a matrix ODE

$$g\dot{\mathbf{M}} = \mathbf{MbM}. \tag{10.3.87}$$

This equation simplifies if written in terms of another matrix $\mathbf{N}(u)$ where

$$\mathbf{N} = -\mathbf{M}^{-1} \tag{10.3.88}$$

for then

$$g\dot{\mathbf{N}} = \mathbf{b}, \tag{10.3.89}$$

and g can be given in terms of $\Delta = \det(\mathbf{N})$ by

$$g^2\Delta = \zeta = \text{const.} \tag{10.3.90}$$

Restricting to three dimensions with $(x^1, x^2, x^3) = (x, y, z)$, the $SU(\infty)$-Toda equation (10.3.46) is given by (10.3.84) with

$$\mathbf{b}(u) = \begin{pmatrix} 1 & 0 & 0 \\ 0 & 1 & 0 \\ 0 & 0 & e^u \end{pmatrix}$$

and as was shown in [158], in this case (10.3.89) can be reduced to the PIII ODE (4.4.11).

For the dKP equation (10.3.52) we have (10.3.84) with

$$\mathbf{b}(u) = \begin{pmatrix} -u & 0 & 1/2 \\ 0 & -1 & 0 \\ 1/2 & 0 & 0 \end{pmatrix}. \tag{10.3.91}$$

(Note that $-\mathbf{b}(u)$ is the inverse of the EW metric.)

Proposition 10.3.6 *Solutions to the dKP equation* (10.3.52) *constant on the central quadric* (10.3.83) *are implicitly given by solutions to PI or PII:*

- *If* $(\mathbf{M}^{-1})_{33} \neq 0$ *then*

$$x^2v - y^2w\,[wv - (\alpha - 1/2)] + \frac{1}{2}t^2\left\{(\alpha - 1/2)^2 + 4\,wv\,[wv - (\alpha - 1/2)] + 2\,v^3\right\}$$
$$+\, xy\,(\alpha - 1/2) - ytv\,(\alpha - 1/2) - 2\,txv^2 = C\,[2wv - (\alpha - 1/2)]^2, \tag{10.3.92}$$

where α is a constant, v is given by

$$v = \frac{1}{2}\dot{w}(u) - w(u)^2 - u,$$

and w is a solution to

$$\frac{d^2w}{du^2} = 8w^3 + 8wu + 4\alpha,$$

which is the rescaled PII ODE (4.4.11).

- *If* $(\mathbf{M}^{-1})_{33} = 0$ *and* $(\mathbf{M}^{-1})_{23} \neq 0$ *then*

$$x^2 + w^2 y^2 - w\left(\frac{\dot{w}^2}{4} - 4w^3\right)t^2 - 4xtw^2 + 2wxy + \left(\frac{\dot{w}^2}{4} - 4w^3\right)yt = C\dot{w}^2, \tag{10.3.93}$$

  where $w(u)$ *satisfies*

$$\frac{d^2 w}{du^2} = 24w^2 + 8u,$$

  which is the rescaled PI ODE (4.4.11).

- *If* $(\mathbf{M}^{-1})_{33} = (\mathbf{M}^{-1})_{23} = 0$ *then*

$$\frac{y^2}{4} + \left[\sin(u)^3\cos(u) - u\sin(u)^2 + \gamma^2\cos(u)^4\right]t^2 - \sin(u)^2 tx - \gamma\cos(u)^2 ty = C\tan(u)^2, \tag{10.3.94}$$

  where γ *is a constant.*

The constant C *can always be set to* 0 *or* 1.

- **Example.** For certain values of α, PII admits particular solutions expressible in terms of 'known' functions. For $\alpha = n \in \mathbb{Z}$ the PII equation possesses rational solutions, and for $\alpha = n + 1/2$ there exists a class of solutions expressible by Airy functions. For example, if $\alpha = 1$, then PII is satisfied by $w = -1/(2u)$. Now $v = -u$, and the coefficients of the quadric become cubic in u:

$$8t^2u^3 + 16xtu^2 + (8x^2 - 4yt)u - (t^2 + 4xy) = C, \quad C = \text{const.}$$

  The three roots of this cubic give three solutions to dKP. A root which yields a real solution is

$$u(x, y, t) = \frac{\sqrt[3]{A + 12t\sqrt{B}}}{12t} + \frac{6yt + 4x^2}{3t\sqrt[3]{A + 12t\sqrt{B}}} - \frac{2x}{3t}, \tag{10.3.95}$$

$$A = 144\,xyt + 64\,x^3 + 108\,t^3 - 108\,tC,$$

$$B = -96\,y^3t - 48\,y^2x^2 + 216\,xyt^2 + 96\,x^3t + 81\,t^4 - 96\,\frac{Cx^3}{t} - 216\,Cxy - 162\,Ct^2 + 81\,C^2.$$

10.4 ASD conformal structures in neutral signature

The relevant Lie group isomorphism in $(+ + - -)$ signature is

$$SO(2, 2) \cong SL(2, \mathbb{R}) \times SL(2, \mathbb{R})/\mathbb{Z}_2. \tag{10.4.96}$$

In the neutral signature the spinor conjugation is a map $\mathbb{S} \to \mathbb{S}$ given by

$$\iota_A = (\alpha, \beta) \to \hat{\iota}_A = (\overline{\alpha}, \overline{\beta}) \tag{10.4.97}$$

(compare (9.2.17)) so that there exists a notion of real spinors. The isomorphism (9.2.15) is replaced by

$$TM \cong \mathbb{S} \otimes \mathbb{S}',$$

where $\mathbb{S}$ and $\mathbb{S}'$ are real rank-two vector bundles.

10.4.1 Conformal compactification

We shall now describe a conformal compactification of the flat neutral metric $\mathbb{R}^{2,2}$. The natural compactification $\overline{\mathbb{R}^{2,2}}$ is a projective quadric in $\mathbb{RP}^5$. To describe it explicitly consider $[\mathbf{x}, \mathbf{y}]$ as homogeneous coordinates on $\mathbb{RP}^5$, and set $\mathcal{Q} = |\mathbf{x}|^2 - |\mathbf{y}|^2$. Here $(\mathbf{x}, \mathbf{y})$ are vectors on $\mathbb{R}^3$ with its natural inner product. The cone $\mathcal{Q} = 0$ is projectively invariant, and the freedom $(\mathbf{x}, \mathbf{y}) \sim (c\mathbf{x}, c\mathbf{y})$, where $c \neq 0$ is fixed to set $|\mathbf{x}| = |\mathbf{y}| = 1$ which is $S^2 \times S^2$. We need to quotient this by the antipodal map $(\mathbf{x}, \mathbf{y}) \to (-\mathbf{x}, -\mathbf{y})$ to obtain the conformal compactification

$$\overline{\mathbb{R}^{2,2}} = (S^2 \times S^2)/\mathbb{Z}_2.$$

Parameterizing the double cover of this compactification by stereographic coordinates we find that the flat metric $|d\mathbf{x}|^2 - |d\mathbf{y}|^2$ on $\mathbb{R}^{3,3}$ yields the metric

$$g_0 = 4\frac{d\zeta d\bar{\zeta}}{(1+\zeta\bar{\zeta})^2} - 4\frac{d\chi d\bar{\chi}}{(1+\chi\bar{\chi})^2} \tag{10.4.98}$$

on $S^2 \times S^2$. To obtain the flat metric on $\mathbb{R}^{2,2}$ we would instead consider the intersection of the zero locus of $\mathcal{Q}$ in $\mathbb{R}^{3,3}$, with a null hypersurface $x_1 - y_1 = 1$.

The metric g_0 is conformally flat and scalar flat, as the scalar curvature is the difference between curvatures on both factors. It is also Kähler with respect to the natural complex structures on $\mathbb{CP}^1 \times \mathbb{CP}^1$ with holomorphic coordinates (ζ, χ).

10.4.2 Curved examples

- **Compact neutral hyper-Kähler metrics.** The only compact four-dimensional Riemannian hyper-Kähler manifolds are the complex torus with the flat metric and $K3$ with a Ricci-flat Calabi–Yau metric.

  To write down explicit examples in neutral signature, consider the following pseudo-hyper-Kähler metric

$$g = d\phi dy - dzdx - \mathcal{Q}(x, y)dy^2, \tag{10.4.99}$$

for an arbitrary function Q. This is the neutral version of the pp-wave metric of general relativity [135], and is a special case of (10.3.65), where the underlying projective structure is flat. It is non-conformally flat for generic Q. Define complex coordinates $z_1 = \phi + iz$ and $z_2 = x + iy$ on $\mathbb{C}^2$. By quotienting the z_1 and z_2 planes by lattices one obtains a product of elliptic curves, a special type of complex torus. If we require Q to be periodic with respect to the z_2 lattice, then (10.4.99) descends to a metric on this manifold.

- **Tod's scalar-flat Kähler metrics on $S^2 \times S^2$.** Consider $S^2 \times S^2$ with the conformally flat metric (10.4.98), that is, the difference of the standard sphere metrics on each factor. Thinking of each sphere as $\mathbb{CP}^1$ and letting ζ and χ be non-homogeneous coordinates for the spheres, this metric is given by (10.4.98). As we have already said, g_0 is scalar flat, indefinite Kähler. The obvious complex structure J with holomorphic coordinates (ζ, χ) gives a closed two-form and $\mathbf{\Omega} := g_0(J., .)$. Moreover g_0 clearly has a high degree of symmetry, since the two-sphere metrics have rotational symmetry. In [161], Tod found deformations of g_0 preserving the scalar-flat Kähler property, by using the Lorentizan version of the expression (10.3.45) for neutral scalar-flat Kähler metrics with symmetry. Take the explicit solution

$$e^u = 4\frac{1 - t^2}{(1 + x^2 + y^2)^2}$$

to the Lorentzian Toda equation (10.3.46) (where $z = it$), which can be obtained by demanding $u = f_1(x, y) + f_2(t)$. There remains a linear monopole equation for V. Setting $W = V(1 - t^2)$ and performing the coordinate transformation $t = \cos\theta$, $\zeta = x + iy$ gives

$$g = 4W\frac{d\zeta d\bar{\zeta}}{(1 + \zeta\bar{\zeta})^2} - Wd\theta^2 - \frac{\sin^2\theta}{W}(d\tau + \eta)^2, \qquad (10.4.100)$$

and W must solve a linear equation. This metric reduces to (10.4.98) for $W = 1$ and $\eta = 0$, with θ, ϕ standard coordinates for the second sphere. Differentiating the linear equation for W and setting $Q = \frac{\partial W}{\partial t}$, one obtains the neutral wave equation

$$\nabla_1^2 Q = \nabla_2^2 Q, \qquad (10.4.101)$$

where $\nabla_{1,2}$ are the Laplacians on the two-spheres, and Q is independent of ϕ, that is, axisymmetric for one of the sphere angles. Equation (10.4.101) can be solved using Legendre polynomials, and one obtains non-conformally flat deformations of (10.4.98) in this way.

- **Ooguri–Vafa metrics.** In [125] Ooguri and Vafa constructed a class of non-compact neutral hyper-Kähler metrics on cotangent bundles of Riemann surfaces with genus ≥ 1 using the heavenly equation formalism (10.2.18), with a $(+ + − −)$ real section of the complexified space-time. Instead of using the real coordinates in (10.2.18) we set

$$w = \zeta, \quad \tilde{w} = \bar{\zeta}, \quad z = ip, \quad \text{and} \quad \tilde{z} = -i\bar{p}, \text{ where } \zeta, p \in \mathbb{C}$$

with $\Omega = i(p\bar{\zeta} - \bar{p}\zeta)$ corresponding to the flat metric. Let Σ be a Riemann surface with a local holomorphic coordinare ζ, such that the Kähler metric on Σ is $h_{\zeta\bar{\zeta}} d\zeta d\bar{\zeta}$. Suppose that p is a local complex coordinate for fibres of the cotangent bundle $T^*\Sigma$. If $\boldsymbol{\Omega}$ is the Kähler form for a neutral metric g then $g_{A\tilde{B}} = \partial_A \partial_{\tilde{B}} \Omega$ for a function Ω on the cotangent bundle. Then the equation

$$\det g_{A\tilde{B}} = -1$$

is equivalent to the first heavenly equation (10.2.18), and gives a Ricci-flat ASD neutral metric.

Suppose that Ω depends only on the globally defined function $X = h^{\zeta\bar{\zeta}} p\bar{p}$, which is the length of the cotangent vector corresponding to p. There is a globally defined holomorphic (2, 0)-form $d\zeta \wedge dp$, which is the holomorphic part of the standard symplectic form on the cotangent bundle, so (ζ, p) are the holomorphic coordinates in the Plebański coordinate system. The heavenly equation reduces to an ODE for $\Omega(X)$, and for solutions of this ODE to exist the metric h must have constant negative curvature, so Σ has genus > 1. In this case one can solve the ODE to find

$$\Omega = 2\sqrt{A^2 + BX} + A \ln \frac{\sqrt{A^2 + BX} - A}{\sqrt{A^2 + BX} + A},$$

where A and B are arbitrary positive constants. The metric g is well behaved when $X \to 0$ (or $p \to 0$), as in this limit $\Omega \to \ln(X)$ and g restricts to $h_{\zeta\bar{\zeta}} d\zeta d\bar{\zeta}$ on Σ and $-h^{\zeta\bar{\zeta}} dp d\bar{p}$ on the fibres. In the limit $X \to \infty$ the metric is flat. To see this one needs to chose a uniformizing coordinate τ on Σ so that h is a metric on the upper half-plane. Then make a coordinate transformation $\zeta_1 = \tau\sqrt{p}$ and $\zeta_2 = \sqrt{p}$. The holomorphic two-form is still $d\zeta_1 \wedge d\zeta_2$, and the Kähler potential $\Omega = i(\zeta_2\bar{\zeta}_1 - \zeta_1\bar{\zeta}_2)\sqrt{B}$ yields the flat metric.

10.5 Twistor theory

Theorem 10.1.2 proved in Section 10.1 motivates the following definition which generalizes the notion of the twistor space of Definition 7.2.1 to the curved case:

Definition 10.5.1 *The twistor space $\mathcal{PT}$ of a holomorphic conformal structure $(M, [g])$ with ASD Weyl curvature is the manifold of α-surfaces in M.*

The twistor space is a three-dimensional complex manifold whose structure is best revealed by exploiting the double fibration picture.

Define the correspondence space $\mathcal{F}$ to be the product $M \times \mathbb{CP}^1$ locally coordinatized by (x^a, λ), where x^a denote the coordinates of a point $p \in M$ and λ is the coordinate on $\mathbb{CP}^1$ that parameterizes the α-surfaces through p in M. We represent $\mathcal{F}$ as the quotient of the primed spin bundle $\mathbb{S}'$ with fibre coordinates $\pi_{A'}$ by the Euler vector field $\Upsilon = \pi^{A'}\partial/\partial\pi^{A'}$. The fibre coordinates are related to λ by $\lambda = \pi_{0'}/\pi_{1'}$. A form with values in the line bundle (see Appendix B) $\mathcal{O}(n)$ on $\mathcal{F}$ can be represented by a homogeneous form κ on the non-projective spin bundle satisfying

$$\Upsilon \lrcorner \kappa = 0 \quad \text{and} \quad \text{Lie}_\Upsilon \kappa = n\kappa.$$

For example, $\pi_{A'} d\pi^{A'}$ descends to an $\mathcal{O}(2)$-valued one-form on $\mathcal{F}$.

The Lax pair on $\mathcal{F}$ arises as the image under the projection $T\mathbb{S}' \longrightarrow T\mathcal{F}$ of the distribution (10.1.1) and the twistor space $\mathcal{PT}$ arises as a quotient of $\mathcal{F}$ by the Lax pair. A twistor function is a function on $\mathcal{F}$ which is constant along the distribution L_A. Similarly a differential form on $\mathcal{F}$ descends to $\mathcal{PT}$ if its Lie derivative along L_A vanishes.

The correspondence space has the alternative definition

$$\mathcal{F} = \mathcal{PT} \times M|_{Z\in L_p} = M \times \mathbb{CP}^1,$$

where L_p is the curve in $\mathcal{PT}$ that corresponds to $p \in M$ and $Z \in \mathcal{PT}$ lies on L_p. This leads to a double fibration

$$M \xleftarrow{r} \mathcal{F} \xrightarrow{q} \mathcal{PT}. \tag{10.5.102}$$

Lemma 10.5.2 *The holomorphic curves $q(\mathbb{CP}^1_p)$ where $\mathbb{CP}^1_p = r^{-1}p$, $p \in M$, have normal bundle $N = \mathcal{O}(1) \oplus \mathcal{O}(1)$.*

Proof The normal bundle $N(L)$ of a submanifold $L \subset \mathcal{PT}$ is defined to be $\cup_{Z\in L} N_Z(L)$, where $N_Z = (T_Z\mathcal{PT})/(T_Z L)$ is a quotient vector space. The double fibration picture allows the identification of the normal bundle with the quotient $r^*(T_p M)/\{\text{span} L_A\}$. In their homogeneous form the Lax operators L_A have weight 1, and the distribution spanned by them is isomorphic to the bundle $\mathbb{C}^2 \otimes \mathcal{O}(-1)$. The definition of the normal bundle as a quotient gives the exact sequence

$$0 \mapsto \mathbb{C}^2 \otimes \mathcal{O}(-1) \mapsto \mathbb{C}^4 \mapsto N \mapsto 0$$

and thus $N = \mathcal{O}(1) \oplus \mathcal{O}(1)$ as the last map, in spinor notation, is given explicitly by $V^{AA'} \mapsto V^{AA'}\pi_{A'}$ clearly projecting onto $\mathcal{O}(1) \oplus \mathcal{O}(1)$. □

The conformal structure $[g]$ on M is encoded in the algebraic geometry of curves in $\mathcal{PT}$ in the following way

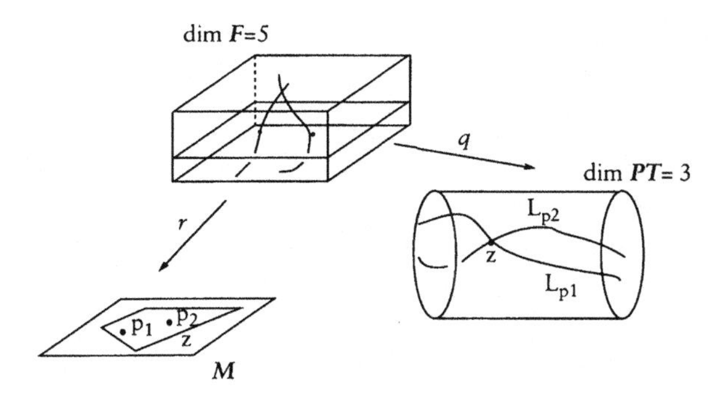

Figure 10.1 *Double fibration*

- Two points p_1 and p_2 in M are null-separated iff the corresponding curves L_{p_1} and L_{p_2} intersect at one point (Figure 10.1).

The point of this definition is that a null geodesic containing p_1 and p_2 lies on a unique α-surface. This is because the tangent vector to any null geodesic must be of the form $\iota^A o^{A'}$, and thus the geodesic is contained in a surface spanned by $o^{A'} \mathbf{e}_{AA'}$.

We have therefore established the first part of the following:

Theorem 10.5.3 (Penrose [131]) *There is a one-to-one correspondence between*

- *Complex ASD conformal structures*
- *Three-dimensional complex manifolds containing a four-parameter family of rational curves with normal bundle $\mathcal{O}(1) \oplus \mathcal{O}(1)$*

Proof To complete the proof we must be able to go in the other direction and reconstruct the ASD conformal structure from its twistor space. Using the Kodaira theorem (Theorem B.3.1) we define M to be the moduli space of $\mathbb{CP}^1$'s with the prescribed normal bundle.

The Kodaira isomorphism (B9) states that a vector at a point p in M corresponds to a holomorphic section of the normal bundle $\mathcal{O}(1) \oplus \mathcal{O}(1)$ of the curve $L_p \cong \mathbb{CP}^1$ in $\mathcal{PT}$. We define a vector in M to be null if this holomorphic section has a zero. Vanishing of a section of $\mathcal{O}(1) \oplus \mathcal{O}(1)$ is a quadratic condition as $V^{AA'}\pi_{A'} = 0$ regarded as a linear system for $\pi_{A'}$ has a solution if $\det(V^{AA'}) = 0$. This gives a conformal structure on M. To prove that this conformal structure is ASD it is enough to show that α-surfaces exist and refer to Theorem 10.1.2. There is a two-parameter family of $\mathcal{O}(1) \oplus \mathcal{O}(1)$ curves through a given point Z of $\mathcal{PT}$ and this defines a surface $Z \subset M$. This surface is totally null with respect to the conformal structure as all points corresponding to curves through Z are null-separated and thus Z is an α- or β-surface. We choose the orientation on M so that it is an α-surface. The twistor space is three-dimensional so there is a three-parameter family of α-surfaces in M and $[g]$ is ASD by Theorem 10.1.2. □

The real ASD conformal structures are obtained by introducing an involution on the twistor space. There are two possibilities leading to the Riemannian and neutral signatures, respectively. In both cases the involutions act on the twistor lines, thus giving rise to maps from $\mathbb{CP}^1$ to $\mathbb{CP}^1$: the antipodal map which in stereographic coordinates is given by $\lambda \to -1/\overline{\lambda}$, or a complex conjugation which swaps the lower and upper hemispheres preserving the real equator. The antipodal map has no fixed points and corresponds to the positive-definite conformal structures. The conjugation corresponds to the neutral case.

- **Euclidean case.** The spinor conjugation (9.2.17) acts on $\mathbb{S}'$ and descends to an involution $\sigma : \mathcal{PT} \to \mathcal{PT}$ such that $\sigma^2 = -\text{Id}$. The twistor curves which are preserved by σ form a real four-parameter family, thus giving rise to a real four-manifold $M_{\mathbb{R}}$. If $Z \in \mathcal{PT}$ then Z and $\sigma(Z)$ are connected by a unique *real curve*. The real curves do not intersect as no two points are connected by a null geodesics in the positive-definite case. Therefore there exists a fibration of the twistor space $\mathcal{PT}$ over a real four-manifold $M_{\mathbb{R}}$. A fibre over a point $p \in M_{\mathbb{R}}$ is a copy of a $\mathbb{CP}^1$. The fibration is not holomorphic, but smooth. In the Atiyah–Hitchin–Singer (AHS) version [10] of the correspondence the twistor space of the positive-definite conformal structure is a real six-dimensional manifold identified with the projective spin bundle $P(\mathbb{S}') \to M_{\mathbb{R}}$. Given a conformal structure $[g]$ on $M_{\mathbb{R}}$ one defines an almost-complex-structure on $P(\mathbb{S}')$ by declaring

$$L_0, \quad L_1, \quad \text{and} \quad \frac{\partial}{\partial \overline{\lambda}}$$

to be the anti-holomorphic vector fields in $T^{0,1}[P(\mathbb{S}')]$. Here L_0 and L_1 are given by (10.1.1).

This almost-complex-structure is integrable in the sense of Theorem 9.3.1 if

$$[T^{0,1}, T^{0,1}] \subset T^{0,1}$$

and this happens iff L_0 and L_1 span an integrable distribution. We have already established that the integrability of L_A is equivalent to ASD of the conformal structure $[g]$. An alternative, but equivalent, way to define the almost-complex-structure on $P(\mathbb{S}')$ is to decompose its tangent space into horisontal and vertical subspaces:

$$T_z\left[P(\mathbb{S}')\right] = H_z \oplus V_z, \quad \text{where } z = (p, [\pi]) \in P(\mathbb{S}')$$

with respect to a connection on $\mathbb{S}'$ induced from a Levi-Civita connection of some $g \in [g]$.

The two-dimensional vector space V_z has a natural complex structure since the fibres of $P(\mathbb{S}') \to M_{\mathbb{R}}$ are Riemann spheres, and for a given $\pi_{A'}$ the almost-complex-structure on H_z is defined by a one-to-one tensor:

$$J_b^a = i\varepsilon_B^A \left[\pi^{A'} \sigma(\pi)_{B'} + \sigma(\pi)^{A'} \pi_{B'} \right],$$

where $\sigma : \mathbb{S}' \to \mathbb{S}'$ and $\sigma(\pi)_{A'} = (\overline{\pi_{1'}}, -\overline{\pi_{0'}})$.

We can summarize all this in the following theorem:

Theorem 10.5.4 (Atiyah–Hitchin–Singer [10]) *The six-dimensional almost-complex-manifold*

$$P(\mathbb{S}') \to M_{\mathbb{R}}$$

parameterizes almost-complex-structures in $(M_{\mathbb{R}}, [g])$*. Moreover* $P(\mathbb{S}')$ *is complex iff* $[g]$ *is ASD.*

- **Neutral case.** The spinor conjugation (10.4.97) allows an invariant decomposition of a spinor into its real and imaginary parts. Recall that the tangent space to an α-surface is spanned by null vectors of the form $\kappa^A \pi^{A'}$ with $\pi^{A'}$ fixed and κ^A arbitrary. A *real α-surface* corresponds to both κ^A and $\pi^{A'}$ being real.

  In general $\pi^{A'} = \mathrm{Re}(\pi^{A'}) + i\mathrm{Im}(\pi^{A'})$, and the correspondence space $\mathcal{F} = P(\mathbb{S}')$ decomposes into two open sets:

$$\mathcal{F}_+ = \{(x^a, [\pi^{A'}]) \in \mathcal{F}; \mathrm{Re}(\pi_{A'})\mathrm{Im}(\pi^{A'}) > 0\} = M_{\mathbb{R}} \times D_+ \quad \text{and}$$

$$\mathcal{F}_- = \{(x^a, [\pi^{A'}]) \in \mathcal{F}; \mathrm{Re}(\pi_{A'})\mathrm{Im}(\pi^{A'}) < 0\} = M_{\mathbb{R}} \times D_-,$$

where $D_\pm$ are two copies of a Poincare disc. These sets are separated by a real correspondence space

$$\mathcal{F}_0 = \{(x^a, [\pi^{A'}]) \in \mathcal{F}; \mathrm{Re}(\pi_{A'})\mathrm{Im}(\pi^{A'}) = 0\} = M_{\mathbb{R}} \times \mathbb{RP}^1.$$

The vector fields (10.1.1) together with the complex structure on the $\mathbb{CP}^1$ give $\mathcal{F}$ the structure of a complex manifold $\mathcal{PT}$ in a way similar to the AHS Euclidean picture: The integrable sub-bundle of $T\mathcal{F}$ is spanned by L_0, L_1, and $\partial_{\overline{\lambda}}$. The distribution (10.1.1) with $\lambda \in \mathbb{RP}^1$ defines a foliation of $\mathcal{F}_0$ with quotient $\mathcal{PT}_0$ which leads to a double fibration:

$$M_{\mathbb{R}} \xleftarrow{p} \mathcal{F}_0 \xrightarrow{q} \mathcal{PT}_0.$$

The twistor space $\mathcal{PT}$ is a union of two open subsets $\mathcal{PT}_+ = (\mathcal{F}_+)$ and $\mathcal{PT}_- = (\mathcal{F}_-)$ separated by a three-dimensional real boundary (*real twistor space*) $\mathcal{PT}_0 := q(\mathcal{F}_0)$.

The *real structure* $\sigma(x^a) = \overline{x}^a$ maps α-surfaces to α-surfaces, and therefore induces an anti-holomorphic involution $\sigma : \mathcal{PT} \to \mathcal{PT}$. The fixed points of this involution correspond to real α-surfaces in $M_{\mathbb{R}}$. There is an $\mathbb{RP}^1$ worth of such α-surfaces through each point of $M_{\mathbb{R}}$. The set of fixed points of σ in $\mathcal{PT}$ is $\mathcal{PT}_0$.

10.5.1 Curvature restrictions

Special conditions on a metric $g \in [g]$ can be encoded into the holomorphic geometry of the twistor space. These conditions involve the canonical bundle $\kappa \to \mathcal{PT}$. This is a holomorphic line bundle of holomorphic three-forms. Restricting κ to a twistor curve $L_p \cong \mathbb{CP}^1$ must therefore be one of the standard line bundles $\mathcal{O}(n)$ (see the Birkhoff–Grothendieck theorem Theorem [B.2.5]). In fact $n = -4$ since

$$\kappa|_{L_p} \cong T^* L_p \otimes \Lambda^2 N(L_p) = \mathcal{O}(-4)$$

as the dual of the normal bundle is $\mathcal{O}(-1) \oplus \mathcal{O}(-1)$ and $T^*\mathbb{CP}^1 = \mathcal{O}(-2)$.

The canonical bundle is related to the bundles over the correspondence space in the following way: Consider a section of κ of the form $bdz^1 \wedge dz^2 \wedge dz^3$, where z^i are local holomorphic coordinates on $\mathcal{PT}$. The pull-back of this three-form to $\mathcal{F}$ is of the form

$$b(\nu \wedge \pi_{A'} D\pi^{A'})(L_0, L_1, \ldots, \ldots, \ldots)$$

where ν is a volume-form on M and

$$D\pi^{A'} = d\pi^{A'} + \Gamma^{A'}{}_{B'}\pi^{B'}.$$

The three-form $(\nu \wedge \pi_{A'} D\pi^{A'})(L_0, L_1, \ldots, \ldots, \ldots)$ is $\mathcal{O}(4)$-valued as the operators L_A are homogeneous of degree one, and $\pi_{A'} D\pi^{A'}$ is homogeneous of degree two. Thus b must take values in $\mathcal{O}(-4)$ for the resulting three-form to be scalar-valued.

Now we shall list additional conditions on $\mathcal{PT}$ characterizing various subclasses of ASD conformal structures.

- A holomorphic fibration $\mu : \mathcal{PT} \to \mathbb{CP}^1$ corresponds to hyper-Hermitian conformal structures [18, 44].
- A preferred section of $\kappa^{-1/2}$ which vanishes at exactly two points on each twistor line corresponds to a scalar-flat Kähler metric in the following way g [137]:

  Given the covariantly constant Kähler form $\mathbf{\Omega}_{ab} = \omega_{A'B'}\varepsilon_{AB}$ one constructs the canonical section by $\omega_{A'B'}\pi^{A'}\pi^{B'}$. Conversely, given a section of $\kappa^{-1/2}$ one pulls it back to $\mathbb{S}'$ where it defines a symmetric spinor $\omega_{A'B'}$ which satisfies the conformally invariant twistor equation

$$\nabla_{A(A'}\omega_{B'C')} = 0.$$

Now pick any metric g in the ASD conformal class $[g]$ and define $\hat{g} = \Omega^2 g$, where

$$2\Omega^{-2} = \omega_{A'B'}\omega^{A'B'}.$$

Then $\hat{g}$ is Kähler with the Kähler form $\Omega^2\omega_{A'B'}\varepsilon_{AB}$.

- A preferred section of $\kappa^{-1/4}$ corresponds to ASD null Kähler g [47].
The details are similar to the scalar-flat Kähler case with the additional complication arising from the fact that the null-Kähler condition does not completely fix the conformal structure. A parallel spinor $\iota_{A'}$ which defines the null-Kähler structure gives rise to a section $\pi \cdot \iota$. Conversely, admitting such a section is equivalent to having a solution of the twistor equation:[2]

$$\nabla_{A(A'}\iota_{B')} = 0. \tag{10.5.103}$$

Therefore

$$\nabla_{AA'}\iota_{B'} = \varepsilon_{A'B'}\alpha_A \tag{10.5.104}$$

for some α_A. Choose a representative in $[g]$ with $R = 0$. Contracting (10.5.104) with $\nabla^A{}_{C'}$ and using the spinor identity (9.2.26) gives

$$\nabla^A{}_{C'}\nabla_{AA'}\iota_{B'} = C_{A'B'C'D'}\iota^{D'} - \frac{1}{12}R\varepsilon_{C'(B'}\iota_{A')} = 0 = \varepsilon_{A'B'}\nabla^A{}_{C'}\alpha_A,$$

so α_A is a solution to the neutrino equation. It can be written in terms of a potential

$$\alpha_A = \iota^{A'}\nabla_{AA'}\phi \tag{10.5.105}$$

because the integrability conditions $\iota^{A'}\iota^{B'}\nabla^A{}_{A'}\alpha_A = \alpha_A\iota^{A'}\nabla^A{}_{A'}\iota^{B'}$ are satisfied. Here ϕ is a function which satisfies

$$\nabla^a\nabla_a\phi + \nabla_a\phi\nabla^a\phi = 0 \tag{10.5.106}$$

as a consequence of the neutrino equation. Consider a conformal rescaling

$$\hat{g} = \Omega^2 g, \quad \hat{\varepsilon}_{A'B'} = \Omega\varepsilon_{A'B'}, \quad \hat{\iota}_{A'} = \Omega\iota_{A'}, \quad \hat{\iota}^{A'} = \iota^{A'}, \quad \text{and}$$
$$\hat{R} = R + \frac{1}{4}\Omega^{-1}\Box\Omega.$$

The twistor equation (10.5.103) is conformally invariant as $\hat{\nabla}_A^{(A'}\hat{\iota}^{B')} = \Omega^{-1}\nabla_A{}^{(A'}\iota^{B')} = 0$. Choose $\Omega \in \ker\Box$ so that $\hat{R} = 0$. Let $\Upsilon_a = \Omega^{-1}\nabla_a\Omega$. Then

$$\hat{\nabla}_{AA'}\hat{\iota}^{B'} = \nabla_{AA'}\iota^{B'} + \varepsilon_{A'}{}^{B'}\Upsilon_{AC'}\iota^{C'} = \varepsilon_{A'}{}^{B'}\left[\iota^{C'}\nabla_{AC'}(\phi + \ln\Omega)\right],$$

[2] Lorentzian metrics admitting a solution to this equation have been found in [105].

where we used (10.5.104) and (10.5.105). Notice that, as a consequence of (10.5.106), $e^{\phi} \in \ker \Box$ and we can choose $\ln\Omega = -\phi$, and

$$\hat{\nabla}_{AA'}\hat{\iota}^{B'} = 0. \qquad (10.5.107)$$

We can still use the residual gauge freedom and add to ϕ an arbitrary function constant along $\iota^{A'}e_{AA'}$. This means (10.5.107) is invariant under a conformal rescaling by functions constant along the leaves of the congruence defined by $\hat{\iota}^{A'}$. Such conformal transformations do not change $\hat{R} = 0$.

- A holomorphic fibration $\mu : \mathcal{PT} \to \mathbb{CP}^1$ together with the non-degenerate $\mathcal{O}(2)$-valued two-form on each fibre of μ (where $\mathcal{O}(2)$ denotes the pull-back bundle from the base of μ) corresponds to a hyper-Kähler metric [10, 79, 131]. (We shall present a proof of this result in next section.)
- A holomorphic one-form τ and a holomorphic three-form ρ such that $\tau \wedge d\tau = 2\Lambda\rho$ and τ is non-zero when contracted with any vector tangent to a twistor curve correspond to an Einstein metric with non-zero cosmological constant Λ [79, 170].

10.5.2 ASD Ricci-flat metrics

Below we shall give the details of the correspondence in the ASD Ricci-flat (hyper-Kähler) case which is relevant to gravitational instantons and the heavenly equations.

Theorem 10.5.5 (Penrose [131]) *There is a one-to-one correspondence between solutions (M, g) to the ASD Ricci-flat equations and three-dimensional complex manifolds $\mathcal{PT}$ with the following structures:*

1. *A projection $\mu : \mathcal{PT} \longrightarrow \mathbb{CP}^1$*
2. *A four-parameter family of sections of μ with normal bundle $\mathcal{O}(1) \oplus \mathcal{O}(1)$*
3. *A non-degenerate two-form Σ on the fibres of μ, with values in the pull-back from $\mathbb{CP}^1$ of $\mathcal{O}(2)$*

Proof Given an ASD Ricci-flat metric g construct the twistor space $\mathcal{PT}$ corresponding to the conformal structure $[g]$ as in Theorem 10.5.3. There exists a covariantly constant spin-frame on $\mathbb{S}'$ (see Theorem 9.3.3) so the equation

$$\nabla_{AA'}\pi_{B'} = 0$$

has solutions. This gives a holomorphic fibration $\mu : \mathcal{PT} \longrightarrow \mathbb{CP}^1$. The base space of this fibration has a one-form $\tau = \varepsilon_{A'B'}\pi^{A'}d\pi^{B'}$, where $\varepsilon_{A'B'}$ is related to the metric g by (9.2.16). Corollary 9.3.4 guarantees the existence of a basis $\Sigma^{A'B'}$ of closed SD two-forms. Let

$$\Sigma(\lambda) = \pi_{A'}\pi_{B'}\Sigma^{A'B'}$$

be a two-form on $\mathbb{S}'$. It is homogeneous of degree two in $\pi_{A'}$, and (for each fixed value of $\pi_{A'}$) it Lie derives along the twistor distribution (10.1.1), which follows because $\Sigma = \nu(L_0, L_1, \ldots, \ldots)$ and $d\Sigma^{A'B'} = 0$. Thus Σ descends to an $\mathcal{O}(2)$-valued two-form on each fibre of μ.

Conversely, consider the twistor space satisfying the conditions of Theorem 10.5.5. To fix a conformal factor leading to a metric $g \in [g]$ in the ASD conformal class it is sufficient to determine $g(U, V)$ for all null vectors U and V. This determines g on all vectors by bilinearity. Define homogeneous coordinates on $\mathcal{PT}$. These are coordinates on $\mathcal{T}$, the total space of the tautological line bundle $\mathcal{O}(-1)$ pulled back from $\mathbb{CP}^1$ to $\mathcal{PT}$. Let $\pi_{A'}$ be homogeneous coordinates on $\mathbb{CP}^1$ pulled back to $\mathcal{T}$ and let ω^A be local coordinates on $\mathcal{T}$ chosen on a neighbourhood of the fibre $\mu^{-1}\{\pi_{0'} = 0\}$ that are homogeneous of degree one and canonical so that $\Sigma = \varepsilon_{AB} d\omega^A \wedge d\omega^B$. Similarly $\tilde{\omega}^A$ are local coordinates on $\mathcal{T}$ on a neighbourhood of the fibre $\mu^{-1}\{\pi_{1'} = 0\}$.

The section

$$\hat{U} = U^{AA'}\pi_{A'}\frac{\partial}{\partial\omega^A}$$

of the normal bundle N corresponding to a null vector U vanishes at exactly one point in $\mathbb{CP}^1$. Let the sections $U^{AA'}\pi_{A'}$ and $V^{AA'}\pi_{A'}$ of N vanish on $\mathbb{CP}^1$ at points represented by spinors $u^{A'}$ and $v^{A'}$, respectively. Define

$$g(U, V) = \frac{(u \cdot v)\, \Sigma(\hat{U}, \hat{V})}{(\pi \cdot u)(\pi \cdot v)}, \tag{10.5.108}$$

where $u \cdot v = \varepsilon_{A'B'} u^{A'} v^{B'}$. The RHS is homogeneous of degree zero in $\pi_{A'}$ and is defined everywhere on $\mathbb{CP}^1$. It is therefore independent on $\pi_{A'}$ by the Liouville theorem (Theorem B.0.4) and defines a metric on M. To see that the metric is Ricci-flat pull-back Σ to $\mathbb{S}'$ where it satisfies

$$\Sigma(\lambda) \wedge \Sigma(\lambda) = 0 \quad \text{and} \quad d_h \Sigma(\lambda) = 0, \tag{10.5.109}$$

where in the exterior derivative d_h, $\pi_{A'}$ is understood to be held constant. The globality conditions give $\pi_{A'}\pi_{B'}\Sigma^{A'B'}$ for some SD two-forms $(\Sigma^{0'0'}, \Sigma^{0'1'}, \Sigma^{1'1'})$ on M which therefore satisfy

$$d\Sigma^{A'B'} = 0 \quad \text{and} \quad \Sigma^{(A'B'} \wedge \Sigma^{C'D')} = 0.$$

This is the hyper-Kähler condition (9.3.33) with (9.3.29). □

In the AHS picture of Theorem 10.5.4 the three complex structures I_j, $j = 1, 2, 3$, on M give a sphere of complex structures:

$$I_\lambda = u_1 I_1 + u_2 I_2 + u_3 I_3,$$

where $\mathbf{u} = (u_1, u_2, u_3)$ is the unit vector related to $\lambda \in \mathbb{CP}^1$ by stereographic projection

$$(u_1, u_2, u_3) = \left(\frac{1 - |\lambda|^2}{1 + |\lambda|^2}, \frac{\lambda + \overline{\lambda}}{1 + |\lambda|^2}, i\frac{\lambda - \overline{\lambda}}{1 + |\lambda|^2} \right).$$

The complex structure on the six-dimensional real manifold $P(\mathbb{S}')$ is $\mathbf{I} = (I_\lambda, I_0)$ where I_0 is rotation by $90°$ on each tangent space $T_\lambda \mathbb{CP}^1 = \mathbb{R}^2$.

10.5.2.1 *Deformation theory*

Here we shall describe one way of obtaining complex three-manifolds satisfying the assumptions of Theorems 10.5.3 and 10.5.5.

Cover $\mathcal{PT}$ by two sets, U and $\tilde{U}$ with $|\lambda| < 1 + \epsilon$ on U and $|\lambda| > 1 - \epsilon$ on $\tilde{U}$ with (ω^A, λ) coordinates on U and $(\tilde{\omega}^A, \lambda^{-1})$ on $\tilde{U}$. The twistor space $\mathcal{PT}$ is then determined by the transition functions

$$\tilde{\omega}^B = \tilde{\omega}^B(\omega^A, \pi_{A'}) \tag{10.5.110}$$

on $U \cap \tilde{U}$ which preserves the fibrewise two-form, $d\omega^A \wedge d\omega_A|_{\lambda=\text{const}} = d\tilde{\omega}^A \wedge d\tilde{\omega}_A|_{\lambda=\text{const}}$. To obtain a non-trivial transition function we can deform the patching of twistor space $\mathbb{CP}^3 - \mathbb{CP}^1$ from Definition 7.2.1 which corresponds to the flat conformal structure. The Kodaira theorem (Theorem B.3.1) guarantees that the deformations preserve the four-parameter family of curves, and thus the deformed twistor space still gives rise to a four-dimensional manifold M.

Infinitesimal deformations are given by elements of $H^1(\mathcal{PT}, \mathbf{\Theta})$, where $\mathbf{\Theta}$ denotes the space (strictly speaking the sheaf of germs [83, 175]) of holomorphic vector fields. Let

$$Y = f^A(\omega^B, \pi_{B'}) \frac{\partial}{\partial \omega^A}$$

be a vector field on the overlap $U \cap \tilde{U}$ defining a class in $H^1(\mathcal{PT}, \mathbf{\Theta})$ that preserves the fibration $\mathcal{PT} \mapsto \mathbb{CP}^1$. The corresponding infinitesimal deformation is given by

$$\tilde{\omega}^A(\omega^A, \pi_{A'}, t) = (1 + tY)(\omega^A) + O(t^2). \tag{10.5.111}$$

From the globallity of $\Sigma(\lambda) = d\omega^A \wedge d\omega_A$ it follows that Y is a Hamiltonian vector field with a Hamiltonian $f \in H^1(\mathcal{PT}, \mathcal{O}(2))$ with respect to the symplectic structure Σ. A finite deformation is given by integrating

$$\frac{d\tilde{\omega}^B}{dt} = \varepsilon^{AB} \frac{\partial f}{\partial \tilde{\omega}^A} \tag{10.5.112}$$

from $t = 0$ to 1.

10.5.2.2 *Heavenly equations*

The heavenly equations (10.2.18) and (10.2.22) arise from choosing a special parameterization of rational curves in the twistor space. Below we shall focus on the second heavenly form.

Choose a constant spinor $o_{A'} \in \mathbb{CP}^1$. Pull back the twistor coordinates to the correspondence space $\mathcal{F}$ and define four coordinates on M by

$$x^{AA'} := \left.\frac{\partial \omega^A}{\partial \pi_{A'}}\right|_{\pi_{A'}=o_{A'}} = \begin{pmatrix} y & w \\ -x & z \end{pmatrix}$$

where the derivative is along the fibres of $\mathcal{F}$ over M. Thus the curve $L_p \subset \mathcal{PT}$ corresponding to $p \in M$ is parameterized by choosing a two-dimensional fibre of $\mu : \mathcal{PT} \to \mathbb{CP}^1$ and defining $x^{A1'} = (w, z)$ to be the coordinates of the initial point of the curve, and $x^{A0'} = (y, -x)$ to be the tangent vector to the curve.

This can alternatively be expressed in affine coordinates on $\mathbb{CP}^1$ by expanding the coordinates ω^A pulled back to $\mathcal{F}$ in powers of $\lambda = \pi_{0'}/\pi_{1'}$. Set $P = \omega^0/\pi_{1'}$ and $Q = \omega^1/\pi_{1'}$. Then

$$P = w + \lambda y + p_2 \lambda^2 + p_3 \lambda^3 + \cdots \quad \text{and} \qquad (10.5.113)$$
$$Q = z - \lambda x + q_2 \lambda^2 + q_3 \lambda^3 + \cdots,$$

where p_i and q_i are functions of $x^{AA'}$. The symplectic two-form Σ on the fibres of μ, when pulled back to the spin bundle, has an expansion in powers of λ that truncates at order three by globality and homogeneity, so that

$$\Sigma(\lambda) = (\pi_{1'})^2\, dP \wedge dQ = \pi_{A'}\pi_{B'}\Sigma^{A'B'},$$

where $\Sigma^{A'B'}$ are SD two-forms on M and the relations (10.5.109) hold.

If we express the forms in terms of $x^{AA'}$, the closure condition is satisfied identically, whereas the truncation condition will give rise to equations on the p_i, q_i allowing one to express them in terms of a function $\Theta(x^{AA'})$ and to field equations on Θ as follows: To deduce the existence of Θ observe that the vanishing of the coefficient of λ^3 in Σ gives

$$\begin{aligned} 0 &= dw \wedge dq_3 + dy \wedge dq_2 - dz \wedge dp_3 + dx \wedge dp_2 \\ &= -d(q_3 dw + q_2 dy - p_3 dz + p_2 dx). \end{aligned}$$

Therefore locally there exists a function $\Theta(w, z, x, y)$ such that

$$p_2 = -\frac{\partial \Theta}{\partial x}, \quad p_3 = \frac{\partial \Theta}{\partial z}, \quad q_2 = -\frac{\partial \Theta}{\partial y}, \quad \text{and} \quad q_3 = -\frac{\partial \Theta}{\partial w}. \qquad (10.5.114)$$

Now

$$\Sigma = dw \wedge dz + \lambda(dx \wedge dw + dy \wedge dz) + \lambda^2 \left[dx \wedge dy - dw \wedge d(\Theta_y) + dz \wedge d(\Theta_x)\right],$$

and the second heavenly equation (10.2.22) arises from the vanishing of the coefficient of λ^4 in $\Sigma \wedge \Sigma$.

- **Example.** This example is modified from [83] to allow the application of the heavenly formalism. Take the Hamiltonian $f \in H^1(\mathcal{PT}, \mathcal{O}(2))$ defining the deformation (10.5.112) to be

$$f = \frac{(\omega^0)^4}{4\pi_{0'}\pi_{1'}}.$$

The deformation equations (10.5.112)

$$\frac{d\omega^0}{dt} = 0 \quad \text{and} \quad \frac{d\omega^1}{dt} = \frac{(\omega^0)^3}{\pi_{0'}\pi_{1'}}$$

integrate to

$$\tilde{\omega}^0 = \omega^0 \quad \text{and} \quad \tilde{\omega}^1 = \omega^1 + t\frac{(\omega^0)^3}{\pi_{0'}\pi_{1'}}.$$

Therefore ω^0 gives a global holomorphic function on $\mathbb{CP}^1$ homogeneous of degree one, so by the Liouville theorem (Theorem B.2.4) it must be linear, that is,

$$\omega^0 = \pi_{0'}y + \pi_{1'}w$$

for some complex numbers (y, w). Substituting this in the formula for $\tilde{\omega}^1$ gives an expression homogeneous of degree one:

$$\tilde{\omega}^1 - \frac{(\pi_{1'})^2}{\pi_{0'}}tw^3 - 3\pi_{1'}tw^2y - 3\pi_{0'}twy^2 = \omega^1 + \frac{(\pi_{0'})^2}{\pi_{1'}}ty^3.$$

The LHS is holomorphic around $\pi_{A'} = (1, 0)$ and the RHS is holomorphic around $\pi_{A'} = (0, 1)$. Thus, again applying the Liouville theorem, we deduce that this expression defines a linear function, say $\pi_{1'}z - \pi_{0'}x$, for some complex numbers (z, x). Rearranging gives

$$\omega^1 = \pi_{1'}z - \pi_{0'}x - \frac{(\pi_{0'})^2}{\pi_{1'}}ty^3.$$

The four complex numbers (w, z, x, y) parameterize the family of curves, and serve as local coordinates on the ASD Ricci-flat four-manifold M.

We can now read off the functions (P, Q) from the formula (10.5.113):

$$P = w + \lambda y \quad \text{and} \quad Q = z - \lambda x - \lambda^2 ty^3. \qquad (10.5.115)$$

Comparing this with (10.5.114) gives the second heavenly potential

$$\Theta = \frac{ty^4}{4}.$$

Finally the ASD Ricci-flat metric (10.2.13) is

$$g = dwdx + dzdy - 3ty^2dw^2. \qquad (10.5.116)$$

The ASD Weyl curvature in the second heavenly formalism is given by

$$C_{ABCD} = \frac{\partial^4\Theta}{\partial x^A \partial x^B \partial x^C \partial x^D}, \qquad (10.5.117)$$

where $x^A = (y, -x)$. Therefore our simple example is of Petrov–Penrose type N and has constant curvature – the only non-vanishing component being $C_{0000} = 6t$, where t is a constant deformation parameter.

We shall re-derive the expression for the metric using the Penrose's original prescription presented in the proof of Theorem 10.5.5 without referring to the heavenly formalism. In this approach the conformal structure on the moduli space of lines parameterized by $x^a = (w, z, x, y)$ is calculated by determining the quadratic condition for a section of the normal bundle to a twistor line to vanish. The sections of the normal bundle to the curve (10.5.115) correspond to tangent vectors and sections with one zero will determine null vectors and therefore the conformal structure. Now take the variation of $P(\lambda)$ and $Q(\lambda)$ for a small change δx^a to obtain

$$\delta P = \delta w + \lambda \delta y = 0 \quad \text{and} \quad \delta Q = \delta z - \lambda \delta x - 3\lambda^2 t y^2 \delta y = 0.$$

Substituting $\lambda = -\delta w/\delta y$ from the first expression to the second and multiplying the resulting expression by δy we find that the conformal structure is represented by the metric (10.5.116). The conformal factor now needs to be determined from (10.5.108) to ensure that the resulting metric is Ricci-flat. We find that this conformal factor is a constant in agreement with the calculation based on the heavenly formalism.

- **Example.** The second heavenly equation (10.2.22) with $\Theta_z = 0$ can be expressed as

$$d\Theta_x \wedge dx \wedge dy + dw \wedge d\Theta_x \wedge d\Theta_y = 0. \qquad (10.5.118)$$

Introduce $p := \Theta_x$ and perform a Legendre transform

$$F(p, y, w) := px(w, y, p) - \Theta(w, y, x(w, y, p)).$$

Then $x = F_p$, $\Theta_y = -F_y$ and (10.5.118) yields the wave equation [65] (which in Riemannian signature is the Laplace equation on $\mathbb{R}^3$):

$$F_{pw} + F_{yy} = 0. \qquad (10.5.119)$$

Implicit differentiation gives

$$\Theta_{yy} = -F_{yy} + \frac{F_{py}}{F_{pp}}, \quad \Theta_{xy} = -\frac{F_{py}}{F_{pp}}, \quad \text{and} \quad \Theta_{xx} = \frac{1}{F_{pp}},$$

and so (with the help of (10.2.13) and (10.5.119))

$$g = F_{pp}(\frac{1}{4}dy^2 + dwdp) - \frac{1}{F_{pp}}(dz - \frac{F_{pp}}{2}dy + F_{py}dw)^2$$
$$= V(\frac{1}{4}dy^2 + dwdp) - V^{-1}(dz + A)^2, \qquad (10.5.120)$$

where $V = F_{pp}$ and $A = F_{py}dw - (F_{pp}/2)dy$ satisfy the monopole equation (9.4.36) which follows from (10.5.119). Thus (10.5.120) is the complexified Gibbons–Hawking metric (9.4.35).

The twistor description is as follows: The vanishing of Θ_z implies that the whole series (10.5.113) for ω^0 truncates at second order. Thus the twistor space admits a global holomorphic function of degree two given by $\pi_{1'}\omega^0$ (i.e. $\mathcal{PT}$ fibres holomorphicaly over the total space of the line bundle $\mathcal{O}(2)$), and this is the Hamiltonian with respect to Σ, for the holomorphic vector field corresponding to the tri-holomorphic Killing field $\partial_z = K^{AA'}\partial_{AA'}$ on M. Conversely, given a tri-holomorphic symmetry, the tri-holomorphicity condition means that its lift to the spin bundle M is horizontal and so on twistor space, the corresponding holomorphic vector field is tangent to the fibres of μ. It also preserves Σ and so is Hamiltonian with Hamiltonian given by a homogeneity degree-two global function. We can choose ω^0 to be this preferred section divided by $\pi_{1'}$ so that the series for ω^0 terminates after λ^2.

Substituting the Legendre transform into (10.5.113) yields

$$P = w + \lambda y - \lambda^2 p \quad \text{and}$$
$$Q = z - \lambda F_p + \lambda^2 F_y + \lambda^3 F_w + \cdots,$$

where $F = F(w, y, p)$. With the definition $\Sigma = d\omega^0 \wedge d\omega^1|_{\lambda=\text{const}}$ (10.5.119) follows from $\Sigma \wedge \Sigma = 0$. The basis of SD two-forms can be read off from $\Sigma = \Sigma^{A'B'}\pi_{A'}\pi_{B'}$:

$$\Sigma^{0'0'} = -dz \wedge dp + dy \wedge dF_p - dw \wedge dF_y, \quad \Sigma^{0'1'} = dz \wedge dy + dw \wedge dF_p,$$
$$\text{and} \quad \Sigma^{1'1'} = dz \wedge dw,$$

and these determine the metric above.

This example is a starting point to constructing ASD Ricci-flat metric without Killing vectors. One assumes that the twistor space admits a holomorphic fibration over a total space of the line bundle $\mathcal{O}(2k)$ for some $k > 2$. In this case the series for ω^0 truncates after $2k + 1$ terms, and the function F in the series for ω^1 is a solution of a system of overdetermined but consistent PDEs generalizing the Laplace equation. See [14, 49, 106] for details.

10.5.2.3 *Recursion operator and twistor functions*

Given a solution $\phi_0 \in \mathcal{W}_g$ to the background-coupled wave equation (10.2.24), define, for $i \in \mathbb{Z}$, a hierarchy of linear fields, $\phi_i \equiv R^i\phi_0$. Define a function on the correspondence space by $\Psi = \sum_{-\infty}^{\infty} \phi_i \lambda^i$ and observe that the recursion equations (10.2.27) are equivalent to $L_A\Psi = 0$. Thus Ψ is a function on the twistor space $\mathcal{PT}$. Conversely every solution of $L_A\Psi = 0$ defined on a neighbourhood of $|\lambda| = 1$ can be expanded in a Laurent series in λ with the coefficients forming a series of elements of $\mathcal{W}_g$ related by the recursion operator. It is clear that a series corresponding to $R\phi_0$ is the function $\lambda^{-1}\Psi$, thus we define $R\Psi = \Psi/\lambda$.

We can in this way build coordinate charts on twistor space from those on M arising from the choices in the heavenly equations. Put $\omega_0^A = w^A = (w, z)$; the surfaces of constant ω_0^A are twistor surfaces. We have that $\nabla^A{}_{0'}\omega_0^B = 0$ so that in particular $\nabla_{A1'}\nabla^A{}_{0'}\omega_0^B = 0$ and if we define $\omega_i^A = R^i\omega_0^A$ then we can choose $\omega_i^A = 0$ for negative i. We define

$$\omega^A = \sum_{i=0}^{\infty} \omega_i^A \lambda^i. \tag{10.5.121}$$

We can similarly define $\tilde{\omega}^A$ by $\tilde{\omega}_0^A = \tilde{w}^A$ and choose $\tilde{\omega}_i^A = 0$ for $i > 0$. Note that ω^A and $\tilde{\omega}^A$ are solutions of L_A holomorphic around $\lambda = 0$ and $\lambda = \infty$, respectively, and they can be chosen so that they extend to a neighbourhood of the unit disc and a neighbourhood of the complement of the unit disc and can therefore be used to provide a patching description (10.5.110) of the twistor space.

The recursion operator acts on linearized perturbations of the ASD Ricci-flat equations. Under the twistor correspondence, these correspond to linearized holomorphic deformations of (part of) $\mathcal{PT}$. Consider the infinitesimal version of (10.5.112) given by

$$\delta\tilde{\omega}^A = \frac{\partial \delta f}{\partial \tilde{\omega}_A}. \tag{10.5.122}$$

If the ASD Ricci-flat metric is determined by a solution Θ to the second heavenly equation (10.2.22) then δf is a linearized deformation of the twistor space corresponding to $\delta\Theta \in \mathcal{W}_g$. The recursion operator acts on linearized deformations as follows:

Proposition 10.5.6 *Let R be the recursion operator defined by* (10.2.27). *Its twistor counterpart is the multiplication operator*

$$R\,\delta f = \frac{\pi_{1'}}{\pi_{0'}}\delta f = \lambda^{-1}\delta f. \tag{10.5.123}$$

(Note that R acts on δf without ambiguity; the ambiguity in boundary condition for the definition of R on space-time is absorbed into the choice of explicit representative for the cohomology class determined by δf.)

Proof Pull back δf to the primed spin bundle $\mathbb{S}'$ on which it is a coboundary so that

$$\delta f(\pi_{A'}, x^a) = h(\pi_{A'}, x^a) - \tilde{h}(\pi_{A'}, x^a), \tag{10.5.124}$$

where h and $\tilde{h}$ are holomorphic on U and $\tilde{U}$, respectively (here we abuse notation and denote by U and $\tilde{U}$ the open sets on the spin bundle that are the preimage of U and $\tilde{U}$ on twistor space). A choice for the splitting (10.5.124) is given by (compare (B8))

$$h = \frac{1}{2\pi i}\oint_\Gamma \frac{(\pi^{A'}o_{A'})^3}{(\rho^{C'}\pi_{C'})(\rho^{B'}o_{B'})^3}\delta f(\rho_{E'})\rho_{D'}d\rho^{D'} \quad \text{and} \tag{10.5.125}$$

$$\tilde{h} = \frac{1}{2\pi i}\oint_{\tilde{\Gamma}} \frac{(\pi^{A'}o_{A'})^3}{(\rho^{C'}\pi_{C'})(\rho^{B'}o_{B'})^3}\delta f(\rho_{E'})\rho_{D'}d\rho^{D'}.$$

Here $\rho_{A'}$ are homogeneous coordinates of $\mathbb{CP}^1$ pulled back to the spin bundle. The contours Γ and $\tilde{\Gamma}$ are homologous to the equator of $\mathbb{CP}^1$ in $U \cap \tilde{U}$ and are such that $\Gamma - \tilde{\Gamma}$ surrounds the point $\rho_{A'} = \pi_{A'}$.

The functions h and $\tilde{h}$ are homogeneous of degree two in $\pi_{A'}$ and do not descend to $\mathcal{PT}$, whereas their difference does so that

$$\pi^{A'}\nabla_{AA'}h = \pi^{A'}\nabla_{AA'}\tilde{h} = \pi^{A'}\pi^{B'}\pi^{C'}\Sigma_{AA'B'C'}, \tag{10.5.126}$$

where the first equality shows that the LHS is global with homogeneity degree three and implies the second equality for some $\Sigma_{AA'B'C'}$ which will be the third potential for a linearized ASD Weyl spinor. $\Sigma_{AA'B'C'}$ is in general defined modulo terms of the form $\nabla_{A(A'}\gamma_{B'C')}$ but this gauge freedom is partially fixed by choosing the integral representation above; h vanishes to third order at $\pi_{A'} = o_{A'}$ and direct differentiation, using $\nabla_{AA'}\delta f = \rho_{A'}\delta f_A$ for some δf_A, gives $\Sigma_{AA'B'C'} = o_{A'}o_{B'}o_{C'}\nabla_{A0'}\delta\Theta$ where

$$\delta\Theta = \frac{1}{2\pi i}\oint_\Gamma \frac{\delta f}{(\rho^{B'}o_{B'})^4}\rho_{D'}d\rho^{D'}. \tag{10.5.127}$$

This is consistent with the Plebański gauge choices leading to (10.2.22). The condition

$$\nabla_{A(D'}\Sigma^{A}{}_{A'B'C')} = 0$$

follows from (10.5.126) which, with the Plebański gauge choice, implies $\delta\Theta \in \mathcal{W}_g$. Thus we obtain a twistor integral formula for the linearization of the second heavenly equation.

Now recall formula (10.2.27) defining R. Let $R\delta f$ be the twistor function corresponding to $R\delta\Theta$ by (10.5.127). The recursion relations yield

$$\oint_\Gamma \frac{R\delta f_A}{(\rho^{B'}o_{B'})^3}\rho_{D'}d\rho^{D'} = \oint_\Gamma \frac{\delta f_A}{(\rho^{B'}o_{B'})^2(\rho^{B'}\iota_{B'})}\rho_{D'}d\rho^{D'}$$

so $R\delta f = \lambda^{-1}\delta f$. □

10.5.2.4 *Hidden symmetry algebra*

The ASD Ricci-flat equations in the Plebański forms (10.2.18) or (10.2.22) have a residual coordinate symmetry. This consists of area-preserving diffeomorphisms in the w^A coordinates together with some extra transformations that depend on whether one is reducing to the first or second form. By regarding the infinitesimal forms of these transformations as linearized perturbations and acting on them using the recursion operator, the coordinate (passive) symmetries can be extended to give 'hidden' (active) symmetries of the heavenly equations. Formulae (10.5.127) and (10.5.123) can be used to recover the relations of the hidden symmetry algebra of the heavenly equations.

Let V be a volume-preserving vector field on M. Define $\delta^0_V \mathrm{e}_{AA'} := [V, \mathrm{e}_{AA'}]$, where $\mathrm{e}_{AA'}$ is a null tetrad of the metric. This is a pure gauge transformation corresponding to the addition of $\mathrm{Lie}_V g$ to the space-time metric and preserves the field equations. Note that

$$[\delta^0_V, \delta^0_W]\mathrm{e}_{AA'} := \delta^0_{[V,W]}\mathrm{e}_{AA'}.$$

Once a Plebański coordinate system and reduced equations have been obtained, the reduced equation will not be invariant under all the SDiff(M) transformations, where SDiff(M) is the group of volume-preserving diffeomorphisms of M. The second form (10.2.22) will be preserved if we restrict ourselves to transformations which preserve the SD two-forms $\Sigma^{1'1'} = dw_A \wedge dw^A$ and $\Sigma^{0'1'} = dx_A \wedge dw^A$. The conditions $\mathrm{Lie}_V\Sigma^{1'1'} = \mathrm{Lie}_V\Sigma^{0'1'} = 0$ imply that V is given by

$$V = \frac{\partial h}{\partial w_A}\frac{\partial}{\partial w^A} + \left(\frac{\partial k}{\partial w_A} - x^B\frac{\partial^2 h}{\partial w_A \partial w^B}\right)\frac{\partial}{\partial x^A},$$

where $h = h(w^A)$ and $k = k(w^A)$. The four-manifold M is now viewed as a cotangent bundle $M = T^*\Sigma^2$ with w^A being coordinates on a two-dimensional complex manifold Σ^2. The full SDiff(M) symmetry breaks down to the semi-direct product of SDiff(Σ^2), which acts on M by a Lie lift, with $\Gamma(\Sigma^2, \mathcal{O})$ which acts on M by translations of the zero section by the exterior derivatives of functions on Σ^2. Let $\delta_V\Theta$ correspond to $\delta^0_V\mathrm{e}_{AA'}$ by

$$\delta^0_V \mathrm{e}_{A1'} = \frac{\partial^2 \delta_V\Theta}{\partial x^A \partial x^B}\frac{\partial}{\partial x_B}.$$

This gives the pure gauge elements in the second heavenly equation. These symmetries take a solution to an equivalent solution. The recursion operator can be used to define an algebra of 'hidden symmetries' that take one solution to a different one as follows: Let $\delta_V^0\Theta$ be a pure gauge which also satisfies $\Box_g \delta_V^0\Theta = 0$. We set

$$\delta_V{}^i\Theta := R^i\delta_V\Theta \in \mathcal{W}_g.$$

Proposition 10.5.7 *Generators of the hidden symmetry algebra of the second heavenly equation satisfy the relation*

$$[\delta_V{}^i, \delta_W{}^j] = \delta_{[V.W]}{}^{i+j}. \tag{10.5.128}$$

Proof Let $\delta_V^i f$ be the twistor function corresponding to $\delta_V^i\Theta$ (by (10.5.127)) treated as an element of $\Gamma(U \cap \tilde{U}, \mathcal{O}(2))$ rather than $H^1(\mathcal{PT}, \mathcal{O}(2))$. Define $[\delta_V^i, \delta_W^j]$ by

$$[\delta_V^i, \delta_W^j]\Theta := \frac{1}{2\pi i}\oint \frac{\{\delta_V^i f, \delta_W^j f\}}{(\pi_{0'})^4}\pi_{A'}d\pi^{A'},$$

where the Poisson bracket is calculated with respect to a canonical Poisson structure on $\mathcal{PT}$. From Proposition 10.5.123 it follows that

$$[\delta_V^i, \delta_W^j]\Theta = \frac{1}{2\pi i}\oint \lambda^{-i-j}\frac{\{\delta_V f, \delta_W f\}}{(\pi_{0'})^4}\pi_{A'}d\pi^{A'} = R^{i+j}\delta_{[V.W]}\Theta$$

as required. □

10.5.2.5 *Hierarchies*

The twistor space $\mathcal{PT}_n$ for a solution to the hierarchy (10.2.29) associated to the Lax system (10.2.30) on $\mathcal{N}$ is obtained by factoring the correspondence space $\mathcal{N} \times \mathbb{CP}^1$ by the twistor distribution L_{Ai}. One can repeat the steps leading to the proof of Theorem 10.5.5 to show that $\mathcal{PT}_n$ is a three-dimensional complex manifold which holomorphically fibres over $\mathbb{CP}^1$ such that the sections of this fibration have normal bundle $\mathcal{O}(n) \oplus \mathcal{O}(n)$ and there exists a non-degenerate two-form Σ on the fibres of $\mu : \mathcal{PT}_n \to \mathbb{CP}^1$, with values in the pull-back from $\mathbb{CP}^1$ of $\mathcal{O}(2n)$.

One can then find the twistor spaces for the four-dimensional hyper-Kähler slices given by $x^{Ai} = \text{const}, i \geq 2$ by taking a sequence of $n-1$ blow-ups of points in the fibre over $o_{A'} \in \mathbb{CP}^1$, the choice of point in the fibre to blow up at the $(n-i+1)$th blowup corresponding precisely to the choice of the values of x^{Ai}. See [45, 49] for details of this construction.

10.5.3 Twistor theory and symmetries

In Section 10.3 we discussed the appearance of EW structures and projective structures in the cases of a non-null and null conformal Killing vectors, respectively. In both cases there is a twistor correspondence which arises as a symmetry reduction of Theorem 10.5.3.

Given a four-dimensional holomorphic ASD conformal structure, its twistor space is the space of α-surfaces. A conformal Killing vector preserves the conformal structure, so preserves α-surfaces, giving rise to a holomorphic vector field on the twistor space. If the Killing vector is non-null then the vector field on twistor space $\mathcal{PT}$ is non-vanishing. This is because a non-null Killing vector is transverse to any α-surface. In this case one can quotient the three-dimensional twistor space by the induced vector field, and it can be shown [93] that the resulting two-dimensional complex manifold contains $\mathbb{CP}^1$'s with normal bundle $\mathcal{O}(2)$.

Theorem 10.5.8 (Hitchin [79]) *There is a one-to-one correspondence between solutions to EW equations* (10.3.41) *and two-dimensional complex manifolds admitting a three-parameter family of rational curves with normal bundle* $\mathcal{O}(2)$.

In this twistor correspondence the points of $\mathcal{W}$ correspond to rational $\mathcal{O}(2)$ curves in the complex surface $\mathcal{Z}$ and points in $\mathcal{Z}$ correspond to totally geodesics null surfaces in $\mathcal{W}$. The conformal structure $[h]$ arises as we define the null vectors at p in $\mathcal{W}$ to be the sections of the normal bundle $N(L_p)$ which vanish at some point to second order. A section of $\mathcal{O}(2)$ has the form $V^{A'B'}\pi_{A'}\pi_{B'}$, thus the vanishing condition $(V^{0'1'})^2 - V^{0'0'}V^{1'1'}$ is quadratic. To define the connection D we define a direction at $p \in \mathcal{W}$ to be a one-dimensional space of sections of $\mathcal{O}(2)$ which vanish at two points Z_1 and Z_2 in L_p. The one-dimensional family of $\mathcal{O}(2)$ curves in $\mathcal{Z}$ passing through Z_1 and Z_2 gives a geodesic curve in $\mathcal{W}$ in a given direction. In the limiting case $Z_1 = Z_2$ these geodesics are null with respect to $[h]$ in agreement with (10.3.40).

The dispersionless integrable systems described in Section 10.3.1 can be encoded in the twistor correspondence of Theorem 10.5.8 if the twistor space admits some additional structures. The coordinate equivalence classes of solutions to the $SU(\infty)$-Toda equations correspond to twistor spaces with a preferred section of $\kappa^{-1/2}$, where κ is the canonical bundle of $\mathcal{Z}$ (see [104] for details). The solutions to dKP correspond to $\mathcal{Z}$ with a preferred section of $\kappa^{-1/4}$ (see [46]). Finally the solutions to the Diff(S^1) equation correspond to $\mathcal{Z}$ which holomorphically fibre over $\mathbb{CP}^1$ (see [51]).

If the Killing vector is null then the induced vector field on the twistor space $\mathcal{PT}$ vanishes on a hypersurface. This is because at each point, the Killing vector is tangent to a single α-surface. Hence it preserves a foliation by α-surfaces, and vanishes at the hypersurface in twistor space corresponding to this foliation. However, one can show [57] that it is possible to continue the vector field on twistor space to a one-dimensional distribution $\hat{K}$ that is nowhere vanishing. Quotienting $\mathcal{PT}$ by this distribution gives a two-dimensional complex manifold $\mathcal{Z}$ containing $\mathbb{CP}^1$'s with normal bundle $\mathcal{O}(1)$, and we can make use of another result of Hitchin:

Theorem 10.5.9 (Hitchin [79]) *There is a one-to-one correspondence between two-dimensional projective structures and two-dimensional complex manifolds admitting a two- parameter family of rational curves with normal bundle* $\mathcal{O}(1)$.

In this correspondence the points of the projective structure U correspond to rational curves in a complex surface $\mathcal{Z}$, and the geodesics of the projective structure correspond to points in $\mathcal{Z}$. Two points in U are connected by a geodesic iff the corresponding rational curves in $\mathcal{Z}$ intersect at one point.

The twistorial version of the correspondence described in Section 10.3.2 is illustrated by the Figure 10.2. In M, a one parameter family of β-surfaces is shown, each of which intersects a one-parameter family of α-surfaces s, also shown. The β-surfaces correspond to a projective structure geodesic in U, shown at the bottom left.

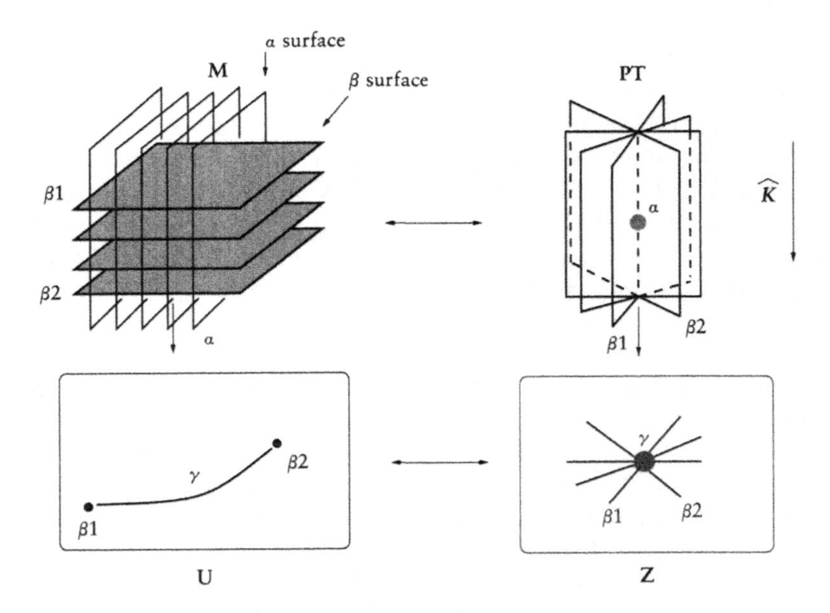

Figure 10.2 *Relationship between M, U, $\mathcal{PT}$, and $\mathcal{Z}$*

The β-surfaces in M correspond to surfaces in $\mathcal{PT}$. These surfaces intersect at the dotted line, which corresponds to the one-parameter family of α-surfaces in M. When we quotient $\mathcal{PT}$ by $\hat{K}$ to get $\mathcal{Z}$, the surfaces become twistor lines in $\mathcal{Z}$, and the dotted line becomes a point at which the twistor lines intersect, this is shown on the bottom right. This family of twistor lines intersecting at a point corresponds to the geodesic of the projective structure. See [29, 57] for the details of this correspondence.

Exercises

1. Show that the curvature scalar of a four-dimensional null-Kähler structure vanishes. [Hint: Differentiate the relation $\nabla_{AA'}\iota_{B'} = 0$ where $\iota_{A'}$ is the parallel spinor defining the null-Kähler structure.]

   Impose the ASD Ricci-flat condition on the null-Kähler structure (10.2.13) and show that the function Θ satisfies the second heavenly equation (10.2.22).

   Calculate the ASD Weyl spinor C_{ABCD} of (10.2.13) in terms of Θ and deduce that SD null-Kähler structures are given by (10.2.13) with $\Theta = \Pi^{ABC}x_A x_B x_C$ where $x_A = (x, y)$ and Π^{ABC} are arbitrary functions of (w, z).
2. Show that the vector fields

$$\tau_+ = -x^2\frac{\partial}{\partial x}, \quad \tau_- = \frac{\partial}{\partial x}, \quad \text{and} \quad \tau_0 = 2x\frac{\partial}{\partial x}$$

   generate the Lie algebra $\mathfrak{sl}(2, \mathbb{R})$. Use the KdV Lax pair (8.1.2) with the matrices replaced by vector fields to obtain solutions to the Diff(S^1) equation (10.3.53) out of solutions to the KdV.

   [Hint: Proceed by analogy with the procedure leading to (10.3.59).]
3. Show that all solutions to the Laplace equation in $\mathbb{R}^n$

$$\sum_{i=1}^{n} \frac{\partial^2 V}{\partial x^{i\,2}} = 0$$

   which are constant on central quadrics are of the form

$$V = \int \frac{dH}{\sqrt{(H-\beta_1)(H-\beta_2)\cdots(H-\beta_n)}},$$

   where

$$\sum_{i=1}^{n} \frac{x^{i\,2}}{H-\beta_i} = C,$$

   and $C, \beta_1, \beta_2, \ldots, \beta_n$ are constants.

4. Consider a deformation of the twistor space given by a Hamiltonian $f = f(\omega^0, \pi_{A'})$ homogeneous of degree two and independent of ω^1, and show that it leads to a family of ASD *pp*-waves generalizing the metric (10.5.116) and given by

$$g = dwdx + dzdy + F(w, y)dw^2,$$

where F is an arbitrary function of two variables.

5. Find the ASD Ricci-flat metric corresponding to a deformed twistor space with a deformation Hamiltonian

$$f = \frac{(\pi_{0'})^4}{\omega^0\omega^1}.$$

[Hint: Show that the deformation equations integrate to

$$\tilde{\omega}^0 = \exp[t(\pi_{0'})^4 Q^{-2}]\omega^0 \quad \text{and} \quad \tilde{\omega}^1 = \exp[-t(\pi_{0'})^4 Q^{-2}]\omega^1, \qquad (10.5.129)$$

where $Q = \omega^0\omega^1$ restricts to $\alpha_{A'}\beta_{B'}\pi^{A'}\pi^{B'}$ on each twistor curve for some $\alpha_{A'}$ and $\beta_{A'}$. Obtain the splitting $Q^{-2}(o \cdot \pi)^4 = \tilde{h} - h$ where, setting $\partial_\alpha := o^{A'}\frac{\partial}{\partial\alpha^{A'}}$ and $\partial_\beta := o^{A'}\frac{\partial}{\partial\beta^{A'}}$,

$$\tilde{h} = 2(\pi \cdot o)\partial_\alpha\partial_\beta\left[\frac{\alpha \cdot o}{(\alpha \cdot \beta)(\pi \cdot \alpha)}\right] \quad \text{and} \quad h = 2(\pi \cdot o)\partial_\alpha\partial_\beta\left[\frac{\beta \cdot o}{(\alpha \cdot \beta)(\pi \cdot \beta)}\right].$$

The twistor curves are now given by

$$\omega^0 = (\gamma \cdot \pi)e^{ht} \quad \text{and} \quad \omega^1 = (\delta \cdot \pi)e^{-ht},$$

for some spinors $\gamma_{A'}$ and $\delta_{A'}$.]

APPENDIX A

Manifolds and Topology

The first six chapters of this book are intended to give an elementary introduction to the subject and the reader is expected only to be familiar with basic real and complex analysis, algebra, and dynamics as covered in the undergraduate syllabus. In particular no knowledge of differential geometry is assumed. One obvious advantage of this approach is that the book is suitable for advanced undergraduate students.

The disadvantage is that the discussion of Hamiltonian formalism and continuous groups of transformations in earlier chapters used phrases like 'spaces coordinatized by (p, q)', 'open sets in $\mathbb{R}^n$', or 'groups whose elements smoothly depend on parameters' instead calling these object by their real name – manifolds. The first part of this appendix is intended to fill this gap. The second part of the appendix contains the discussion of homotopy groups and topological degree needed in Chapters 5–7.

Definition A.0.1 *An n-dimensional smooth manifold is a set M together with a collection of open sets U_α called the coordinate charts such that*

- *The open sets U_α labelled by a countable index α cover M.*
- *There exist one-to-one maps $\phi_\alpha : U_\alpha \to V_\alpha$ onto open sets in $\mathbb{R}^n$ such that for any pair of overlapping coordinate charts the maps*

$$\phi_\beta \circ \phi_\alpha^{-1} : \phi_\alpha(U_\alpha \cap U_\beta) \longrightarrow \phi_\beta(U_\alpha \cap U_\beta)$$

  are smooth (i.e. infinitely differentiable) functions from $\mathbb{R}^n$ to $\mathbb{R}^n$. (Figure A.1)

Thus a manifold is a topological space together with additional structure which makes local differential calculus possible. The space $\mathbb{R}^n$ itself is of course a manifold which can be covered by one coordinate chart.

- **Example.** A less trivial example is the unit sphere

$$S^n = \{\mathbf{r} \in \mathbb{R}^{n+1}, |\mathbf{r}| = 1\}.$$

  To verify that this is indeed a manifold, cover S^n by two open sets $U_1 = U$ and $U_2 = \widetilde{U}$:

$$U = S^n/\{0, \dots, 0, 1\} \quad \text{and} \quad \widetilde{U} = S^n/\{0, \dots, 0, -1\},$$

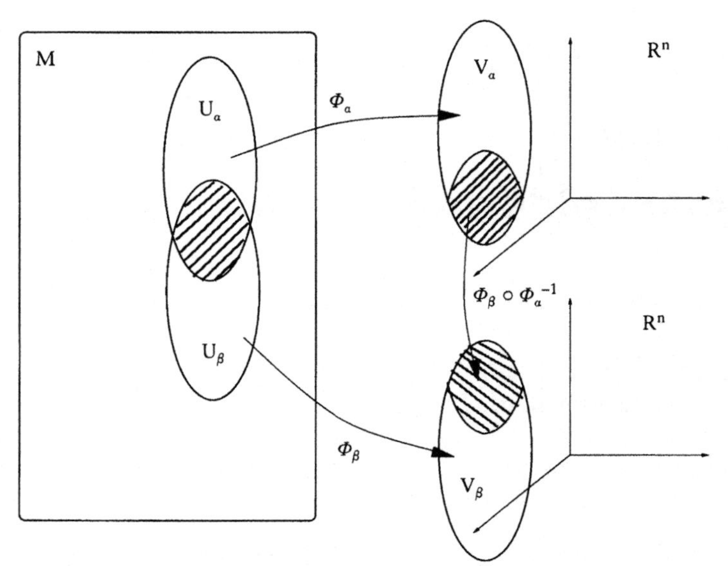

Figure A.1 *Manifold*

and define the local coordinates by stereographic projections

$$\phi(r_1, r_2, \ldots, r_{n+1}) = \left(\frac{r_1}{1 - r_{n+1}}, \ldots, \frac{r_n}{1 - r_{n+1}}\right) = (x_1, \ldots, x_n) \in \mathbb{R}^n \quad \text{and}$$

$$\tilde{\phi}(r_1, r_2, \ldots, r_{n+1}) = \left(\frac{r_1}{1 + r_{n+1}}, \ldots, \frac{r_n}{1 + r_{n+1}}\right) = (x_1, \ldots, x_n) \in \mathbb{R}^n.$$

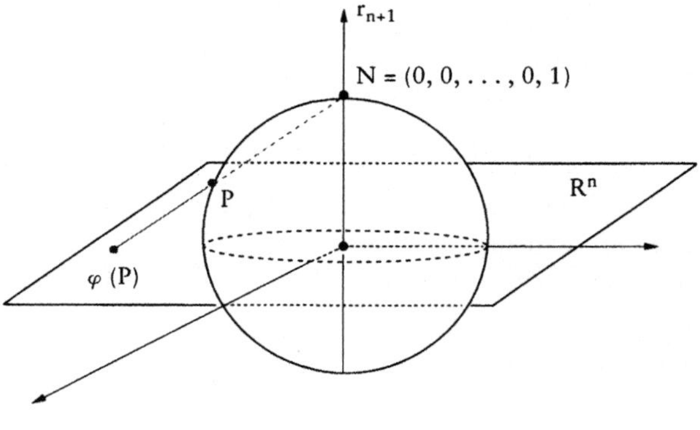

Using

$$\frac{r_k}{1 + r_{n+1}} = \left(\frac{1 - r_{n+1}}{1 + r_{n+1}}\right) \frac{r_k}{1 - r_{n+1}}, \quad k = 1, \ldots, n,$$

where $r_{n+1} \neq \pm 1$ shows that on the overlap $U \cap \tilde{U}$ the transition functions

$$\phi \circ \tilde{\phi}^{-1}(x_1, \ldots, x_n) = \left(\frac{x_1}{x_1^2 + \cdots + x_n^2}, \ldots, \frac{x_n}{x_1^2 + \cdots + x_n^2} \right)$$

are smooth.

The Cartesian product of manifolds is also a manifold. For example, the n-torus arising in the Arnold–Liouville theorem (Theorem 1.2.2) is the Cartesian product of n one-dimensional spheres.

Another way to obtain interesting manifolds is to define them as surfaces in $\mathbb{R}^n$ by a vanishing condition for a set of functions. If $f_1, \ldots, f_k : \mathbb{R}^n \to \mathbb{R}$ then the set

$$M_f := (x \in \mathbb{R}^n, \quad f_i(x) = 0, \quad i = 1, \ldots, k) \tag{A1}$$

is a manifold if the rank of the $k \times n$ matrix of gradients ∇f_i is constant in a neighborhood of M_f in $\mathbb{R}^n$. If this rank is maximal and equal to k then $\dim M_f = n - k$. The manifold axioms can be verified using the implicit function theorem. For example, the sphere S^{n-1} arises this way with $k = 1$ and $f_1 = 1 - |\mathbf{x}|^2$. There is a theorem which says that every manifold arises as some surface in $\mathbb{R}^n$ for sufficiently large n. If the manifold is m-dimensional then n is at most $2m + 1$. This useful theorem is now nearly forgotten – differential geometers like to think of manifolds as abstract objects defined by a collections of charts as in Definition A.0.1.

A map between smooth manifolds $f : M \to \tilde{M}$, where $\dim M = n$ and $\dim \tilde{M} = \tilde{n}$, is called smooth if it is smooth in local coordinates. This means that the maps

$$\tilde{\phi}_\beta \circ f \circ \phi_\alpha{}^{-1} : \mathbb{R}^n \longrightarrow \mathbb{R}^{\tilde{n}}$$

are smooth maps in the ordinary sense. Here (U_α, ϕ_α) and $(\tilde{U}_\beta, \tilde{\phi}_\beta)$ are coordinate charts for M and $\tilde{M}$, respectively.

Let $\gamma : \mathbb{R} \to M$ be a smooth curve such that

$$\gamma(0) = p \in M \quad \text{and} \quad \frac{d\gamma(\varepsilon)}{d\varepsilon}|_{\varepsilon=0} = V \in T_pM,$$

where V is a vector tangent to γ at p and the tangent space T_pM consists of all tangent vectors to all possible curves through p. If $\dim M = n$, the tangent space is an n-dimensional vector space. The collection of all tangent spaces as p varies in M is called the tangent bundle $TM = \cup_{p \in M} T_pM$. The tangent bundle is a manifold of dimension $2n$.

A smooth map f between two manifolds induces, for each $p \in M$, a smooth map between tangent spaces

$$f_* : T_pM \longrightarrow T_{f(p)}\tilde{M}$$

such that

$$f_*(V) = \frac{df(\gamma(\varepsilon))}{d\varepsilon}|_{\varepsilon=0}. \tag{A2}$$

This map is called the tangent map. It depends smoothly on a point $p \in M$ and thus it extends to the tangent bundle TM. If V^i are components of the vector field

V with respect to the natural basis $\{\partial/\partial x^i\}$ then

$$(f_* V)^i = V^j \frac{\partial y^i}{\partial x^j}.$$

The Lie derivative of a vector W along a vector V is defined as

$$\mathrm{Lie}_V W = \lim_{\epsilon \to 0} \frac{W(p) - \gamma(\epsilon)_* W(p)}{\epsilon}, \tag{A3}$$

where $\gamma(\epsilon)$ is the one-parameter group of transformations generated by V. Thus, using the Leibniz rule

$$\mathrm{Lie}_V(f) = V(f), \quad \mathrm{Lie}_V(W) = [V, W], \quad \text{and} \quad \mathrm{Lie}_V(\omega) = d(V \lrcorner\, \omega) + V \lrcorner\, (d\omega),$$

where f, W, and ω are a function, a vector field and a one-form, respectivelly, and $\lrcorner$ is a contraction of a differential form with a vector field.

A.1 Lie groups

We can now give a proper definition of a Lie group:

Definition A.1.1 *A Lie group G is a group and, at the same time, a smooth manifold such that the group operations*

$$G \times G \to G, \quad (g_1, g_2) \to g_1 g_2, \quad \textit{and} \quad G \to G, \quad g \to g^{-1}$$

are smooth maps between manifolds.

- **Example.** The general linear group $G = GL(n, \mathbb{R})$ is an open set in $\mathbb{R}^{n^2}$ defined by the condition $\det g \neq 0, g \in G$. It is therefore a Lie group of dimension n^2. The special orthogonal group $SO(n)$ is defined by (A1), where the $n(n+1)/2$ conditions in $\mathbb{R}^{n^2}$ are

$$gg^{\mathrm{T}} - \mathbf{1} = 0, \quad \det g = 1.$$

  The determinant condition just selects a connected component in the set of orthogonal matrices, so it does not count as a separate condition. It can be shown that the corresponding matrix of gradients has constant rank and thus $SO(n)$ is an $[n(n-1)/2]$-dimensional Lie group.

In Chapter 4 a Lie algebra $\mathfrak{g}$ was defined as a vector space with an antisymmetric bilinear operation which satisfies the Jacobi identity (4.2.6).

A Lie algebra of a Lie group G is the tangent space to G at the identity element, $\mathfrak{g} = T_e G$ with the Lie bracket defined by a commutator of vector fields at e. For any $g \in G$ define left translation L_g using the group multiplication

$$L_g : G \longrightarrow G \quad \text{and} \quad L_g(h) = gh.$$

The tangent mapping (A2) maps $T_e G = \mathfrak{g}$ to $T_g G$ and each element $V \in \mathfrak{g}$ corresponds to a vector field $(L_g)_* V$ on the group manifold. Theses vector fields are called left-invariant. The Lie bracket of two left-invartiant vector fields is again left-invariant as

$$[(L_g)_*(V), (L_g)_*(W)] = (L_g)_*[V, W]_{\mathfrak{g}}$$

from the properties of the tangent map (A2). (The bracket on the LHS is the Lie bracket of two vector fields. The symbol $[\,,\,]_{\mathfrak{g}}$ on the RHS is the bracket in the Lie algebra $\mathfrak{g}$.) Therefore the elements of $\mathfrak{g}$ can be represented by global vector fields on G, and Lie groups are paralizable as they globally admit $\dim(G)$ non-vanishing vector fields which are left translations of vectors in $\mathfrak{g}$.

Let $L_\alpha, \alpha = 1, \ldots, \dim \mathfrak{g}$ be a basis of left-invariant vector fields such that

$$[L_\alpha, L_\beta] = f_{\alpha\beta}^{\gamma} L_\gamma,$$

and let σ^α be the dual basis of one-forms such that $L_\alpha \lrcorner\, \sigma^\beta = \delta_\alpha^\beta$. The identity

$$d\omega(V, W) = V[\omega(W)] - W[\omega(V)] - \omega([V, W])$$

with $V = L_\alpha$, $W = L_\beta$, and $\omega = \sigma^\gamma$ gives

$$d\sigma^\alpha + \frac{1}{2} f_{\beta\gamma}^{\alpha} \sigma^\beta \wedge \sigma^\gamma = 0.$$

If G is a matrix group the one-form

$$g^{-1} dg$$

is called the Maurer–Cartan one-from on G. The Maurer–Cartan one-form is invariant under left multiplication of g by a constant group element. This one-form takes its values in $\mathfrak{g}$, as for any smooth curve $g(s)$ in G we have

$$g^{-1}(s) g(s + \varepsilon) = 1 + \varepsilon g^{-1} \frac{dg}{ds}|_{\varepsilon=0} + O(\varepsilon^2),$$

so $g^{-1}(dg/ds)$ is a tangent vector to G at g, and so it is an element of $\mathfrak{g}$. Thus we can write

$$g^{-1} dg = \sigma^\alpha T_\alpha,$$

where T_α are matrices spanning $\mathfrak{g}$ and σ^α are left-invariant one-forms on G. The metric

$$h = -\mathrm{Tr}(g^{-1} dg\, g^{-1} dg) = -\mathrm{Tr}(T_\alpha T_\beta) \sigma^\alpha \sigma^\beta$$

is the left-invariant metric on the Lie group.

The right-invariant one-forms $\tilde{\sigma}^\alpha$ are defined by

$$g^{-1} dg = T_\alpha \tilde{\sigma}^\alpha,$$

and the right-invariant vector fields R_α are defined by the duality $R_\alpha \lrcorner\, \tilde{\sigma}^\beta = \delta_\alpha^\beta$. They satisfy

$$[R_\alpha, L_\beta] = 0 \quad \text{and} \quad [R_\alpha, R_\beta] = -f_{\alpha\beta}^{\gamma} R_\gamma.$$

- **Example.** The left-invariant one-forms σ^α on the Lie group $SO(3)$ can be explicitly given in terms of Euler angles:

$$\sigma^1 = \cos\psi \, d\theta + \sin\psi \sin\theta \, d\phi, \quad \sigma^2 = -\sin\psi \, d\theta + \cos\psi \sin\theta \, d\phi, \quad \text{and}$$
$$\sigma^3 = d\psi + \cos\theta \, d\phi.$$

- **Example.** Consider the three-dimensional Heisenberg group (4.1.1) with the corresponding Lie algebra (4.1.3). We have

$$g^{-1} = \begin{pmatrix} 1 & -m_1 & -m_3 + m_1 m_2 \\ 0 & 1 & -m_2 \\ 0 & 0 & 1 \end{pmatrix}$$

so

$$\begin{aligned} g^{-1} dg &= T_\alpha \sigma^\alpha \\ &= T_1 dm_1 + T_2 dm_2 + T_3 (dm_3 - m_1 dm_2). \end{aligned}$$

This gives

$$d\sigma^1 = 0, \quad d\sigma^2 = 0, \quad \text{and} \quad d\sigma^3 = -\sigma^1 \wedge \sigma^2.$$

The left-invariant metric

$$\begin{aligned} h &= (\sigma^1)^2 + (\sigma^2)^2 + (\sigma^3)^2 \\ &= dm_1^2 + dm_2^2 + (dm_3 - m_1 dm_2)^2 \end{aligned}$$

has a Kaluza–Klein interpretation: Its geodesics, when projected to the (m_1, m_2)-plane, are trajectories of a particle moving in a uniform magnetic field $F = dm_1 \wedge dm_2$ with a potential $A = m_1 dm_2$. The first integral $\dot{m}_3 - m_1 \dot{m}_2 = \text{const}$ of the geodesic motion corresponds to charge conservation.

If G is a transformation group of some manifold X, then the elements of $\mathfrak{g}$ can also be represented by vector fields on X. If $\rho : G \times X \longrightarrow X$ then for any $V \in \mathfrak{g}$ we define $\rho(V)$ to be a vector in X by demanding that its flow coincides with a one-parameter subgroup $e^{\varepsilon V}$ of G in X. This induces a Lie algebra homomorphism form $\mathfrak{g}$ to the Lie algebra of vector fields on X, that is,

$$[\rho(V), \rho(W)] = \rho([V, W]_{\mathfrak{g}}), \qquad V, W \in \mathfrak{g}.$$

We will usually omit the reference to the map ρ, and denote the vectors in $\mathfrak{g}$ and the corresponding vector fields in TX by the same symbol.

- **Example.** The group $SO(3, \mathbb{R})$ acts on $\mathbb{R}^3$ as the group of rotations. The action $\mathbf{x} \longrightarrow A\mathbf{x}$ is infinitesimaly generated by three vector fields

$$X_c = \frac{1}{2} \varepsilon_{abc} x^a \frac{\partial}{\partial x^b}.$$

The group preserves the Euclidean distance, and so its action descends to the two-sphere $S^2 \subset \mathbb{R}^3$ given by $|\mathbf{x}| = 1$. The S^2-volume form $\omega = d(\cos\theta) \wedge d\psi$ is preserved by this action, and the corresponding vector fields are Hamiltonian, that is, $X_c \lrcorner\, \omega = -dh_c$ with Hamiltonians $h_a : S^2 \longrightarrow \mathbb{R}$:

$$h_1 = \sin\theta \sin\psi, \quad h_2 = -\sin\theta\cos\psi, \quad \text{and} \quad h_3 = \cos\theta$$

such that

$$\{h_a, h_b\} = \frac{1}{2}\varepsilon_{abc} h_c. \tag{A4}$$

Proof of the first part of Arnold–Liouville's Theorem 1.2.2. The gradients ∇f_k are independent, thus the set

$$M_f := \{(p, q) \in M;\ f_k(p, q) = c_k\},$$

where $c_1, c_2, \ldots, c_n$ are constant defines a manifold of dimension n. Let $\xi^a = (p, q)$ be local coordinates on M such that the Poisson bracket is

$$\{f, g\} = \omega^{ab}\frac{\partial f}{\partial \xi^a}\frac{\partial g}{\partial \xi^b}, \quad a, b = 1, 2, \ldots, 2n,$$

where ω is the constant antisymmetric matrix

$$\begin{pmatrix} 0 & 1_n \\ -1_n & 0 \end{pmatrix}.$$

The vanishing of the Poisson brackets $\{f_j, f_k\} = 0$ implies that each Hamiltonian vector field

$$X_{f_k} = \omega^{ab}\frac{\partial f_k}{\partial \xi^b}\frac{\partial}{\partial \xi^a}$$

is orthogonal (in the Euclidean sense) to any of the gradients $\partial_a f_j, a = 1, \ldots, 2n,\ j, k = 1, \ldots, n$. The gradients are perpendicular to M_f, thus the Hamiltonian vector fields are tangent to M_f. They are also commuting as

$$[X_{f_j}, X_{f_k}] = -X_{\{f_j, f_k\}} = 0,$$

so the vectors generate an action of the abelian group $\mathbb{R}^n$ on M. This action restricts to an $\mathbb{R}^n$ action on M_f. Let $p_0 \in M_f$, and let Γ be the lattice consisting of all vectors in $\mathbb{R}^n$ which fix p_0 under the group action. Then Γ is a discrete subgroup of $\mathbb{R}^n$ and (by an intuitively clear modification of the orbit-stabiliser theorem) we have

$$M_f = \mathbb{R}^n/\Gamma.$$

Assuming that M_f is compact, this quotient space is diffeomorphic to a torus T^n.

□

In fact this argument shows that we get a torus for any choice of the constants c_k. Thus, varying the constants, we find that the phase-space M is foliated by n-dimensional tori.

A.2 Degree of a map and homotopy

Definition A.2.1 *Let M_1 and M_2 be oriented, compact D-dimensional manifolds without boundary, and let ω be a volume-form on M_2. A degree deg(f) of a smooth map $f : M_1 \to M_2$ is given by*

$$\int_{M_1} f^*\omega = [deg(f)] \int_{M_2} \omega. \tag{A5}$$

Rescaling ω by a constant does not change the degree, and neither does choosing a different volume-form because

$$\int_{M_2} \omega = \int_{M_2} \omega'$$

implies that $\omega - \omega' = d\alpha$ for some $(D-1)$-form α (the D-dimensional cohomology of M_2 is one-dimensional) and

$$f^*\omega' = f^*\omega - f^*d\alpha = f^*\omega - d(f^*\alpha)$$

gives

$$\int_{M_1} (f^*\omega' - f^*\omega) = 0$$

by application of Stokes' theorem. We conclude that the deg(f) depends only on f.

There is another useful way of calculating degree by counting a number of pre-images. Let $y \in M_2$ be a generic point, that is, the set $f^{-1}(y) = \{x; f(x) = y\}$ is finite, and the Jacobian $J(f) \neq 0$ (recall that if $x \in U$ has local coordinates x^i, and $y \in f(U)$ has local coordinates y^i, then $J = \det(\partial y^i/\partial x^j)$ if $y^i = y^i(x^1, \ldots, x^D)$).

Proposition A.2.2 *deg(f) is the integer given by*

$$deg(f) = \sum_{x \in f^{-1}(y)} sign[J(x)]. \tag{A6}$$

Proof Let $f^{-1}(y) = \{x_\alpha\}$, where α is a discrete index with possibly infinite range. The set of critical values of f has measure zero (by the Sard theorem [43]), and such points do not contribute to the integral. Clearly (A6) is an integer, but it may depend on the choice of y. Choose a neighbourhood V of y and U_α of x_α for some fixed value of α such that $f : U_\alpha \to V$ is one-to-one and onto. Let ω have support on $\prod f(U_\alpha) \subset V$ (i.e. $\omega = 0$ outside V). Therefore

$$\int_{M_1} f^*(\omega) = \sum_\alpha \int_{U_\alpha} f^*\omega.$$

Now change coordinates from x^j to y^j near x_α. Then

$$\int_{U_1} f^*\omega = \left(\int_V \omega\right) \text{sign}[J(x_\alpha)],$$

where we have used

$$\omega = \rho(y)dy^1 \wedge \cdots \wedge dy^D$$

and

$$\int f^*\omega = \int \rho[f(x)]J\,dx^1 \wedge \cdots \wedge dx^D = \Big[\int \rho(y)dy^1 \wedge \cdots \wedge dy^D\Big]\mathrm{sign}(J)$$

as the last change of variables from x to y introduces the term $|J|^{-1}$. We obtain similar relations on all other open sets U_α, which proves (A6). □

- **Example.** $f : S^1 \to S^1$. In this case the topological degree is called the winding number. Formula (A6) gives

$$N = \sum_{\theta: f(\theta) = f_0} \mathrm{sign}(df/d\theta).$$

If the graph of f looks like

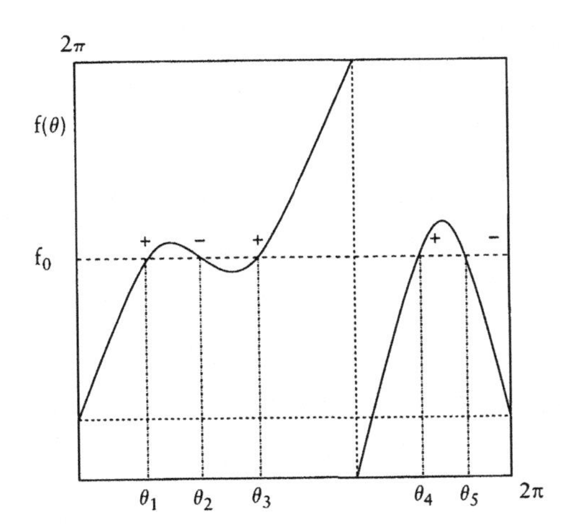

then applying (A6) we find

$$N = 1 - 1 + 1 + 1 - 1 = 1.$$

Alternatively the degree can be calculated from the definition (A5) which yields

$$N = (\mathrm{vol}_{S^1})^{-1} \int_{S^1} df = \frac{1}{2\pi} \int_0^{2\pi} \frac{df}{d\theta} d\theta.$$

If we think of S^1 as the unit circle $|z| = 1$ in the complex plane then every map of degree k is homotopic to $f(z) = z^k$.

- **Example.** For any $k \in \mathbb{Z}$ there exist smooth maps from S^n to S^n of degree k. Let $f : S^{n-1} \longrightarrow S^{n-1}$ be a smooth map which takes a unit vector $\mathbf{n}$ in $\mathbb{R}^n$ to another such vector $f(\mathbf{n})$. We can construct the suspension $\Sigma f : S^n \longrightarrow S^n$ of f by the following prescription:

$$(\Sigma f)(\hat{\mathbf{n}}) = (\cos\theta, \sin\theta \; f(\mathbf{n})), \quad \text{for} \quad \hat{\mathbf{n}} = (\cos\theta, \sin\theta \; \mathbf{n}) \in S^n,$$

so that (Σf) coincides with f on the equator of S^n which is the intersection of S^n with the plane $\cos\theta = 0$. The suspension does not change the degree.

Now take any map $f : S^1 \longrightarrow S^1$, and suspend it $n-1$ times to get the required map. For example, in spherical polar coordinates on S^2 the map $(\theta, \phi) \longrightarrow (\theta, k\phi)$ has degree k.

- **Example.** Let $f : S^2 \longrightarrow S^2$ be given by $f^a = f^a(x^i) \in \mathbb{R}^3, |f| = 1$, where x^i are local coordinates on S^2. Then

$$\deg(f) = \frac{1}{\mathrm{vol}(S^2)} \int \frac{1}{2} \varepsilon^{abc} f^a df^b \wedge df^c = \frac{1}{8\pi} \int \varepsilon^{ij} \varepsilon^{abc} f^a \partial_i f^b \partial_j f^c d^2x, \qquad \text{(A7)}$$

as the area of the unit two-sphere is 4π and the relation between the area form in spherical polars and Cartesian coordinates is

$$\sin\theta d\theta \wedge d\phi = r^{-3}(xdy \wedge dz + ydz \wedge dy + zdx \wedge dy),$$

where $0 \le \theta < \pi, 0 \le \phi < 2\pi$ and

$$x = r \sin\theta \cos\phi, \quad y = r \sin\theta \sin\phi, \quad \text{and} \quad z = r\cos\theta.$$

- **Example.** Let $f : X \longrightarrow SU(2) = S^3$, where X is a closed three-manifold. Then

$$\deg(f) = \frac{1}{24\pi^2} \int_X \mathrm{Tr}[(f^{-1}df)^3], \qquad \text{(A8)}$$

as the volume of S^3 is $2\pi^2$.

A.2.1 Homotopy

A smooth (continuous) homotopy of a map $f : M \longrightarrow N$ is a smooth (continuous) map

$$F : M \times [0, 1] \longrightarrow N$$

such that $F[x, 0] = f(x)$ for all $x \in M$. Each map $f_t(x) := F(x, t)$ is said to be homotopic to f. The notion of homotopy is an equivalence relation and the maps can be classified into homotopy classes.

Let $\pi_1(M, x_0)$ be the set of homotopy classes of loops, that is, maps f from $[0, 1]$ to M which are based at x_0 in the sense that $f(0) = f(1) = x_0$. This space and its group structure which we are just about to describe do not depend on the choice of the base point x_0 if M is path-connected. We will therefore often write $\pi_1(M)$. The space $\pi_1(M)$ consists of oriented loops, that is, copies of S^1 in M. It has the strucure of a group. A composition of two loops f_1, and f_2 is a loop $f_1 * f_2$ obtainded by following f_1 by f_2:

$$(f_1 * f_2)(t) = \begin{cases} f_1(2t) & 0 \le t \le \frac{1}{2} \\ f_2(2t-1) & \dfrac{1}{2} \le t \le 1. \end{cases}$$

The inversion of f is the loop $\hat{f}(t) = f(1-t)$. The group $\pi_1(M)$ is called the fundamental group of M. By definition, M is simply connected if this group is trivial.

- **Example.** $\pi_1(S^1) = \mathbb{Z}$. Consider a continuous map $f : [0, 2\pi] \longrightarrow S^1$. The base condition is $f(0) = 0$, and the continuity implies $f(2\pi) = 2\pi k$, where $k \in \mathbb{Z}$. Two maps f_1 and f_2 with the same k are homotopic by the relation

$$f_\tau = (1-\tau) f_1 + \tau f_2$$

as $f_\tau(2\pi) = 2\pi k$.

The higher homotopy groups $\pi_k(M)$ generalize $\pi_1(M)$ replacing $[0, 1]$ by a k-dimensional closed disc $D^k = [0, 1]^k$. More precisely an element $\pi_k(M, x_0)$ is a homotopy class of maps $D^k \longrightarrow M$ sending the boundary S^{k-1} of D^k of the disc to the point x_0. The group operation is introduced as follows:

$$(f_1 * f_2)(t_1, \dots, t_k) = \begin{cases} f_1(t_1, \dots, t_{k-1}, 2t_k) & 0 \le t_k \le \frac{1}{2} \\ f_2(t_1, \dots, t_{k-1}, 2t_k - 1) & \frac{1}{2} \le t_k \le 1, \end{cases}$$

where we have regarded S^k as a quotient space of a cube $[0, 1]^k$ obtained by collapsing the boundary of the cube to a point. Using the last coordinate to define the product is immaterial, and one gets the same group operation using any other coordinate. The group $\pi_k(M)$ is abelian if $k > 1$. We list various results about homotopy groups without proofs [21, 43]:

$$\pi_n(S^n) = \mathbb{Z}, \quad \pi_3(S^2) = \mathbb{Z}, \quad \pi_{n+1}(S^n) = \mathbb{Z}_2 \quad \text{for} \quad n > 2,$$

$$\pi_k(S^n) = \{0\} \quad \text{for} \quad k < n, \quad \pi_k(S^n) = \pi_{k+1}(S^{n+1}) \quad \text{for} \quad k < 2n - 1.$$

In particular the last two relations imply that S^n is simply connected for $n > 1$.

The formulae (A5) and (A6) allow us to compute the degree of a smooth map. The case of particular interest is $M_1 = M_2 = S^n$. How about maps which are merely continuous? Any continuous map from S^n to itself is homotopic to some smooth map [43], which allows us to define the degree of a continuous map f to be degree of any smooth map homotopic to f. This means that two continuous maps from S^n to itself are homotopic iff they have the same degree, and so the degree is an effective way of computing the homotopy class of a map. One consequence important in soliton theory is that a map from S^n to itself of degree 0 is homotopic to a constant.

In Section 8.2.3 we will need the following result:

Proposition A.2.3 *Let g_1 and g_2 be maps from S^3 to $U(n)$ and let $g_1 g_2 : S^3 \longrightarrow U(n)$ be given by*

$$g_1 g_2(x) := g_1(x) g_2(x), \quad x \in S^3,$$

where the product on the RHS is the point-wise group multiplication. Then

$$[g_1 g_2] = [g_1] + [g_2], \tag{A9}$$

where

$$[g] = \frac{1}{24\pi^2}\int_{S^3} Tr[(g^{-1}dg)^3].$$

Proof This result holds because

$$\mathrm{Tr}\{[(g_1g_2)^{-1}d(g_1g_2)]^3\} = \mathrm{Tr}[(g_1{}^{-1}dg_1)^3 + (g_2^{-1}dg_2)^3] + d\beta,$$

where β is a two-form, and so $d\beta$ integrates to 0 by Stokes' theorem. This was explicitly demonstrated by Skyrme [147] in the case of $SU(2)$.

Rather than exhibiting the exact form of β we shall use the following general argument. The higher homotopy groups $\pi_d(G)$ of a Lie group G are abelian, and the group multiplication in G induces the addition in the homotopy groups: if g_1 and g_2 are maps from S^d to G then the homotopy class of the map $g_1g_2 : S^d \longrightarrow G$ defined by the group multiplication is the sum of homotopy classes of g_1 and g_2. The proof of this is presented for example in [21] and essentially follows the proof that the fundamental group of a topological group is abelian. Now $\pi_3(G) = \mathbb{Z}$ for any compact simple Lie group. If $G = SU(2)$ this result just reproduces the calculation done by Skyrme as two continuous maps from S^3 to itself are homotopic iff they have the same topological degree.

A.2.2 Hermitian projectors

We shall give a more detailed computation of the second homotopy group and the associated topological charge for the complex projective space $\mathbb{CP}^{n-1}$. We shall need this in Section 8.2. It is convenient to choose a map and perform calculation in a local framework. We can represent complex directions in $\mathbb{C}^n$, which are the elements of $\mathbb{CP}^{n-1}$, by vectors in $\mathbb{C}^n$ with their first component fixed to 1. Then the map f defined by

$$\mathbb{CP}^{n-1} \ni (1, f_1, \ldots, f_{n-1}) \longrightarrow (f_1, \ldots, f_{n-1}) \in \mathbb{C}^{n-1} \tag{A10}$$

belongs to the maximal holomorphic atlas of $\mathbb{CP}^{n-1}$. The results do not depend on the choice of this map. Define the Hermitian projector to be the matrix R given by

$$R = \frac{q \otimes q^\dagger}{q^\dagger q}, \quad \text{where} \quad q = \begin{pmatrix} 1 \\ f_1 \\ \vdots \\ f_{n-1} \end{pmatrix}. \tag{A11}$$

It satisfies $R^2 = R$.

Now consider maps $R : \mathbb{R}^2 \to \mathbb{CP}^{n-1}$ which extend from $\mathbb{R}^2$ to the conformal compactification S^2. The topological charge of R is defined by

$$N = -\frac{i}{2\pi}\int_{\mathbb{R}^2} \mathrm{Tr}(R[R_x, R_y])dxdy = -\frac{1}{8\pi}\int_{\mathbb{R}^2} R^*\mathbf{\Omega}, \tag{A12}$$

where

$$\mathbf{\Omega} = -4i\, \partial\bar{\partial} \ln(1 + \sum_{l=1}^{n} |f_l|^2) = -4i \frac{\delta_{jk}\left(1 + \sum\limits_{l=1}^{n-1} |f_l|^2\right) - f_j \bar{f}_k}{\left(1 + \sum\limits_{l=1}^{n-1} |f_l|^2\right)^2} df_k \wedge d\bar{f}_j \qquad \text{(A13)}$$

is the Kähler form of the Fubini–Study metric on $\mathbb{CP}^{n-1}$ and $R^*\mathbf{\Omega}$ denotes its pull-back. The first expression for N given in (A12) is often more convenient for calculations, while the second clarifies the topological character. The equality (A12) is proved by establishing that in the chosen map both expressions give

$$-i \int_{\mathbb{R}^2} \sum_{k,j=1}^{n-1} \frac{\delta_{kj}(1 + \sum\limits_{l=1}^{N-1} |f_l|^2) - \bar{f}_k f_j}{(1 + \sum\limits_{l=1}^{n-1} |f_l|^2)^2} \frac{\partial(f_k, \bar{f}_j)}{\partial(x, y)} dxdy. \qquad \text{(A14)}$$

APPENDIX B

Complex analysis

This appendix introduces some elements of the theory of complex manifolds and holomorphic vector bundles used in the twistor constructions as well as in other parts of the book.

A function $f : U \to \mathbb{C}$ defined on an open set $U \subset \mathbb{C}$ is complex differentiable at $\lambda \in U$ if

$$\lim_{h \to 0} \frac{f(\lambda + h) - f(\lambda)}{h}$$

exists. A function which is complex differentiable at any point of U is called *holomorphic* on U. Holomorphic functions satisfy the Cauchy–Riemann equations: if $\lambda = x + iy$ then

$$\frac{\partial f}{\partial \overline{\lambda}} = 0, \quad \text{where} \quad \frac{\partial}{\partial \overline{\lambda}} = \frac{1}{2}\left(\frac{\partial}{\partial x} + i\frac{\partial}{\partial y}\right).$$

If f is holomorphic inside and on an anticlockwise-oriented closed contour $\Gamma \subset \mathbb{C}$, then the value of f at any point λ inside Γ is given by the Cauchy integral formula

$$f(\lambda) = \frac{1}{2\pi i} \oint_\Gamma \frac{f(\xi)}{\xi - \lambda} d\xi.$$

It will become clear that the need for the rather sophisticated complex analysis arises from the need to overcome two classical results: the Liouville theorem and the maximum modulus theorem.

Theorem B.0.4 (Liouville theorem) *Let the function $f : \mathbb{C} \to \mathbb{C}$ be holomorphic on the whole complex plane. If f is bounded then f is constant.*

Proof The holomorphic function f admits the Taylor expansion around $\lambda = 0$ convergent on $\mathbb{C}$

$$f(\lambda) = \sum_{m=0}^{\infty} a_m \lambda^m$$

with the coefficients a_m are determined by the Cauchy integral formula

$$a_m = \frac{1}{2\pi i} \oint_{C_r} \frac{f(\xi)}{\xi^{m+1}} d\xi$$

where C_r is a circle of radius r centred at 0. If there exist $M > 0$ such that $|f(\lambda)| \leq M$ for all λ then the estimate

$$|a_m| \leq \frac{1}{2\pi}\oint_{C_r}\frac{|f(\xi)|}{|\xi^{m+1}|}|d\xi| \leq \frac{1}{2\pi}\int_0^{2\pi}\frac{M}{r^m}d\theta \leq \frac{M}{r^m}$$

holds. The statement now follows from taking a limit $r \to \infty$. Thus $a_m = 0$ if $m > 0$ and $f = a_0$ is a constant. □

The next closely related result asserts that if a modulus of a holomorphic function attains a maximum on an open set, then the function is necessarily constant.

Theorem B.0.5 (Maximum modulus theorem) *Let the function f be holomorphic on an open disc with centre $\lambda \in \mathbb{C}$ and radius R, and such that $|f(\xi)| \leq |f(\lambda)|$ for all ξ inside the disc. Then f is constant.*

Proof Let $C_{(r,\lambda)}$ be a circle of radius $r < R$ centreed at λ. Setting $\xi = \lambda + re^{i\theta}$ on the circle, the Cauchy integral formula gives

$$f(\lambda) = \frac{1}{2\pi}\int_0^{2\pi} f(\lambda + re^{i\theta})d\theta.$$

Taking the absolute values and using the assumptions of the theorem yields

$$|f(\lambda)| \leq \frac{1}{2\pi}\int_0^{2\pi} |f(\lambda + re^{i\theta})|d\theta \leq |f(\lambda)|$$

and so

$$\int_0^{2\pi}[|f(\lambda)| - |f(r + \lambda e^{i\theta})|]d\theta = 0.$$

The integrand is continuous and non-negative so it must vanish for all $r < R$ and $\theta \in [0, 2\pi]$. Thus $|f(\lambda)| = |f(\xi)|$ for all ξ inside the disc and f has constant modulus. The Cauchy–Riemann equations now yield

$$(\partial_\lambda f)\overline{f} = 0$$

and hence f is a constant. □

B.1 Complex manifolds

An n-dimensional complex manifold M is defined as in Definition A.0.1 where the transition maps between local coordinate systems $\mathbb{C}^n$ are required to be holomorphic. In a neighbourhood U of each point, there exist local holomorphic coordinates z^a $(a = 1, 2, \ldots, n)$ such that in the intersection $U \cap \widetilde{U}$ the transition maps $\tilde{z}^a(z^1, \ldots, z^n)$ between two coordinate systems z^a and $\tilde{z}^a$ satisfy the Cauchy–Riemann (CR) equations $\partial\tilde{z}^a/\partial\overline{z}^b = 0$ and the non-degeneracy condition $\det(\partial\tilde{z}^a/\partial z^b) \neq 0$.

Holomorphic functions, vector fields, and other tensors can be defined as in the real case by adding the requirement that the components should depend holomorphically on the coordinates.

- **Example. Riemann sphere.** The two-dimensional sphere is a one-dimensional complex manifold with local coordinates defined by stereographic projection. Let $(u_1, u_2, u_3) \in S^2$. Define two open subsets covering S^2:

$$U = S^2 - \{(0,0,1)\} \quad \text{and} \quad \tilde{U} = S^2 - \{(0,0,-1)\},$$

and introduce complex coordinates λ and $\tilde{\lambda}$ on U and $\tilde{U}$, respectively, by

$$\lambda = \frac{u_1 + iu_2}{1 - u_3} \quad \text{and} \quad \tilde{\lambda} = \frac{u_1 - iu_2}{1 + u_3}.$$

The domain of λ is the whole sphere less the north pole; the domain of $\tilde{\lambda}$ is the whole sphere less the south pole. On the overlap $U_0 \cap U_1$ we have $\tilde{\lambda} = 1/\lambda$ which is a holomorphic function.

- **Example. Projective spaces.** The n-dimensional projective space $\mathbb{CP}^n$ is the quotient of $\mathbb{C}^{n+1}$ by the equivalence relation

$$(Z^0, Z^1, \ldots, Z^n) \sim (cZ^0, cZ^1, \ldots, cZ^n) \quad \text{for} \quad c \in \mathbb{C}^*.$$

The homogeneous coordinates Z^α label the points uniquely, up to an overall non-zero complex scaling factor. The complex manifold structure on $\mathbb{CP}^n$ is introduced by using the inhomogeneous coordinates. On the open set U_α in which $Z^\alpha \neq 0$, we define z^a $(a = 1, \ldots, n)$ by

$$z^0 = Z^0/Z^\alpha, \ldots, \quad z^{\alpha-1} = Z^{\alpha-1}/Z^\alpha, \quad z^{\alpha+1} = Z^{\alpha+1}/Z^\alpha, \quad \ldots, \quad z^n = Z^n/Z^\alpha.$$

We shall use a special notation for the complex projective line $\mathbb{CP}^1$. The homogeneous coordinates are denoted by $Z^A = (Z^0, Z^1)$, the two set covering is $U_0 = U$ and $U_1 = \tilde{U}$, and $\lambda = Z^1/Z^0$ and $\tilde{\lambda} = 1/\lambda$ are inhomogeneous coordinates in U and $\tilde{U}$, respectively.

Definition B.1.1 *A holomorphic map $f : M \to \widetilde{M}$ is a continuous map such that for each coordinate chart $\phi_\alpha : U_\alpha \to \mathbb{C}^n$ on M and $\tilde{\phi}_\beta : \tilde{U}_\beta \to \mathbb{C}^{\tilde{n}}$ on $\widetilde{M}$, $\tilde{\phi}_\beta \circ f \circ \phi_\alpha^{-1}$ is holomorphic.*

Two complex manifolds M and $\widetilde{M}$ are isomorphic if they are diffeomorphic by a bi-holomorphic diffeomorphism, that is, if f and f^{-1} are holomorphic in local holomorphic coordinates. When $n = 1$, the projective space is isomorphic to the Riemann sphere since we can take $\lambda = Z^1/Z^0$ and $\tilde{\lambda} = Z^0/Z^1$ as the two local coordinates (in the southern and northern hemispheres). An explicit isomorphism between S^2 and $\mathbb{CP}^1$ is given by

$$(u_1, u_2, u_3) \longrightarrow [1 - u_3, u_1 + iu_2]. \tag{B1}$$

A holomorphic map $f : M \to \mathbb{C}$ is called a holomorphic function on M.

Theorem B.1.2 *If M is connected and compact, the only holomorphic functions on M are the constants.*

Proof This follows from the maximum modulus theorem: $|f|$ has to have a maximum on the compact space M, but in a coordinate neighbourhood of this point $f \circ \phi_\alpha^{-1}$ is a holomorphic function whose modulus has a maximum in the interior of an open set in $\mathbb{C}^n$. Therefore $f \circ \phi_\alpha^{-1}$ is constant by the maximum modulus theorem (Theorem B.0.5), and so f is constant. □

B.2 Holomorphic vector bundles and their sections

To obtain a useful theory of functions on compact complex manifold we have to introduce holomorphic line (and vector) bundles and their sections. Roughly speaking, a vector bundle over a manifold is a collection of vector spaces, one at each point of the manifold

Definition B.2.1 *A holomorphic vector bundle of rank k over a complex manifold M is a complex manifold E, and a holomorphic projection $\pi : E \to M$ such that*

- *For each $z \in M$, $\pi^{-1}(z)$ is a k-dimensional complex vector space.*
- *Each point $z \in M$ has a neighbourhood U_α and a homeomorphism χ_α such that the diagram*

$$\begin{array}{ccc} \pi^{-1}(U_\alpha) & \overset{\chi_\alpha}{\cong} & U_\alpha \times \mathbb{C}^k \\ & \pi \searrow \quad \swarrow & \\ & U_\alpha & \end{array}$$

  is commutative.
-
$$F_{\alpha\beta} := \chi_\beta \circ {\chi_\alpha}^{-1} : U_\alpha \cap U_\beta \to \mathrm{GL}(k, \mathbb{C})$$

  is a holomorphic map to the space of invertible $k \times k$ matrices.

Remarks

- Rank-one vector bundles are called line bundles.
- We call $F_{\alpha\beta}$ a transition function, or a patching matrix. On the overlap $U_\alpha \cap U_\beta$ of two local trivializations we have

$$(z, u) \sim (z, u') = \big(z, F_{\alpha\beta}(z)u\big), \qquad z \in U_\alpha \cap U_\beta.$$

  One way that we can specify a holomorphic vector bundle is by giving its transition maps $F_{\alpha\beta}$ between the open sets of some open cover U_α (α in some indexing set). These are holomorphic matrix-valued functions on the intersections with the following properties:

$$F_{\alpha\alpha} = 1, \quad F_{\alpha\beta} F_{\beta\alpha} = 1, \quad F_{\alpha\beta} F_{\beta\gamma} F_{\gamma\alpha} = 1, \quad \text{no summation.}$$

The last relation is called the cocyle property, and holds on each triple intersection:

$$U_\alpha \cap U_\beta \cap U_\gamma.$$

It is not hard to see that every holomorphic vector bundle can be represented in this way.

- There is a certain non-uniqueness in the definition of transition functions. Let E be given by (χ_α, U_α) and $\widetilde{E}$ by $(\tilde{\chi}_\alpha, U_\alpha)$. Put $H_\alpha = \chi_\alpha \circ (\tilde{\chi}_\alpha)^{-1} \in \text{Map}(U_\alpha, GL(k, \mathbb{C}))$. Then clearly

$$\widetilde{F}_{\alpha\beta} = (H_\beta)^{-1} F_{\alpha\beta} H_\alpha. \tag{B2}$$

We shall regard E and $\widetilde{E}$ as equivalent if their transition functions satisfy (B2).
- All the algebraic operations on vector spaces can be extended to holomorphic bundles.

The product $E = M \times \mathbb{C}^k$ is called a trivial vector bundle.

Lemma B.2.2 *The bundle is trivial, iff there exist holomorphic splitting matrices*

$$H_\alpha : U_\alpha \to \text{GL}(k, \mathbb{C})$$

such that

$$F_{\alpha\beta} = H_\beta H_\alpha^{-1}. \tag{B3}$$

Proof If (B3) holds, then $F_{\alpha\beta}$ is equivalent to Id by the relation (B2), so E is trivial.
If E given by (U_α, χ_α) is trivial, then there exists a bi-holomorphism $\chi : E \to M \times \mathbb{C}^k$. We define $H_\alpha = \chi_\alpha \circ (\chi|_\alpha)^{-1}$, and note that

$$H_\beta^{-1} F_{\alpha\beta} H_\alpha = \text{Id}.$$

□

Definition B.2.3 *A holomorphic section of a vector bundle E over M is a holomorphic map $s : M \to E$ such that $\pi \circ s = \text{id}_M$.*

The local description is given by a collection of holomorphic maps $s_\alpha : U_\alpha \to \mathbb{C}^k$

$$z \longrightarrow (z, s_\alpha(z)), \quad \text{for} \quad z \in U_\alpha$$

with the transition rule $s_\beta(z) = F_{\alpha\beta}(z) s_\alpha(z)$.

In the case of trivial bundles Lemma B.2.2 implies that we can specify global sections by putting $s_\alpha(z) = H_\alpha(z)\xi$ for some constant vector ξ.

We denote the space of holomorphic sections over $U \subset M$ by $\Gamma(U, E)$ (or $H^0(U, L)$ if $E = L$ is a line bundle). An important theorem (which we shall not prove) is that $\Gamma(M, E)$ is finite-dimensional if M is compact.

- **Example. Tautological bundle.** Consider $M = \mathbb{CP}^1$ with the usual coordinate patches U_0 and U_1. First define a tautological line bundle

$$\mathcal{O}(-1) = \{(\lambda, (Z^0, Z^1)) \in \mathbb{CP}^1 \times \mathbb{C}^2 | \lambda = Z^1/Z^0\}.$$

If we represent the Riemann sphere as the projective line, as above, then we have the projection $\mathbb{C}^2 \to \mathbb{CP}_1$. The fibre above the point with coordinate $[Z]$ is the one-dimensional line cZ through the origin in $\mathbb{C}^2$ containing the the point (Z^0, Z^1). Define the trivializations

$$\chi_0([Z], c(Z^0, Z^1)) = ([Z], cZ^0) \in U_0 \times \mathbb{C} \quad \text{and}$$
$$\chi_1([Z], c(Z^0, Z^1)) = ([Z], cZ^1) \in U_1 \times \mathbb{C},$$

so that

$$cZ^1 = \frac{Z^1}{Z^0} cZ^0 = F_{01}\, cZ^0$$

giving the transition function $F_{01} = \lambda$.

- **Example.** Other line bundles can be obtained by algebraic operations:

$$\mathcal{O}(-n) = \mathcal{O}(-1)^{\otimes n}, \quad \mathcal{O}(n) = \mathcal{O}(-n)^*, \quad \text{and} \quad \mathcal{O} = \mathcal{O}(-1) \otimes \mathcal{O}(1), \quad n \in \mathbb{N}.$$

The transition function for $\mathcal{O}(n)$ is $F = \lambda^{-n}$ on $U_0 \cap U_1 \cong \mathbb{C}^*$. A global holomorphic section of this line bundle is given by functions s and $\tilde{s}$ on $\mathbb{C}$ related by

$$s(\lambda) = \lambda^n \tilde{s}(\tilde{\lambda})$$

on the overlap $\mathbb{C}^*$. Expanding these functions as power series in their respective local coordinates and using the fact that $\tilde{\lambda} = \lambda^{-1}$, we get

$$\sum_0^\infty a_m \lambda^m = \lambda^n \sum_0^\infty \tilde{a}_m \lambda^{-m}.$$

Equating coefficients, we find that $\tilde{a}_m = a_m = 0$ for $m > n$ and $\tilde{a}_0 = a_n$, $\tilde{a}_1 = a_{n-1}$, etc. Thus the global sections are given by polynomials

$$\sum_0^n a_m \lambda^m$$

of degree less than or equal to n, and hence the space of holomorphic sections is $(n+1)$-dimensional, that is,

$$H^0(\mathbb{CP}^1, \mathcal{O}(n)) = 0 \quad (n < 0) \quad \text{and} \quad \dim H^0(\mathbb{CP}^1, \mathcal{O}(n)) = n + 1 \quad (n \geq 0). \tag{B4}$$

The transition relation implies that

$$s(\lambda)(Z^0)^n = \tilde{s}(\tilde{\lambda})(Z^1)^n,$$

so we can define a function $G : \mathbb{C}^2 \to \mathbb{C}$ by

$$G(Z) = s(\lambda)(Z^0)^n \quad \text{for} \quad [Z] \in U_0,$$
$$= \tilde{s}(\tilde{\lambda})(Z^1)^n \quad \text{for} \quad [Z] \in U_1.$$

Then G is homogeneous of degree n. Conversely any holomorphic function on $\mathbb{C}^2$ homogeneous of degree n gives rise to a section of $\mathcal{O}(n)$. A global holomorphic section of $\mathcal{O}(n)$, $n \geq 0$, is same as a global function on $\mathbb{C}^2$ homogeneous of degree n (a polynomial). Thus we get an extension of Liouville theorem:

Theorem B.2.4 *A holomorphic function on $\mathbb{C}^2$ homogeneous of degree $n > 0$ is of the form*

$$f(Z) = \phi_{AB\cdots C} Z^A Z^B \cdots Z^C$$

for some symmetric constant 'spinor' $\phi_{AB\cdots C}$.

- **Example. Tangent and cotangent bundles.** Holomorphic vector fields on $\mathbb{CP}^1$ are sections of the holomorphic tangent bundle $T\mathbb{CP}^1$. Holomorphic one-forms are sections of the holomorphic cotangent bundle $T^*\mathbb{CP}^1$. Observe that

$$\frac{\partial}{\partial \lambda} = -\lambda^{-2} \frac{\partial}{\partial \tilde{\lambda}} \quad \text{and} \quad d\lambda = -\lambda^2 d\tilde{\lambda}.$$

We absorb the minus signs into the local trivializations, and deduce that

$$T\mathbb{CP}^1 = \mathcal{O}(2) \quad \text{and} \quad T^*\mathbb{CP}^1 = \mathcal{O}(-2).$$

Therefore a global section of $T\mathbb{CP}^1$ is of the form

$$(u\lambda^2 + x\lambda + v)\frac{d}{d\lambda}, \tag{B5}$$

for $(u, x, v) \in \mathbb{C}^3$ and there are no global section of $T^*\mathbb{CP}^1$.

- **Example.** To any rank-k vector bundle E we assign a line bundle $\det E := \Lambda^k(E)$. It has transition functions $\det F_{\alpha\beta}$. Define the canonical bundle K of M to be $\det(T^*M)$. This is the line bundle of holomorphic volume forms. On $\mathbb{CP}^1$ we have an isomorphism $K \cong \mathcal{O}(-2)$.

The theorem of Grothendieck states that all holomorphic line bundles over a rational curve are equivalent to $\mathcal{O}(n)$ for some n. In fact more is true

Theorem B.2.5 (Birkhoff–Grothendieck) *A rank-k holomorphic vector bundle $E \rightarrow \mathbb{CP}^1$ is isomorphic to a direct sum of line bundles $\mathcal{O}(m_1) \oplus \cdots \oplus \mathcal{O}(m_k)$ for some integers m_i.*

This theorem is proved for example in [121].

Therefore for a rank k vector bundle the transition matrix

$$F : \mathbb{C}^* \rightarrow \mathrm{GL}(k, \mathbb{C})$$

can be written in the form

$$F = \tilde{H} \operatorname{diag}(\lambda^{-m_1}, \ldots, \lambda^{-m_k}) H^{-1},$$

where $H : U \rightarrow \mathrm{GL}(k, \mathbb{C})$ and $\tilde{H} : \tilde{U} \rightarrow \mathrm{GL}(k, \mathbb{C})$ are holomorphic.

- **Example.** Let E_t be a one-parameter family of rank-two vector bundles over $\mathbb{CP}^1$ determined by a patching matrix

$$F = \begin{pmatrix} \lambda & t \\ 0 & \lambda^{-1} \end{pmatrix}.$$

For $t = 0$, F is already in the form given by the Birkhoff–Grothendieck theorem, with $H = \tilde{H} = 1$, $m_1 = -1$, and $m_2 = 1$. But for $t \neq 0$, we have

$$F = \begin{pmatrix} 0 & t \\ -t^{-1} & \lambda^{-1} \end{pmatrix} \begin{pmatrix} 1 & 0 \\ t^{-1}\lambda & 1 \end{pmatrix}, \tag{B6}$$

so that $m_1 = m_2 = 0$ by Lemma B.2.2. Therefore $E_0 \simeq \mathcal{O}(1) \oplus \mathcal{O}(-1)$, but E_t is the trivial bundle for $t \neq 0$. This is an example of 'jumping': as t changes through 0, the holomorphic structure of the bundle changes discontinuously, in spite of the fact that the bundles E_t are all the same (and all trivial) from the topological point of view.

B.3 Čech cohomology

An element of the first cohomology group of $E \to M$ relative to the cover U_α of M is a map that assigns a holomorphic section $f_{\alpha\beta} \in \Gamma(U_\alpha \cap U_\beta)$ to each non-empty intersection such that

$$f_{\alpha\beta} + f_{\beta\alpha} = 0 \quad \text{and} \quad f_{\alpha\beta} + f_{\beta\gamma} + f_{\gamma\alpha} = 0.$$

Two such maps f and f' are equivalent iff

$$f'_{\alpha\beta} - f_{\alpha\beta} = h_\alpha - h_\beta,$$

(the RHS is called a co-boundary) where h_α is a holomorphic section of E over U_α. The first cohomology group is a quotient of the additive group $\{f_{\alpha\beta}\}$ by this relation. (The definition is in fact independent of the covering chosen [121].) A more concrete definition can be used if it is possible to choose an open cover consisting on two pseudo-convex sets U and $\tilde{U}$:

$$H^1(M, E) = \frac{\Gamma(U \cap \tilde{U}, E)}{\Gamma(U, E) + \Gamma(\tilde{U}, E).}$$

We will find $H^1(\mathbb{CP}^1, \mathcal{O}(k))$. This is the space of functions f_{01} holomorphic on $U_0 \cap U_1$ and homogeneous of degree k in coordinates $[Z^0, Z^1]$, modulo coboundaries. In a trivialization over $U_0 = U$ f_{01} is represented by a holomorphic function f on $\mathbb{C}^*$. In the trivialization over $U_1 = \tilde{U}$, f_{01} is represented by $\lambda^{-k} f$. For $k \geq -1$ we can write

$$f = \sum_{-\infty}^{\infty} f_i \lambda^i = \lambda^k \tilde{h} - h,$$

where

$$h = -\sum_{0}^{\infty} f_i \lambda^i \quad \text{and} \quad \tilde{h} = \sum_{1}^{\infty} f_{-i} \lambda^{-i-k}$$

are holomorphic in U and $\tilde{U}$, respectively. Therefore $f \sim 0$, and the first cohomology group vanishes. The splitting is not unique unless $k = -1$, as we are free to set the first $k+1$ terms in h to 0 by modifying $\tilde{h}$.

If $k \leq -2$ we have $f = \lambda^k \tilde{h} - h + q$, where

$$-h = \sum_{0}^{\infty} f_i \lambda^i, \quad \tilde{h} = \sum_{-k}^{\infty} f_{-i} \lambda^{-i-k}, \quad \text{and} \quad q = \sum_{1}^{-k-1} f_{-i} \lambda^{-i}.$$

The functions h and $\tilde{h}$ are holomorphic in U and $\tilde{U}$, respectively, therefore $f \sim q$, and the class of f is uniquely determined by the coefficients $f_{-1}, \ldots, f_{k+1}$. We conclude that

$$H^1(\mathbb{CP}^1, \mathcal{O}(k)) = \begin{cases} 0 & \text{for } k > -2 \\ \mathbb{C}^{-k-1} & \text{for } k \leq -2. \end{cases} \tag{B7}$$

Therefore $f \in H^1(\mathbb{CP}^1, \mathcal{O}(-1))$ can be split uniquely, as $f = \tilde{h} - h$. We shall often use the explicit form of this splitting:

$$h = \frac{1}{2\pi i} \oint_{\Gamma} \frac{f(\zeta)}{\lambda - \zeta} d\zeta \quad \text{and} \quad \tilde{h} = \frac{1}{2\pi i} \oint_{\tilde{\Gamma}} \frac{f(\zeta)}{\lambda - \zeta} d\zeta, \tag{B8}$$

where ζ is an affine coordinate on $\mathbb{CP}^1$ (Figure B.1). The contours Γ and $\tilde{\Gamma}$ are homologous to the equator of $\mathbb{CP}^1$ in $U \cap \tilde{U}$ and are such that $\Gamma - \tilde{\Gamma}$ surrounds the point $\lambda = \zeta$. We see that $f = \tilde{h} - h$ follows from the Cauchy's integral formula.

Given $f \in H^1(\mathbb{CP}^1, \mathcal{O}(k))$, $k > -1$, we may divide it by a homogeneous polynomial of degree $k+1$, and apply (B8) to the quotient. The non-uniqueness of this procedure (the choice of the polynomial) is measured by $H^0(\mathbb{CP}^1, \mathcal{O}(k+1))$.

B.3.1 Deformation theory

Let L be a complex submanifold of a complex manifold $\mathcal{Z}$. The normal bundle $N(L) \to L$ is defined to be $\cup_{\zeta \in L} N_\zeta(L)$ where $N_\zeta = (T_\zeta \mathcal{Z})/(T_\zeta L)$ is a quotient vector space.

The following result of Kodaira underlies the twistor approach to curved geometries. Let $\mathcal{Z}$ be a complex manifold of dimension $d+r$. A pair $(\mathcal{F}, M)$ is called *a complete analytic family of compact submanifolds of* $\mathcal{Z}$ *of dimension* d if

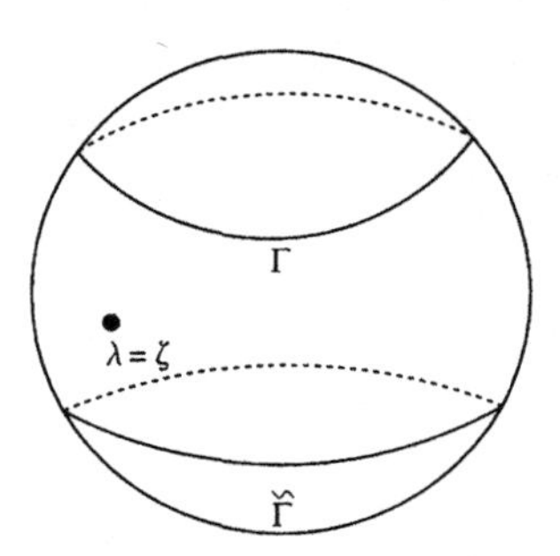

Figure B.1 *Splitting formula*

- $\mathcal{F}$ is a complex analytic submanifold of $\mathcal{Z} \times M$ of co-dimension r with the property that for each $t \in M$ the intersection $L_t \times t := \mathcal{F} \cap (\mathcal{Z} \times t)$ is a compact submanifold of $\mathcal{Z} \times t$ of dimension d.
- There exists an isomorphism

$$T_t M \simeq H^0(L_t, N_t) \tag{B9}$$

where $L_t \subset \mathcal{Z}$ is submanifold of $\mathcal{Z}$ and $N_t \longrightarrow L_t$ is the normal bundle of L_t in $\mathcal{Z}$.

Theorem B.3.1 (Kodaira [95]) *Let L be a d-dimensional complex compact submanifold of a complex manifold $\mathcal{Z}$, and let N be the normal bundle of L in $\mathcal{Z}$. If $H^1(L, N) = 0$ then there exists a complete analytic family of compact submanifolds $(\mathcal{F}, M)$ such that $L = L_{t_0}$ for some $t_0 \in M$.*

In Chapter 10 we apply the above theorem to the situation when $\mathcal{Z}$ is a twistor space and $L = \mathbb{CP}^1$. Roughly speaking, the moduli space M is the 'arena' of differential geometry and integrable systems. One way to analyse such moduli spaces is to consider infinitesimal deformations and to exponentiate them.

APPENDIX C

Overdetermined PDEs

C.1 Introduction

This appendix treats geometric approaches to DEs, both ODEs and PDEs. Geometry in this context means that certain results do not depend on coordinate choices made to write down a DE, and also that structures like connection and curvature are associated to DEs.

The subject can get very technical but we shall take a low-technology approach. This means that sometimes, for the sake of explicitness, a coordinate calculation will be performed instead of presenting an abstract coordinate-free argument. We shall also skip some proofs, and replace them by examples illustrating the assumptions and applications. The proofs can be found in [23] (see also [90] and [151]).

Given a system of DEs it is natural to ask the following questions:

- Are there any solutions?
- If yes, how many?
- What data is sufficient to determine a unique solution?
- How to construct solutions?

These are all local questions, that is, we are only interested in a solution in a small neighbourhood of a point in a domain of definition of dependent variables. We shall mostly work in the smooth category, except when a specific reference to the Cauchy–Kowalewska theorem is made. This theorem holds only in the real analytic category.

Problem 1. Consider an ODE

$$\frac{du}{dx} = F(x, u) \tag{C1}$$

where F and $\partial_u F$ are continuous[1] in some open rectangle

$$U = \{(x, u) \in \mathbb{R}^2, a < x < b, c < u < d\}.$$

The Picard theorem states that for all $(x_0, u_0) \in U$ there exists an interval $I \subset \mathbb{R}$ containing x_0 such that there is a unique function $u : I \to \mathbb{R}$ which satisfies (C1)

[1] In fact it is sufficient if F is Lipschitz.

and such that $u(x_0) = u_0$. We say that the general solution to this first-order ODE depends on one constant. The unique solution in Picard theorem arises as a limit

$$u(x) = \lim_{n\to\infty} u_n(x)$$

of a uniformly convergent sequence of functions $\{u_n(x)\}$ defined iteratively by

$$u_{n+1}(x) = u_0 + \int_{x_0}^{x} F(t, u_n(t))dt.$$

One can treat a system of n first-order ODEs with n unknowns in the same way: The unique solution depends on n constants of integration.

The conditions in Picard theorem always need to be checked and should not be taken for granted. For example, the ODE

$$\frac{du}{dx} = u^{1/2}, \qquad u(0) = 0$$

has two solutions in any neighbourhood of $(0, 0)$: $u(x) = 0$ and $u(x) = x^2/4$.

More geometrically, the solutions to (C1) are curves tangent to a vector field

$$X = \frac{\partial}{\partial x} + F(x, u)\frac{\partial}{\partial u}.$$

The Picard theorem states that the tangent directions always fit together to form a curve.

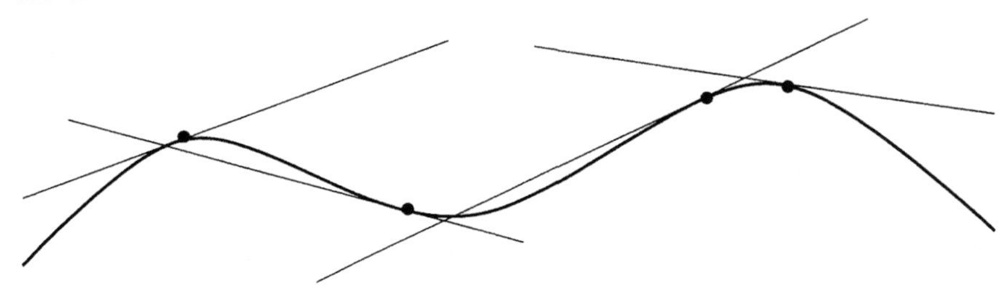

One can rephrase this in a language of differential forms. The one-form annihilated by X is (a multiple of) $\theta = du - Fdx$ and a parameterized curve $x \to (x, u(x))$ is an integral curve of (C1) if θ (or any of its multiples) vanishes on this curve. In general, if θ is a k-form on a manifold M the submanifold $S \subset M$ is an integral of θ if $f^*(\theta) = 0$, where $f : S \to M$ is an immersion.

We aim to reformulate systems of DEs as the vanishing of a set of differential forms (in general of various degree). This gives a coordinate invariant formulation of DEs as exterior differential systems (EDSs), and allows a discussion of the dimension of integral manifold.

Problem 2. Consider a system of PDEs

$$u_x = A(x, y, u) \quad \text{and} \quad u_y = B(x, y, u), \tag{C2}$$

where $u_x = \partial_x u$, etc. Both derivatives of u are determined at each point $(x, y, u) \in \mathbb{R}^3$ where A, B, A_u, and B_u are continuous. This gives rise to a two-dimensional plane spanned by two vectors

$$X_1 = \frac{\partial}{\partial x} + A\frac{\partial}{\partial u} \quad \text{and} \quad X_2 = \frac{\partial}{\partial y} + B\frac{\partial}{\partial u}.$$

Do these planes fit together to form a solution surface in a neighbourhood of (say) $(0, 0, u_0) \in \mathbb{R}^3$? Let us try two successive applications of Picard theorem.

- Set $y = 0$, $u(0, 0) = u_0$. The Picard theorem guarantees the existence of the unique $\tilde{u}(x)$ such that

$$\frac{d\tilde{u}}{dx} = A(x, 0, \tilde{u}), \qquad \tilde{u}(0) = u_0.$$

- Consider $\tilde{u}(x)$ and hold x fixed, regarding it as a parameter. Picard theorem gives the unique $u(x, y)$ such that

$$\frac{du}{dy} = B(x, y, u), \qquad u(x, 0) = \tilde{u}(x).$$

We have therefore constructed a function $u(x, y)$ but it may not satisfy the original PDE (C2) which is overdetermined and requires that the compatibility condition

$$(u_x)_y = (u_y)_x$$

holds. Expanding the mixed partial derivatives yields

$$A_y - B_x + A_u B - B_u A = 0. \tag{C3}$$

Do we need more compatibility conditions arising from differentiating (C3) and using (C2) to get rid of u_x, u_y ? The answer is no. This follows from the Frobenius theorem which we are going to prove in Section C.2 (the LHS of (C3) is the obstruction to the vanishing of the commutator $[X_1, X_2]$). If (C3) holds then solving the pair of ODEs gives the solution surface depending on one constant.

What happens if (C3) does not hold?

- If u does not appear in (C3) then (C3) is a curve in $\mathbb{R}^2$ and there is no solution in an open set containing $(0, 0, u_0)$.
- If (C3) gives an implicit algebraic relation between (x, y, u), then solve this relation to get a surface $(x, y) \to (x, y, u(x, y))$. This may or may not be a solution to the original pair of PDEs (C2). In particular the initial condition may not be satisfied.

This simple example raises a number of questions. How should we deal with more complicated compatibility conditions? When can we stop cross-differentiating? Theorems C.3.1 and C.3.2 proved in Section C.3 and more generally the Cartan test (Theorem C.6.5) discussed in Section C.6 give some of the answers.

Problem 3. Consider a system of linear PDEs

$$u_x = \alpha u + \beta v, \quad u_y + v_x = \gamma u + \delta v, \quad \text{and} \quad v_y = \epsilon u + \phi v, \tag{C4}$$

where $\alpha, \beta, \ldots, \phi$ are some functions of (x, y) defined on an open set $U \subset \mathbb{R}^2$. This is an overdetermined system as there are three equations for two unknowns, but (unlike the system (C2)) it is not overdetermined enough, as the partial derivatives are not specified at each point. Therefore we cannot start the process of building the solution surface as we cannot specify the tangent planes. One needs to use the process of *prolongation* and introduce new variables for unknown derivatives hoping to express derivatives of these variables using the (differential consequences of) the original system. In our case it is enough to define

$$w = u_y - v_x$$

(there are other choices, e.g. $w = u_y$, but the solution surface will not depend on the choices made). Now

$$u_y = \frac{1}{2}(\gamma u + \delta v + w) \quad \text{and} \quad v_x = \frac{1}{2}(\gamma u + \delta v - w),$$

and we can impose the compatibility conditions

$$(u_y)_x = (u_x)_y, \quad \text{and} \quad (v_y)_x = (v_x)_y.$$

These conditions will lead to expressions

$$w_x = \ldots \quad \text{and} \quad w_y = \ldots,$$

where $(\ldots)$ denote terms linear in (u, v, w). The system is now closed as first derivatives of (u, v, w) are determined at each point thus specifying a family of two-dimensional planes in $\mathbb{R}^5$. Do these two planes fit in to form a solution surface

$$(x, y) \longrightarrow (x, y, u(x, y), v(x, y), w(x, y))$$

in $\mathbb{R}^5$? Not necessarily, as there are more compatibility conditions to be imposed (e.g. $(w_x)_y = (w_y)_x$). These additional conditions will put restrictions of the functions $(\alpha, \beta, \ldots, \phi)$. In Section C.4 we shall see how to deal with the prolongation procedure systematically.

This simple example of prolongation arises naturally in the geometry of surfaces. Assume you are given a metric (a first fundamental form) on a surface

$$g = Edx^2 + 2F dxdy + Gdy^2.$$

Does there exist a one-form $K = udx + vdy$ such that the Killing equations

$$\nabla_{(i} K_{j)} = 0 \quad \text{and} \quad x^i = (x, y)$$

are satisfied, where ∇ is the Levi-Civita connection of g? Expanding the Killing equations in terms of the Christoffel symbols leads to the system (C4) where the six functions $(\alpha, \beta, \ldots, \phi)$ are given in terms of E, F, G, and their derivatives. The consistency conditions for the prolonged system to admit non-zero solutions give differential constraints on E, F, and G. These constraints can be expressed in tensor form as differential invariants of the metric g. In Section C.4.1 we shall discuss an approach to constructing such invariants and find necessary and sufficient conditions of a metric g to admit a Killing vector.

C.2 Exterior differential system and Frobenius theorem

Definition C.2.1 *An EDS is a pair* $(M, \mathcal{I})$ *where* M *is a smooth manifold and* $\mathcal{I} \subset \Omega^*(M)$ *is a graded differential ideal in a ring of differential forms that is closed under exterior differentiation:*

$$d\theta \in \mathcal{I} \quad \textit{if} \quad \theta \in \mathcal{I}.$$

For example, the set of forms

$$\{dy - pdx,\ dp \wedge dx,\ dx\}$$

gives EDS where $M = \mathbb{R}^3$. We shall use the following notation: $\mathcal{I}^k = \mathcal{I} \cap \Omega^k(M)$ is a set of all forms of degree k in $\mathcal{I}$. The evaluation of a form θ at $x \in M$ will be denoted θ_x and $\mathcal{I}_x$ will denote the evaluation of all forms in $\mathcal{I}$ at x.

One way to present an EDS is by specifying the set of differential generators

$$<\theta^1, \ldots, \theta^n>_{\text{diff}} := \{\gamma_1 \wedge \theta^1, \ldots, \gamma_n \wedge \theta^n, \beta_1 \wedge d\theta^1, \ldots, \beta_n \wedge d\theta^n\},$$

where γ and β are arbitrary differential forms. We shall assume that none of the generators are zero-forms (i.e. functions). Otherwise we shall restrict the EDS to submanifolds on which these functions vanish. An EDS whose generators are one-forms is called a Pfaffian system.

We shall also use the notation

$$<\theta^1, \ldots, \theta^n>_{\text{alg}} := \{\gamma_1 \wedge \theta^1, \ldots, \gamma_n \wedge \theta^n\}$$

to denote the set of forms generated algebraically by exterior multiplication.

Definition C.2.2 *An integral manifold of* $\mathcal{I}$ *is a submanifold* $f : S \to M$ *such that* $f^*(\theta) = 0$ *for all* $\theta \in \mathcal{I}$.

In particular S is an integral submanifold of $\mathcal{I} = <\theta^1, \ldots, \theta^n>_{\text{diff}}$ iff $f^*(\theta^i) = 0$.

- **Example.** A system of N first-order ODEs

$$\frac{du^\alpha}{dx} = F^\alpha(x, u^1, \ldots, u^N), \qquad \alpha = 1, \ldots, N$$

  is modelled by the EDS $\mathcal{I}$ generated by N one-forms $<du^\alpha - F^\alpha dx>_{\text{diff}}$ on an open set in $\mathbb{R}^{N+1}$. The integral manifolds of this EDS are integral curves of the vector field

$$X = \frac{\partial}{\partial x} + \sum_{\alpha=1}^{N} F^\alpha \frac{\partial}{\partial u^\alpha}$$

  which annihilates all forms in $\mathcal{I}$.
- **Example.** The pair of PDEs $u_x = A(x, y, u)$ and $u_y = B(x, y, u)$ is modelled by an ideal generated by one one-form

$$\mathcal{I} = <du - Adx - Bdy>.$$

The vectors $\partial_x + A\partial_u$ and $\partial_y + B\partial_u$ annihilating this one-form are tangent to the integral surface if one exists. There are no integral surfaces if the compatibility (C3) does not hold.

Two EDSs $(M, \mathcal{I})$ and $(\hat{M}, \hat{\mathcal{I}})$ are equivalent if there exist a diffeomorphism such that

$$f : M \longrightarrow \hat{M} \quad \text{and} \quad f^*(\hat{\mathcal{I}}) = \mathcal{I}.$$

This notion can be applied to determine whether two systems of DEs are equivalent and, in particular, to linearize some DEs. In the next two examples we shall use the following notation: if $\theta = \theta_j(x^1, \ldots, x^i)\,dx^j$ then $\hat{\theta} = \theta_j(\hat{x}^1, \ldots, \hat{x}^i)\,d\hat{x}^j$ where x^j and $\hat{x}^j$ are local coordinates on M and $\hat{M}$, respectively.

- **Example.** Consider the Monge–Ampere equation

$$u_{xx}u_{yy} - u_{xy}^2 = 1,$$

where $u = u(x, y)$. This non-linear equation is modelled by the EDS

$$<\theta^1 = du - pdx - qdy, \theta^2 = dp \wedge dq - dx \wedge dy>_{\text{diff}} \qquad \text{(C5)}$$

on $\mathbb{R}^5$. In particular, it is not a Pfaffian system. Consider $f : \mathbb{R}^5 \to \mathbb{R}^5$ given by

$$f(x, y, u, p, q) = (\hat{x}, \hat{y}, \hat{u}, \hat{p}, \hat{q}) := (x, q, u - qy, p, -y).$$

We verify that

$$f^*(\hat{\theta}^1) = d\hat{u} - \hat{p}d\hat{x} - \hat{q}d\hat{y} = du - pdx - qdy \quad \text{and}$$
$$f^*(\hat{\theta}^2) = d\hat{p} \wedge d\hat{q} - d\hat{x} \wedge d\hat{y} = dy \wedge dp + dq \wedge dx.$$

The integral manifolds of the pulled-back ideal are

$$du - pdx - qdy = 0 \quad \text{and} \quad dy \wedge dp + dq \wedge dx = 0.$$

Vanishing of the one-form gives $p = u_x, q = u_y$, and vanishing of the two-form gives the linear Laplace equation

$$u_{xx} + u_{yy} = 0.$$

Some care needs to be taken with this example: We have established a one-to-one correspondence between integral surfaces of the Laplace equation and the Monge–Ampere equation, but not between solutions as some integral surfaces may have $dx \wedge dy = 0$.

- **Example.** A similar procedure can be used to reduce the general four-dimensional Ricci-flat Kähler metric with a tri-holomorphic Killing vector to the Gibbons–Hawking form where the non-linear Ricci-flat condition reduces to the Laplace equation on $\mathbb{R}^3$. Consider a Kähler metric in an open ball in $\mathbb{C}^2$ with local holomorphic coordinates (w, z) given in terms of the (non-holomorphic) Kähler potential $\Omega : \mathbb{C}^2 \longrightarrow \mathbb{R}$:

$$g = \Omega_{w\bar{w}}dw\,d\bar{w} + \Omega_{w\bar{z}}dw\,d\bar{z} + \Omega_{z\bar{w}}dz\,d\bar{w} + \Omega_{z\bar{z}}dz\,d\bar{z}. \qquad \text{(C6)}$$

The Ricci-flat condition on g gives a non-linear Monge–Ampere equation on Ω (compare (9.3.34)):

$$\Omega_{w\bar{w}}\Omega_{z\bar{z}} - \Omega_{w\bar{z}}\Omega_{z\bar{w}} = 1. \tag{C7}$$

Assume that this metric admits the Killing vector[2] $K = i(\partial_w - \partial_{\bar{w}})$. The Killing equations yield $K(\Omega) = 0$ and the Monge–Ampere equation reduces to

$$\Omega_{vv}\Omega_{z\bar{z}} - \Omega_{vz}\Omega_{v\bar{z}} = 1,$$

where $\Omega = \Omega(z, \bar{z}, v)$ and $v = i(\bar{w} - w) \in \mathbb{R}$. This non-linear PDE is modelled by the EDS generated by

$$<\theta^1 = d\Omega - pdv - qdz - \bar{q}d\bar{z},\ \theta^2 = dq \wedge dp \wedge dz - dz \wedge d\bar{z} \wedge dv>_{\text{diff}}$$

together with the independence condition $dz \wedge d\bar{z} \wedge dv \neq 0$ on an open set in $\mathbb{R}^7$. Consider

$$f(\Omega, z, v, p, q) = (\hat{\Omega}, \hat{z}, \hat{v}, \hat{p}, \hat{q}) = (\Omega - pv, z, p, -v, q).$$

Vanishing of the forms

$$f^*(\hat{\theta}^1) = d\Omega - pdv - qdz - \bar{q}d\bar{z} \quad \text{and} \quad f^*(\hat{\theta}^2) = -dq \wedge dv \wedge dz - dz \wedge d\bar{z} \wedge dp$$

gives the Laplace equation on $\mathbb{R}^3$

$$\Omega_{vv} + \Omega_{z\bar{z}} = 0.$$

In this derivation we assumed non-vanishing of $d\hat{z} \wedge d\bar{\hat{z}} \wedge d\hat{p}$. If this three-form vanishes then $\hat{\Omega}$ is linear in $\hat{v}$ and the Monge–Ampere equations (with hats over all variables) implies that the resulting metric $\hat{g}$ is flat.

Exercise. Implement the change of coordinates at the level of $\hat{g}$ given by (the hatted version of) (C6) to show that it is equivalent to the Gibbons–Hawking form

$$g = Vd\mathbf{x}^2 + V^{-1}(d\tau + A)^2,$$

where $\hat{z} = x + iy$, $\hat{w} = (\tau + iv)/2$, the coordinates (x, y, v, τ) are real, $\mathbf{x} = (x, y, p)$, and (A, V) are a one-form and a harmonic function which satisfy (9.4.36)

$$*_3 dV = dA$$

as a consequence of the Laplace equation (here $*$ is the Hodge operator on $\mathbb{R}^3$ with its flat Euclidean metric).

We shall now prove the existence theorem of integral manifolds which applies to ideals generated by one-forms.

[2] In fact using the freedom $\Omega \to \Omega + \kappa + \bar{\kappa}$ where $\kappa = \kappa(w, z)$ is holomorphic and redefining the holomorphic coordinates $(w, z) \to (\hat{w}(w, z), \hat{z}(w, z))$ one can show that this is the most general form of a Killing vector with which Lie derives the Kähler form and the holomorphic two-form $dw \wedge dz$.

Theorem C.2.3 (Frobenius – Version 1) *Let $\mathcal{I}$ be a differential ideal which is algebraically generated by one-forms $\theta^1, \ldots, \theta^{n-r}$ on some n-dimensional manifold M such that*

$$d\theta^i = \sum_{j=1}^{n-r} \gamma^i_j \wedge \theta^j \tag{C8}$$

for some one-forms γ^i_j (so that $\mathcal{I}$ is closed). In any sufficiently small neighbourhood of a point where θ^i are linearly independent there exists a coordinate system $(y^1, \ldots, y^n)$ such that $\mathcal{I}$ is generated by $dy^{r+1}, \ldots, dy^n$ and the maximal, r-dimensional integral manifolds are

$$y^{r+1} = \text{const}, \; y^{r+2} = \text{const}, \quad \ldots, \quad y^n = \text{const}.$$

Proof Let $W_x = \text{span}(\theta^i|_x) \subset T^*_x M$ and let $W_x{}^\perp \subset T_x M$ be an r-dimensional subspace of vectors annihilating $(\theta^i)_x$.

We shall follow the proof given in [23] and proceed by induction with respect to r. If $r = 1$ then $W^\perp$ is spanned by one vector field X. The Picard existence theorem for ODEs implies the existence of a local coordinate system[3] $y^1, \ldots, y^n$ such that $X = \partial/\partial y^1$. Therefore $W_x = \text{span}\,(dy^2, \ldots, dy^n)$ and we are done. Note that no integrability condition is needed for existence of integral curves so we did not have to use (C8) which in fact holds identically if $r = 1$.

Now assume that $r > 1$ and suppose that the theorem holds for $r - 1$ (which is to say that it holds for $(n - r + 1)$ one-forms). Let x^i be local coordinates such that the set of one-forms $\mathcal{I}' := \{\theta^1, \ldots, \theta^{n-r}, dx^r\}$ is linearly independent. The forms $\theta^1, \ldots, \theta^{n-r}$ satisfy the closure condition (C8) and so this condition is also satisfied by the generators of $\mathcal{I}'$. Therefore, by the inductive hypothesis, there exist coordinates $y^1, \ldots, y^n$ such that $dy^r, \ldots, dy^n$ span $\mathcal{I}'$ and so $x^r = x^r(y^r, \ldots, y^n)$. Assume, without loss of generality, that $\partial x^r/\partial y^r \neq 0$ (no summation!) and solve the relation

$$dx^r = \frac{\partial x^r}{\partial y^r} dy^r + \sum_{i=1}^{n-r} \frac{\partial x^r}{\partial y^{r+i}} dy^{r+i}$$

for dy^r. The one-forms θ^i are in the span of $dy^r, \ldots, dy^n$. Therefore, substituting for dy^r, we get

$$\theta^i = b^i dx^r + \sum_{j=1}^{n-r} a^i{}_j dy^{j+r}, \quad i = 1, \ldots, n - r.$$

The forms θ^i and dx^r are linearly independent so the matrix $(a^i{}_j)$ is non-singular, or otherwise $\sum_i V_i(\theta^i - b^i dx^r) = 0$ for some $V \in \ker(a)$. Thus $a^{-1}\boldsymbol{\theta}$ gives a new set

[3] To see it set $X = \partial/\partial y^1$ at $x = (0, 0, \ldots, 0)$. Then, there is a unique integral curve through each point $(0, a^2, \ldots, a^n)$. If a point x lies on the integral curve through this point we can use $(y^2, \ldots, y^n)$ as the last $(n - 1)$ coordinates of x and the time interval it takes the curve to get to x as the first coordinate.

of generators

$$\tilde{\theta}^i = dy^{r+i} + p^i dx^r, \quad i = 1, \ldots, n-r.$$

The closure condition (C8) gives

$$d\tilde{\theta}^i = dp^i \wedge dx^r = \sum_{k=1}^{r-1} \frac{\partial p^i}{\partial y^k} dy^k \wedge dx^r = 0 \mod \tilde{\theta}^i$$

(recall that dy^r is a combination of dx^r and dy^{r+i} so it does not appear in the summation). Therefore

$$p^i = p^i(y^r, y^{r+1}, \ldots, y^n)$$

and the $(n-r)$ forms $\theta^1, \ldots, \theta^{n-r}$ satisfy the Frobenius condition (C8) in $(n-r+1)$ coordinates. This case corresponds to $r = 1$ and was dealt with at the beginning of the proof. □

We shall now give two more formulations of the Frobenius theorem. One in terms of vector fields and one in terms of overdetermined PDEs.

Recall that a distribution $\mathcal{D}$ of vector fields on a manifold M (or distribution for short) is a sub-bundle of a tangent bundle TM. At each point $x \in M$ it consists of $k(x)$ linearly independent vector fields, where $k(x) \leq \dim M$ is an integer.[4] A distribution is integrable if a Lie bracket of any two vector fields in $\mathcal{D}$ belongs to $\mathcal{D}$. The integrability conditions is often written as $[\mathcal{D}, \mathcal{D}] \subset \mathcal{D}$. Thus, the integral manifolds in the Frobenius theorem are leaves of r-dimensional foliation of M by a distribution $W^{\perp} := \cup_x W_x^{\perp} \subset TM$.

Assume that the Frobenius condition (C8) holds and extend the ideal $\mathcal{I}$ to a basis

$$\theta^1, \ldots, \theta^{n-r}, \theta^{n-r+1}, \ldots, \theta^n$$

of $T_x^* M$ so that

$$d\theta^i = \frac{1}{2} \sum_{j,k=1}^{n} C^i_{jk} \theta^j \wedge \theta^k, \qquad i = 1, \ldots, n$$

for some C^i_{jk}. The closure condition (C8) is equivalent to

$$C^m_{pq} = 0, \qquad m = 1, \ldots, n-r, \quad p, q = (n-r+1), \ldots, n.$$

Define the dual basis X_i of $T_x M$ by

$$df = \sum_{i=1}^{n} X_i(f)\theta^i,$$

where f is any function on M. Differentiating this relation gives

$$0 = d^2 f = \sum_{i,j} X_j[X_i(f)]\theta^j \wedge \theta^i + \frac{1}{2} \sum_{i,j,k} X_i(f) C^i_{jk} \theta^j \wedge \theta^k,$$

[4] The distributions need not have constant rank. For example, a distribution $\{\partial_x, \partial_x + z\partial_y\}$ in $\mathbb{R}^3$ has rank two if $z \neq 0$ and rank one otherwise.

and finally

$$[X_p, X_q] = -\sum_s C^s_{pq} X_s, \qquad p, q, s = (n-r+1), \ldots, n,$$

where the vectors $\{X_{n-r+1}, \ldots, X_n\}$ span the distribution $W^\perp$. However the same distribution is spanned by $\{\partial/\partial y^1, \ldots, \partial/\partial y^r\}$ which gives

Theorem C.2.4 (Frobenius – Version 2) *Let $\{X_{n-r+1}, \ldots, X_n\}$ be an r-dimensional distribution on M such that*

$$[X_p, X_q] = -C^s_{pq} X_s, \qquad p, q, s = (n-r+1), \ldots, n. \tag{C9}$$

In any sufficiently small neighbourhood of a point where X_i are linearly independent there exists a coordinate system $(y^1, \ldots, y^n)$ such that

$$span\{X_{n-r+1}, \ldots, X_n\} = span\{\partial/\partial y^1, \ldots, \partial/\partial y^r\}.$$

For the last formulation of the Frobenius theorem consider a system of PDEs

$$\frac{\partial u^\rho}{\partial x^i} = \psi^\rho_i(x, u), \qquad i = 1, \ldots, n, \quad \rho = 1, \ldots, N, \tag{C10}$$

where $u : \mathbb{R}^n \longrightarrow \mathbb{R}^N$. We want to construct a solution through each point

$$(x^1, \ldots, x^n, u^1, \ldots, u^N) \in \mathbb{R}^{n+N}.$$

This is the same as constructing a foliation of $\mathbb{R}^{n+N}$ by n-dimensional integral surfaces of the ideal generated by

$$<\theta^\rho = du^\rho - \psi^\rho_i dx^i>_{\text{diff}}, \qquad \rho = 1, \ldots, N.$$

The annihilator $W^\perp$ of this ideal is spanned by the vector fields

$$X_i = \frac{\partial}{\partial x^i} + \sum_\rho \psi^\rho_i \frac{\partial}{\partial u^\rho}, \qquad i = 1, \ldots, n.$$

The Frobenius integrability condition

$$[X_i, X_j] = 0$$

gives the necessary and sufficient condition for the existence of the integral manifolds. Note that in this case the commutators must vanish exactly as there is no way of generating $\partial/\partial x^i$ on the RHS of the commutator. Expanding the commutators yields.

Theorem C.2.5 (Frobenius – Version 3) *The necessary and sufficient conditions for the unique solution $u^\alpha = u^\alpha(x)$ to the system (C10) such that $u(x_0) = u_0$ to exist for any initial data $(u_0, x_0) \in \mathbb{R}^{n+N}$ is that the relations*

$$\frac{\partial \psi^\alpha_i}{\partial x^j} - \frac{\partial \psi^\alpha_j}{\partial x^i} + \sum_\beta \left(\frac{\partial \psi^\alpha_i}{\partial u^\beta} \psi^\beta_j - \frac{\partial \psi^\alpha_j}{\partial u^\beta} \psi^\beta_i \right) = 0, \qquad i, j = 1, \ldots, n, \quad \alpha, \beta = 1, \ldots, N \tag{C11}$$

hold.

- **Example.** The one-form

$$\theta = du - A(x, y, u)\,dx - B(x, y, u)\,dy$$

in $\mathbb{R}^3$ satisfies (C8) iff

$$d\theta = \gamma \wedge \theta$$

for some one-form γ, or, equivalently, iff

$$\theta \wedge d\theta = 0.$$

This condition holds iff the compatibility condition (C3) for the pair of overdetermined PDEs $u_x = A$ and $u_y = B$ are satisfied. The Frobenius theorem implies that in this case $\theta = \mu df$ where μ and f are some functions of (x, y, u) and that $f = \text{const}$ is the solution surface in $\mathbb{R}^3$.

- **Example.** Another simple application of the Frobenius theorem is used in general relativity. Any metric g with a Killing vector K on an n-dimensional manifold can locally be written as

$$g = Vh + V^{-1}(d\tau + A)^2,$$

where $(\tau, x^1, \ldots, x^{n-1})$ is a local coordinate system such that $K = \partial/\partial\tau$ and

$$V = V(x), \quad A = A_i(x)\,dx^i, \quad \text{and} \quad h = h_{ij}\,(x)dx^i\,dx^j.$$

Moreover in the twist-free case $K \wedge dK{=}0$ one can redefine the coordinates, the function V and the metric h to set $A = 0$ (we follow the usual abuse of notation and denote the vector K and the one-form $g(K, \ldots)$ by the same symbol).

C.3 Involutivity

Any system of DEs can be rewritten as a system of algebraic equations on a manifold where higher derivatives are regarded as independent variables. This idea is formalized by the apparatus of jet spaces. Let $u : \mathbb{R}^n \longrightarrow \mathbb{R}^N$, so that we can write $u = u^\alpha(x^i)$. The space of k jets $J^k(\mathbb{R}^n, \mathbb{R}^N)$ is the space of Taylor polynomials of u of degree k. It is a smooth manifold of dimension

$$n + N\binom{n+k}{k}$$

with local coordinates

$$\{x^i, u^\alpha, p^\alpha_i, p^\alpha_{ij}, \ldots, p^\alpha_{i_1 i_2 \ldots i_k}\}, \quad \alpha = 1, \ldots, N, \; i = 1, \ldots, n.$$

Any map $u : \mathbb{R}^n \longrightarrow \mathbb{R}^N$ can be lifted to a k graph of u (a section of the jet bundle $J^k(\mathbb{R}^n, \mathbb{R}^N) \to \mathbb{R}^n$) by

$$u^\alpha = u^\alpha(x), \quad p^\alpha_i = \frac{\partial u^\alpha}{\partial x^i}(x), \quad \ldots, \quad p^\alpha_{i_1 i_2 \cdots i_k} = \frac{\partial^k u^\alpha}{\partial x^{i_1}\partial x^{i_2}\cdots\partial x^{i_k}}(x).$$

The system of r kth-order PDEs

$$F^{\rho}\left(x^i, u^{\alpha}, \frac{\partial u^{\alpha}}{\partial x^i}, \dots, \frac{\partial^k u^{\alpha}}{\partial x^{i_1}\partial x^{i_2}\cdots \partial x^{i_k}}\right) = 0, \qquad \rho = 1, \dots, r, \tag{C12}$$

gives a submanifold $M^{(k)}$ of co-dimension r in $J^k(\mathbb{R}^n, \mathbb{R}^N)$ and a k graph of the solution to (C12) is an n-dimensional integral submanifold $S \subset M^{(k)}$ of the ideal associated to (C12) such that $dx^1 \wedge dx^2 \wedge \cdots \wedge dx^n \neq 0$ on S. The $(k+1)$th graph of the solution lies in a manifold $M^{(k+1)} \subset J^{k+1}(\mathbb{R}^n, \mathbb{R}^N)$ called a prolongation of $M^{(k)}$. The manifold $M^{(k+1)}$ is defined as a zero locus

$$F^{\rho} = 0, \quad \frac{dF^{\rho}}{dx^i} = 0, \qquad \rho = 1, \dots, r, \quad i = 1, \dots, n$$

in $J^{k+1}(\mathbb{R}^n, \mathbb{R}^N)$.

For any integer $l \geq 0$ define the family of projections

$$\pi_l : J^{l+1}(\mathbb{R}^n, \mathbb{R}^N) \longrightarrow J^l(\mathbb{R}^n, \mathbb{R}^N)$$

by

$$\pi_l(x^i, u^{\alpha}, p_i^{\alpha}, \dots, p^{\alpha}_{i_1 i_2 \dots i_l}, p^{\alpha}_{i_1 i_2 \dots i_l i_{l+1}}) = (x^i, u^{\alpha}, p_i^{\alpha}, \dots, p^{\alpha}_{i_1 i_2 \dots i_l}).$$

Therefore $\mathrm{Im}[M^{(k+1)}] \subset M^{(k)}$ (this is obvious as $F^{\rho} = 0$ holds on $M^{(k+1)}$) but π_k does not have to be surjective: differentiating the PDEs (C12), mixing partial derivatives, and using (C12) gives rise to new PDEs of order lower than k. So the image of $M^{(k+1)}$ under π_k will in general be a submanifold of $M^{(k)}$ of some non-zero co-dimension. Therefore the k jets of a solutions do not have to extend to $(k+1)$ jets. We keep differentiating and adding lower order conditions restricting $M^{(k)}$. When can we stop this process? The combined system of equations and lower order conditions must be involutive. In general one needs the Cartan test which will be discussed in Section C.6. Theorems C.3.1 and C.3.2 which we will prove in this section answer this question for systems of first-order PDEs (C10):

$$\frac{\partial u^{\rho}}{\partial x^i} = \psi_i^{\rho}(x, u), \qquad i = 1, \dots, n, \quad \rho = 1, \dots, N.$$

If the Frobenius integrability conditions (C11) hold, the general solution of (C10) depends on N arbitrary constants. Otherwise (C11) give a set of algebraic equations

$$F_1(u, x) = 0$$

which must be satisfied by any solution to (C10). Differentiating these equations and eliminating the derivatives of u using (C10) leads to a new set of equations

$$F_2(u, x) = 0.$$

Proceeding in this way we get a sequence of sets of equations

$$F_1(u, x) = 0, \quad F_2(u, x) = 0, \quad F_3(u, x) = 0, \dots .$$

If the system (C10) admits a solution there must be an integer K such that the equations in the set $F_{K+1} = 0$ are satisfied as a consequence of the equations in the first K sets. Otherwise we would obtain more than N independent conditions on $(u^1, \dots, u^N)$ which would imply a relation between the independent variables. In particular we must have $K \leq N$. This proves the 'only if' statement in the following:

Theorem C.3.1 *The system* (C10) *admits solutions iff there exists a positive integer $K \leq N$ such that the set of algebraic equations*

$$F_1 = F_2 = \cdots = F_K = 0$$

is compatible for all $x \in U \subset \mathbb{R}^n$ and that the set $F_{K+1} = 0$ is satisfied identically. If p is the number of independent equations in the first K sets, then the general solution depends on $(N - p)$ arbitrary constants.

Proof It remains to prove the 'if' part. We follow the classical treatment given for example in [167]. Assume that the first K independent sets impose $p < N$ independent conditions

$$G_\nu(u, x) = 0, \qquad \nu = 1, \dots, p. \tag{C13}$$

Therefore

$$\text{rank}\left(\frac{\partial G_\nu}{\partial u^\alpha}\right) = p$$

and, by the implicit function theorem, the relations (C13) can be solved for (say) the first p functions $u^1, \dots, u^p$:

$$u^\lambda = \phi^\lambda(u^{p+1}, \dots, u^N, x), \qquad \lambda = 1, \dots, p.$$

Differentiate this and use (C10) to eliminate the derivatives

$$\psi_i^\lambda - \sum_{\nu=p+1}^{N} \frac{\partial \phi^\lambda}{\partial u^\nu} \psi_i^\nu - \frac{\partial \phi^\lambda}{\partial x^i} = 0.$$

These equations belong to the set $F_{K+1} = 0$ so they hold by assumption. We rewrite the above equations substituting $\psi_i^\lambda = \partial u^\lambda / \partial x^i$ and subtracting

$$\frac{\partial u^\lambda}{\partial x^i} - \psi_i^\lambda - \sum_{\nu=p+1}^{N} \frac{\partial \phi^\lambda}{\partial u^\nu} \left(\frac{\partial u^\nu}{\partial x^i} - \psi_i^\nu\right) = 0$$

so

$$\frac{\partial u^\nu}{\partial x^i} = \overline{\psi}_i^\nu(u^{p+1}, \dots, u^N, x), \tag{C14}$$

where $\nu = p + 1, \dots, N$ and

$$\overline{\psi}_i^\nu = \psi_i^\nu|_{u^\lambda = \phi^\lambda(u^{p+1}, \dots, u^N, x)}.$$

The system (C14) is Frobenius integrable as the consistency belongs to the set

$$F_1 = \cdots = F_K = 0$$

so, by the Frobenius theorem (Theorem C.2.5), there is a solution which involves $(N - p)$ constants. □

In many applications the functions ψ_i^α in (C10) are linear and homogeneous in u^ρ. This allows the following geometric interpretation of the last theorem. Let us write the system of linear homogeneous PDEs

$$\frac{\partial u^\rho}{\partial x^i} = \psi^\rho_{\gamma i}(x) u^\gamma$$

as

$$d\mathbf{u} + \mathbf{\Omega}\mathbf{u} = 0, \tag{C15}$$

where $\mathbf{u} = (u^1, \ldots, u^N)^{\mathrm{T}}$ and $\mathbf{\Omega} = -\psi^\rho_{\gamma i} dx^i$ is a matrix-valued one-form on an open set $U \subset \mathbb{R}^n$. Therefore solutions to (C15) correspond to parallel sections $\mathbf{u} : U \to \mathbb{E}$ of a rank N vector bundle $\mathbb{E} \to U$ with connection $D = d + \mathbf{\Omega}$. Locally the total space of this bundle is an open set in $\mathbb{R}^{n+N}$. To simplify notation let us assume that $n = 2$ and (x^1, x^2) are local coordinates in $U \subset \mathbb{R}^2$.

Differentiating (C15) and eliminating $d\mathbf{u}$ yields $\mathbf{F}\mathbf{u} = 0$, where

$$\begin{aligned}\mathbf{F} &= d\mathbf{\Omega} + \mathbf{\Omega} \wedge \mathbf{\Omega} = (\partial_1 \mathbf{\Omega}_2 - \partial_2 \mathbf{\Omega}_1 + [\mathbf{\Omega}_1, \mathbf{\Omega}_2]) dx^1 \wedge dx^2 \\ &= F dx^1 \wedge dx^2\end{aligned}$$

is the curvature of D. Thus we need

$$F\mathbf{u} = 0, \tag{C16}$$

where $F = F(x^1, x^2)$ is an $N \times N$ matrix. This is the first set of conditions $F_1 = 0$ in Theorem C.3.1. In this case these conditions are just linear homogeneous equations. If $F = 0$ and the connection is flat, there exist N-independent parallel sections. In this case the Frobenius integrability conditions (C11) hold. On the other hand if $\det(F) \neq 0$ then no non-zero parallel sections exists.

In general we want to determine the dimension of the space of parallel sections. To achieve this, differentiate the condition (C16) and use (C15) to obtain

$$0 = dF\mathbf{u} - F\mathbf{\Omega}\mathbf{u} = [(\partial_i F - F\mathbf{\Omega}_i)\mathbf{u}]\, dx^i.$$

Using $F\mathbf{u} = 0$ we rewrite this as

$$(D_i F)\mathbf{u} = 0,$$

where $D_i F = \partial_i F + [\mathbf{\Omega}_i, F]$.

We continue differentiating to produce algebraic matrix equations

$$F\mathbf{u} = 0, \quad (D_i F)\mathbf{u} = 0, \quad (D_i D_j F)\mathbf{u} = 0, \quad (D_i D_j D_k F)\mathbf{u}, \quad \ldots.$$

These are the conditions $F_1 = 0, F_2 = 0, F_3 = 0, \ldots$ in Theorem C.3.1. After K differentiations this leads to $r(K)$ linear equations which we write as

$$\mathcal{F}_K \mathbf{u} = 0,$$

where $\mathcal{F}_K$ is a $r(K)$ by N matrix. We also set $\mathcal{F}_0 = F$. Theorem C.3.1 adapted to (C16) and (C15) tells us when we can stop the process.

Theorem C.3.2 *Assume that the ranks of the matrices $\mathcal{F}_K$, $K = 0, 1, 2, \ldots$, are maximal and constant.*[5] *Let K_0 be the smallest natural number such that*

$$rank\ (\mathcal{F}_{K_0}) = rank\ (\mathcal{F}_{K_0+1}). \tag{C17}$$

If K_0 exists then $rank(\mathcal{F}_{K_0}) = rank(\mathcal{F}_{K_0+k})$ for $k \in \mathbb{N}$ and the space of parallel sections (C15) *of $d + \mathbf{\Omega}$ has dimension $(N - rank(\mathcal{F}_{K_0}))$.*

Thus if the curvature of $(\mathbb{E}, D)$ does not vanish, the non-zero solutions to the system of linear PDEs can exist if the holonomy D lies in some proper subgroup of $GL(N, \mathbb{R})$.

C.4 Prolongation

The theorems presented in the last section apply to systems of first-order PDEs. Given an arbitrary system of PDEs we could aim to represent it as a first-order system on a jet space of higher dimension by introducing new variables for second and higher derivatives. This process will however lead to systems where not all first derivatives are determined (compare the system (C4)) and Theorems C.2.5 and C.3.1 cannot be applied to construct the solution surfaces. The idea of prolongation is to introduce more new variables for unknown derivatives aiming to express derivatives of these variables using the (differential consequences of) the original system. Apriori it is not clear that this process will work (i.e. the process of adding new variables may never terminate). The relevant theorems which state under what circumstances the prolongation works were, in case of linear PDEs, given independently by Spencer, Kuranishi, and Goldschmidt. See chapter 5 of [23] for a complete exposition of these ideas and [20] for a treatment which uses vector bundles and is close to our approach.

Let $P : E_1 \longrightarrow E_2$ be a linear kth-order differential operator between two smooth vector bundles over a manifold M. In local coordinates

$$P(v) = a^{i_1 i_2 \cdots i_k} \frac{\partial^k v}{\partial x^{i_1} \partial x^{i_2} \cdots \partial x^{i_k}} + \cdots,$$

where $(\ldots)$ denote lower order terms. The leading term $a^{i_1 i_2 \cdots i_k}$ transforms as a tensor under the change of coordinates and gives rise to a bundle map called the symbol of P:

$$\sigma(P) : \odot^k \Lambda^1(M) \otimes E_1 \longrightarrow E_2.$$

[5] This can always be achieved by restricting to a sufficiently small neighbourhood of some point $x \in U$.

Thus the symbol is a matrix whose components are polynomials homogeneous of degree k:

$$\sigma(P) = (a^{i_1 i_2 \cdots i_k}\xi_{i_1}\xi_{i_2}\cdots\xi_{i_k})^{\beta}_{\alpha}, \qquad \alpha = 1, \ldots, \operatorname{rank}(E_1), \ \beta = 1, \ldots, \operatorname{rank}(E_2).$$

For any integer $s \geq k$ define the vector spaces

$$V_s := (\odot^k \Lambda^1(M) \otimes E_1) \cap (\odot^{(s-k)} \Lambda^1(M) \otimes \ker[\sigma(P)]).$$

The system (P, E_1, E_2) is said to be of *finite type* if $V_s = 0$ for s sufficiently large. The seminal result of Spencer [150] is that for systems of finite type the equation

$$P(v) = 0$$

is equivalent to a closed system of PDEs of the form (C10), where all partial derivatives of the dependent variables are determined. The criterion for a given system to be of finite type is given in [150], but in practice it can be difficult to implement, as the vector spaces V_s cannot be easily constructed. For systems not of finite type the process of adding new variables and cross-differentiating the equations will never end.

In the last section we explained how to regard a closed linear system as a vector bundle $\mathbb{E}$ with a connection D. In the work of Spencer the bundle $\mathbb{E}$ arises as a direct sum $\oplus_s V_s$. Theorem C.3.2 can be adapted to systems of finite type.

Theorem C.4.1 *For systems of finite type there exists a vector bundle $\mathbb{E} \to M$ with a connection D and a bijection*

$$\{v \in \Gamma(E_1) \ \textit{such that} \ P(v) = 0\} \to \{\mathbf{u} \in \Gamma(\mathbb{E}), \ D\mathbf{u} = 0\}.$$

The dimension of the kernel of P is bounded by the rank of $\mathbb{E}$.

The determined system of equations for $D\mathbf{u} = 0$ is the prolongation of the system $P(v) = 0$. Theorem C.3.2 can now be applied to give an algorithm for calculating the dimension of the kernel of D. In many geometric applications, where P is built out of covariant derivatives for some connection on TM, the bundle with connection $(\mathbb{E}, D)$ is called the tractor bundle [20].

- **Example.** Let (M, g) be an n-dimensional (pseudo) Riemannian manifold and let ∇ be the Levi-Civita connection of g. The Killing equations

$$\nabla_{(i}v_{j)} = 0 \tag{C18}$$

can be put into the framework described in this section with

$$E_1 = \Lambda^1(M) \quad \text{and} \quad E_2 = \Lambda^1(M) \odot \Lambda^1(M).$$

The system (C18) is equivalent to the first-order system

$$\nabla_i v_j = \mu_{ij}$$
$$\nabla_i \mu_{jk} = R_{jki}{}^m v_m,$$

where μ_{ij} is antisymmetric, $R_{jki}{}^m$ is the Riemann curvature of g, and we arrived at the second equation by using $\nabla_{[i}\mu_{jk]} = 0$ and commuting the covariant derivatives on v. We combine (v_i, μ_{ij}) into a section

$$\mathbf{u} = \begin{pmatrix} v_i \\ \mu_{ij} \end{pmatrix}$$

of the vector bundle $\mathbb{E} = \Lambda^1(M) \oplus \Lambda^2(M)$ with connection D:

$$\begin{pmatrix} v_j \\ \mu_{jk} \end{pmatrix} \stackrel{D_i}{\longmapsto} \begin{pmatrix} \nabla_i v_j - \mu_{ij} \\ \nabla_i \mu_{jk} - R_{jki}{}^m v_m \end{pmatrix}. \tag{C19}$$

The solutions of the Killing equation (C18) are in one-to-one correspondence with parallel sections of D. The number of these parallel sections does not exceed

$$\mathrm{rank}(\mathbb{E}) = \mathrm{rank}(\Lambda^1) + \mathrm{rank}(\Lambda^2) = \frac{n(n+1)}{2}.$$

This upper bound is also the dimension of the Lie algebra of the orthogonal group. It is achieved for spaces of constant curvature.

- **Example.** The CR equations

$$u_x = v_y \quad \text{and} \quad u_y = -v_x$$

where u and v are functions of (x, y) are not of finite type (the reader is invited to try first few iterations of the prolongation procedure). In fact no uniqueness result analogous to Theorem C.3.2 is expected to hold. The general solution to the CR equations depends on one holomorphic function of $(x + iy)$ rather than on a finite number of constants.

C.4.1 Differential invariants

The prolongation procedure together with Theorem C.3.2 gives a straightforward algorithm for constructing invariants which obstruct existence of certain geometric structures. We shall look at two examples: a relatively simple (but sufficiently non-trivial!) example of Killing equations in Riemannian geometry and more involved problem of existence of metric connections in a given projective class [26]. Our treatment of the subject is based on restricting the holonomy of a connection of some vector bundle. The more common principal bundle approach (due to Cartan) is used in [23].

- **Question.** Let g be a (pseudo) Riemannian metric on an open set U in $\mathbb{R}^2$. When is g the metric on a surface of revolution?

Any metric on a surface of revolution takes the form

$$g = dx^2 + f(x)dy^2$$

in some coordinates where $f = f(x)$ is a non-vanishing function of one variable. This metric admits the Killing vector $v = \partial/\partial y$. Conversely, the existence of a non-trivial solution to the Killing equations (C18) guarantees the existence of this coordinate system. Therefore an equivalent form of the question is: When does a metric on a surface admit a solution to (C18)? The answer must have been known to the classical differential geometers in the nineteenth century: Darboux states it in his book [36] without proof. We shall give the answer as the vanishing of two weighted scalar invariants constructed out of g: one invariant of order four and one invariant of order five.

The metrics of constant curvature admit three Killing vectors (which is the maximal number). The following theorem (also known to Vladimir Matveev and proved in [101] using different methods) applies to metrics with non-constant curvature.

Theorem C.4.2 *A Riemannian metric g on a surface with non-constant scalar curvature R admits a Killing vector in a neighbourhood of a point $p \in U$ such that $dR \neq 0$ at p iff*

$$I_1 := dR \wedge d(|\nabla R|^2) = 0 \quad and \quad I_2 := dR \wedge d[\Delta(R)] = 0, \tag{C20}$$

where

$$|\nabla R|^2 = g^{ij}\nabla_i R \nabla_j R \quad and \quad \Delta(R) = g^{ij}\nabla_i \nabla_j R.$$

Proof Solutions to the Killing equation (C18) are in one-to-one correspondence with parallel sections of the connection (C19). We want to find necessary and sufficient conditions for the existence of one such section. The prolongation procedure simplifies in two dimensions. Firstly any two-form is a multiple of a (chosen) volume form, thus we can write

$$\mu_{ij} = |g|^{1/2}\varepsilon_{ij}\,\mu,$$

where $|g| = |\det g|$ for some section of the canonical bundle μ. Moreover the Riemann tensor is determined by the scalar curvature R:

$$R_{ijkl} = \frac{R}{2}(g_{ik}g_{jl} - g_{jk}g_{il}).$$

With these simplifications the connection (C19) reduces to a connection D on a rank-three vector bundle $\mathbb{E} \to U$:

$$\begin{bmatrix} v_j \\ \mu \end{bmatrix} \overset{D_i}{\longmapsto} \begin{bmatrix} \nabla_i v_j - |g|^{1/2}\varepsilon_{ij}\,\mu \\ \nabla_i \mu - \frac{1}{2}|g|^{-3/2}\varepsilon_i^{\,j} R\, v_j \end{bmatrix}.$$

Using $\nabla_{[i}\nabla_{j]}\mu = 0$ and eliminating the first derivatives of $\mathbf{u} = (v_i, \mu)^T$ gives

$$(\nabla_i R)v^i = 0, \tag{C21}$$

where $v^i = g^{ij}v_j$. This is the condition (C16) leading to Theorem C.3.2 where the curvature of the connection is given by the 3×3 matrix of rank one:

$$F = \begin{pmatrix} 0 & 0 & 0 \\ 0 & 0 & 0 \\ \nabla^1 R & \nabla^2 R & 0 \end{pmatrix}.$$

Differentiating (C21) or equivalently differentiating the tractor curvature F covariantly with respect to D gives two more conditions:

$$|g|^{3/2}(\nabla^i R)\mu + \varepsilon^{ji}(\nabla_j \nabla_k R)v^k = 0. \tag{C22}$$

Therefore the determinant of a 3×3 matrix

$$\begin{pmatrix} \nabla_1 R & \nabla_2 R & 0 \\ -\nabla_2\nabla_1 R & -\nabla_2\nabla_2 R & |g|^{3/2}\nabla^1 R \\ \nabla_1\nabla_1 R & \nabla_1\nabla_2 R & |g|^{3/2}\nabla^2 R \end{pmatrix} \tag{C23}$$

should vanish for non-zero parallel sections of $(\mathbb{E}, D)$ to exist. Calculating this determinant yields the first obstruction I_1 in (C20). This is the necessary condition for the existence of a Killing vector. Assume that this condition holds. The rank of the matrix (C23) has to be smaller than three. It is equal to zero if the scalar curvature R is constant. In this case the tractor connection is flat. Otherwise, in a neighbourhood of a point where $\nabla_i R \neq 0$, the rank is equal to two and constant. Theorem C.3.2 implies that the sufficient conditions are obtained by demanding that the rank of the 6×3 matrix obtained from the matrix (C23) and the second derivatives of (C21) does not go up and is equal to two. This could a priori lead to three additional obstructions. However only one of them is a new condition and the other two follow as differential consequences of (C21). To see this write the first algebraic obstruction (C21) as

$$\mathbf{V} \cdot \mathbf{u} = 0,$$

where $\mathbf{V} = (\nabla_1 R, \nabla_2 R, 0)$. Let $\mathbf{V}_{ij\dots k}$ denote the vector in $\mathbb{R}^3$ orthogonal to $\mathbf{u}$ which is obtained by eliminating the derivatives of $\mathbf{u}$ from $\partial_i\partial_j \cdots \partial_k(\mathbf{V}\cdot\mathbf{u}) = 0$. Vanishing of the first obstruction (C21) implies the linear dependence condition

$$c\mathbf{V} + c_1\mathbf{V}_1 + c_2\mathbf{V}_2 = 0 \tag{C24}$$

for some functions c, c_1, c_2 on U. Assume that we add one more condition

$$e\mathbf{V} + e_1\mathbf{V}_1 + e_2\mathbf{V}_2 + e_{12}\mathbf{V}_{12} = 0$$

for some functions $(e, \dots, e_{12})$ on U. This gives an obstruction $I_2 := \det(\mathbf{V}, \mathbf{V}_i, \mathbf{V}_{12}) = 0$ where i equals 1 or 2 (there is only one obstruction because of the earlier linear dependence condition). Now differentiating (C24) with respect to x^i and using $\mathbf{V}_{12} = \mathbf{V}_{21}$ which holds modulo lower order terms imply that $\mathbf{V}_{11}$ and $\mathbf{V}_{22}$ are in the span of $\mathbf{V}, \mathbf{V}_1, \mathbf{V}_2$ and no additional conditions need to be added. To write the second obstruction I_2 we could differentiate (C22) and take a determinant of one of the resulting 3×3 matrices. Alternatively we can take the Laplacian of (C21) and eliminate the first derivatives of $\mathbf{u}$. This leads to the linear dependence of dR and $d[\Delta(R)]$ which is equivalent to the vanishing of I_2 in (C20). Both methods lead to obstructions of differential order five in the components of the metric g. The argument presented above shows that the resulting sets of obstructions are equivalent. This completes the proof. □

We shall give one more example using the prolongation procedure and Theorem C.3.2 to produce differential invariants. This time two iterations of the prolongation procedure will be needed to close the system. The aim is to answer the following:

- **Question.** Cover a two-dimensional plane with a family of curves, one curve through each point in each direction. How can you tell whether these curves are the geodesics of some metric?

This is an old problem which goes back at least to the work of Roger Liouville [107] in 1887. The solution was given in [26]. The following discussion summarizes the main results. Assume that the curves are presented as integral curves of a second-order ODE

$$\frac{d^2y}{dx^2} = \Lambda\left(x, y, \frac{dy}{dx}\right).$$

Thus we want to find conditions on the ODE so that its integral curves are unparameterized geodesics of some metric connection. First of all they need to be geodesics of some symmetric connection with Christoffel symbols Γ^k_{ij}. Eliminating the parameter t between the geodesic equations

$$\ddot{x}^i + \Gamma^i_{jk}\dot{x}^j\dot{x}^k \sim \dot{x}^i, \qquad x^i = x^i(t)$$

with $(x^1, x^2) = (x, y)$ yields a second-order ODE of the form

$$\frac{d^2y}{dx^2} = A_0(x, y) + A_1(x, y)\frac{dy}{dx} + A_2(x, y)\left(\frac{dy}{dx}\right)^2 + A_3(x, y)\left(\frac{dy}{dx}\right)^3, \tag{C25}$$

where

$$A_0 = -\Gamma^2_{11}, \quad A_1 = \Gamma^1_{11} - 2\Gamma^2_{12}, \quad A_2 = 2\Gamma^1_{12} - \Gamma^2_{22}, \quad \text{and} \quad A_3 = \Gamma^1_{22}.$$

Conversely, any ODE of the form (C25) defines an equivalence class of connections which share the same unparameterized geodesics. Thus

$$\frac{\partial^4 \Lambda}{\partial(y')^4} = 0$$

is the first necessary condition for metricity of paths. One can check that this condition is invariant under the coordinate transformations $(x, y) \to (\hat{x}(x, y), \hat{y}(x, y))$.

Now assume that there exists a (pseudo) Riemannian metric

$$g = Edx^2 + 2F\,dxdy + Gdy^2$$

such that the functions $A_0, \ldots, A_3$ arise from the Levi-Civita connection Γ^k_{ij} of g. Following R, Liouville [107] introduce the 2×2 matrix

$$\sigma^{ij} = \begin{pmatrix} \psi_1 & \psi_2 \\ \psi_2 & \psi_3 \end{pmatrix},$$

where

$$E = \psi_1/\Delta, \quad F = \psi_2/\Delta, \quad G = \psi_3/\Delta, \quad \text{and} \quad \Delta = (\psi_1\psi_3 - {\psi_2}^2)^2.$$

Calculating the Levi-Civita connection in terms of the ψ's shows that the integral curves of the ODE (C25) are metrizable on a neighbourhood of a point $x \in U$ iff there exists σ^{ij} such that $\det(\sigma)$ does not vanish at x and following set of equations hold:[6]

$$\begin{aligned}
\frac{\partial \psi_1}{\partial x} &= \frac{2}{3} A_1 \psi_1 - 2 A_0 \psi_2, \\
\frac{\partial \psi_3}{\partial y} &= 2 A_3 \psi_2 - \frac{2}{3} A_2 \psi_3, \\
\frac{\partial \psi_1}{\partial y} + 2 \frac{\partial \psi_2}{\partial x} &= \frac{4}{3} A_2 \psi_1 - \frac{2}{3} A_1 \psi_2 - 2 A_0 \psi_3, \\
\frac{\partial \psi_3}{\partial x} + 2 \frac{\partial \psi_2}{\partial y} &= 2 A_3 \psi_1 - \frac{4}{3} A_1 \psi_3 + \frac{2}{3} A_2 \psi_2.
\end{aligned} \tag{C26}$$

We need to prolong this system and look for integrability conditions, but let us first rewrite the system in more invariant form. Recall that a projective structure on an open set $U \subset \mathbb{R}^2$ is an equivalence class of torsion-free connections $[\Gamma]$. Two connections Γ and $\hat{\Gamma}$ are projectively equivalent if they share the same unparameterized geodesics. The analytic expression for this equivalence is

$$\hat{\Gamma}_{ij}^k = \Gamma_{ij}^k + \delta_i^k \omega_j + \delta_j^k \omega_i, \qquad i, j, k = 1, 2 \tag{C27}$$

for some one-form $\omega = \omega_i dx^i$.

Thus, in the language of projective differential geometry, we are looking for local conditions on a connection Γ_{ij}^k for the existence of a one-form ω_i and a symmetric non-degenerate tensor g_{ij} such that the projectively equivalent connection is the Levi-Civita connection for g_{ij}, that is,

$$\Gamma_{ij}^k + \delta_i^k \omega_j + \delta_j^k \omega_i = \frac{1}{2} g^{kl} \left(\frac{\partial g_{il}}{\partial x^j} + \frac{\partial g_{jl}}{\partial x^i} - \frac{\partial g_{ij}}{\partial x^l} \right).$$

This is an overdetermined system: there are six components in Γ_{ij}^k and five components in the pair (g_{ij}, ω_i).

Let $\Gamma \in [\Gamma]$ be a connection in the projective class. Its curvature is defined by

$$[\nabla_i, \nabla_j] X^k = R_{ijl}{}^k X^l$$

and can be uniquely decomposed as

$$R_{ijl}{}^k = \delta_i^k \mathrm{P}_{jl} - \delta_j^k \mathrm{P}_{il} + \beta_{ij} \delta_l^k \tag{C28}$$

where β_{ij} is skew. In dimensions higher than two there would be another term (the projective Weyl tensor) in this curvature but in two dimensions this vanishes identically.

If we change the connection in the projective class using (C27) then

$$\hat{\mathrm{P}}_{ij} = \mathrm{P}_{ij} - \nabla_i \omega_j + \omega_i \omega_j \quad \text{and} \quad \hat{\beta}_{ij} = \beta_{ij} + 2 \nabla_{[i} \omega_{j]}.$$

If the de Rham cohomology class $[\beta] \in H^2(U, \mathbb{R})$ vanishes then we can set β_{ij} to 0 by a choice of ω_i in (C27). We are looking for a local metrisability condition

[6] Calculating the expressions $A_0, \ldots, A_3$ directly in terms of (E, F, G) and their first derivatives without introducing ψ's would lead to non-linear relations.

on U so we shall assume that this global cohomological obstruction vanishes. The residual freedom in changing the representative of the equivalence class (C27) is given by gradients $\omega_i = \nabla_i f$, where f is a function on U.

Now $P_{ij} = P_{ji}$ and the Ricci tensor of Γ is symmetric. The Bianchi identity implies that Γ is flat on the bundle of volume forms on U. Thus the equivalent way to normalize ∇_i is to require the existence of skew-symmetric ε^{ij} such that

$$\nabla_i \varepsilon^{jk} = 0.$$

We shall use the volume forms to raise and lower indices according to $z_i = \varepsilon_{ij} z^j$ and $z^i = z_j \varepsilon^{ji}$ where $\varepsilon_{ij}\varepsilon^{ik} = \delta_j^k$. Locally, such a volume form is unique up to scale: let us fix one.

With these preliminaries there exists a representative Γ in a projective class such that the linear system (C26) becomes

$$\nabla_{(i}\sigma_{jk)} = 0,$$

where $\sigma_{ij} = \varepsilon_{il}\varepsilon_{jk}\sigma^{kl}$. Its prolongation gives rise to a connection on a rank-six vector bundle $\mathbb{E}$ over U. Specifically, sections of this bundle comprise triples of contravariant tensors $\mathbf{u} = (\sigma^{ij}, \mu^i, \rho)$ with σ^{ij} being symmetric. The connection is given by

$$\begin{bmatrix} \sigma^{jk} \\ \mu^j \\ \rho \end{bmatrix} \xrightarrow{D_i} \begin{bmatrix} \nabla_i \sigma^{jk} - \delta_i^j \mu^k - \delta_i^k \mu^j \\ \nabla_i \mu^j - \delta_i^j \rho + P_{ik}\sigma^{jk} \\ \nabla_i \rho + 2P_{ij}\mu^j - 2Y_{ijk}\sigma^{jk} \end{bmatrix}, \tag{C29}$$

where $Y_{ijk} = \frac{1}{2}(\nabla_i P_{jk} - \nabla_j P_{ik})$ is the Cotton tensor. The curvature of the connection D is obtained from $\nabla_{[i}\nabla_{j]}\rho = 0$. It is a 6×6 matrix of rank one. The first condition analogous to (C16) is

$$5Y_i\mu^i + (\nabla_i Y_j)\sigma^{ij} = 0, \qquad \text{where} \quad Y_k = \varepsilon^{ij} Y_{ijk}.$$

Differentiating this equation twice and eliminating the first derivatives shows that the 6×6 matrix

$$\mathcal{M} = \left[\begin{bmatrix} 0 \\ 5Y_k \\ \nabla_{(j}Y_{k)} \end{bmatrix}, D_i \begin{bmatrix} 0 \\ 5Y_k \\ \nabla_{(j}Y_{k)} \end{bmatrix}, D_{(i}D_{j)} \begin{bmatrix} 0 \\ 5Y_k \\ \nabla_{(k}Y_{l)} \end{bmatrix} \right] \tag{C30}$$

must be singular. Its determinant gives the first obstruction to metrisability of a projective structure. A more detailed calculation shows that the expression for $\det(\mathcal{M})$ involves raising an index 14 times using the volume form ε and gives rise to a projectively invariant section of the 14th power of the canonical bundle

$$\det(\mathcal{M})\,(dx \wedge dy)^{\otimes 14}$$

which gives a projective invariant.

Analysis of the necessary conditions using Theorem C.3.2 leads to higher order obstructions. If $\det(\mathcal{M}) = 0$ and $\mathrm{rank}(\mathcal{M}) = 5$ there will be two additional obstructions of order six in the components of the connection. If $2 < \mathrm{rank}(\mathcal{M}) < 5$ then there is one obstruction of order seven in the rank-four case and of order eight in the rank-three case. If rank($\mathcal{M}$) = 2 there always exists a four-dimensional space of metrics compatible with the projective structure. Finally if rank($\mathcal{M}$) < 2 then Γ is projectively flat. See [26] for details and proofs.

C.5 Method of characteristics

If a differential ideal on M generated by a one one–form θ is closed then the Frobenius theorem (Theorem C.2.3) provides a simple local normal form: There exist functions μ and y on M such that $\theta = \mu\, dy$. The next theorem gives a stronger result and can be applied to the case when the Frobenius conditions do not hold.

Theorem C.5.1 (Pfaff) *Let $(M, \mathcal{I})$ be an EDS such that $\mathcal{I} = <\theta>_{\text{diff}}$ for some non-vanishing one-form θ and let $r \geq 0$ be the smallest integer such that*

$$\theta \wedge d\theta^{r+1} = 0.$$

Set dim $(M) = N$. *For each $x \in M$ such that $\theta \wedge d\theta^r \neq 0$ at x there exists a coordinate system*

$$(v, y^1, \ldots, y^r, q_1, \ldots, q_r, z^{2r+2}, \ldots, z^N)$$

in the neighbourhood of x such that $\mathcal{I} = <dv>$ if $r = 0$ and, if $r > 0$,

$$\mathcal{I} = <dv - q_1 dy^1 - \cdots - q_r dy^r>_{\text{diff}}$$

and moreover

- *There exists a maximal $(N - r - 1)$-dimensional integral manifold of $\mathcal{I}$*

$$v = q_1 = q_2 = \cdots = q_r = 0.$$

- *Any integral manifold near this one depends on one arbitrary function of r variables, $f(y^1, \ldots, y^r)$ and is given by*

$$v = f(y^1, \ldots, y^r) \quad \text{and} \quad q_k = \frac{\partial f}{\partial y^k}(y^1, \ldots, y^r), \quad k = 1, \ldots, r.$$

This theorem is proved in [23]. We shall not reproduce this proof, but instead concentrate on one important application: the method of characteristics.

Consider a single first-order PDE

$$F\left(x^1, \ldots, x^n, u, \frac{\partial u}{\partial x^1}, \ldots, \frac{\partial u}{\partial x^n}\right) = 0. \tag{C31}$$

This PDE defines a co-dimension one-manifold $M \subset J^1(\mathbb{R}^n, \mathbb{R})$ of the $(2n + 1)$-dimensional first jet space $J^1(\mathbb{R}^n, \mathbb{R})$ with coordinates $(x^i, u, p_i := \partial u/\partial x^i)$. If we assume that F is smooth and not all partial derivatives $\partial F/\partial p_i$ vanish at any single point then the implicit function theorem implies that the surface M given by

$$F(x^1, \ldots, x^n, u, p_1, \ldots, p_n) = 0$$

is a smooth manifold. The PDE (C31) is modelled by an EDS $\mathcal{I}$ generated on M by a one-form

$$\theta = du - p_i dx^i.$$

On M the one-forms $\{dx^i, dp_i, du\}$ are linearly dependent as

$$0 = dF = \frac{\partial F}{\partial x^i}dx^i + \frac{\partial F}{\partial p_i}dp_i + \frac{\partial F}{\partial u}du.$$

Moreover $\theta \wedge (d\theta)^n = 0$ and $\theta \wedge (d\theta)^{n-1} \neq 0$. Therefore the Pfaff theorem (Theorem C.5.1) implies the existence of a coordinate system

$$(v, y^1, \ldots, y^{n-1}, q_1, \ldots, q_{n-1}, z)$$

such that

$$\theta = \mu(dv - q_1 dy^1 - \cdots - q_{n-1}dy^{n-1})$$

for some non-vanishing function μ. The vector field

$$\frac{\partial}{\partial z}$$

is a characteristic vector field as it satisfies[7]

$$\frac{\partial}{\partial z} \lrcorner\, \theta = 0 \quad \text{and} \quad \frac{\partial}{\partial z} \lrcorner\, d\theta = 0.$$

Using the original coordinate system we verify that the vector field

$$Z = \frac{\partial F}{\partial p_i}\frac{\partial}{\partial x^i} + p_i \frac{\partial F}{\partial p_i}\frac{\partial}{\partial u} - \left(\frac{\partial F}{\partial x^i} + p_i \frac{\partial F}{\partial u}\right)\frac{\partial}{\partial p_i} \tag{C32}$$

on $J^1(\mathbb{R}^n, \mathbb{R})$ is tangent to the level set $M = F^{-1}(0)$ and satisfies

$$Z \lrcorner\, \theta = 0 \quad \text{and} \quad Z \lrcorner\, d\theta = 0 \mod \theta.$$

Thus, $Z = \nu\, \partial/\partial z$ for some non-vanishing function ν. The initial value problem for the PDE (C31) can now be solved in the following steps:

- The initial data for (C31) is an $(n-1)$-dimensional submanifold Σ of $\mathbb{R}^{n+1}$ given in parametric form by

  $$(s_1, \ldots, s_{n-1}) \longrightarrow (x^i(s), u(s)) \subset \mathbb{R}^{n+1}.$$

  The natural lift of this submanifold to a graph in $J^1(\mathbb{R}^n, \mathbb{R})$ gives an $(n-1)$-dimensional integral manifold $\Sigma \subset M$ of $\mathcal{I}$ that is transverse to Z.
- Construct an n-dimensional integral manifold by solving a system of ODEs to find integral curves of Z (called the characteristic curves) and taking the union of these curves through Σ. If a characteristic curve has a point in common with the graph of a solution, it lies entirely on the graph.

[7] In general Z is a Cauchy characteristic vector field if $Z \lrcorner\, \theta \in \mathcal{I}$ for all $\theta \in \mathcal{I}$.

- In the coordinates of the Pfaff theorem (Theorem C.5.1) the n-dimensional integral manifold is given by

$$v = f(y^1, \ldots, y^{n-1}) \quad \text{and} \quad q_i = \frac{\partial f}{\partial y^i}(y^1, \ldots, y^{n-1})$$

for some function f which should be determined from the initial data.

Consider the special case of quasi-linear PDE (C31) where

$$F\left(x, u, \frac{\partial u}{\partial x^i}\right) = R^i(x, u)\frac{\partial u}{\partial x^i} + S(x, u)$$

and the Cauchy characteristic vector field (C32) is

$$Z = R^i \frac{\partial}{\partial x^i} + p_i R^i \frac{\partial}{\partial u} - \left[p_i \frac{\partial R^i}{\partial x^j} + \frac{\partial S}{\partial x^j} + p_j \left(\frac{\partial R_i}{\partial u} p_i + \frac{\partial S}{\partial u} \right) \right] \frac{\partial}{\partial p_j}.$$

The classical treatment of this quasi-linear problem does not use the jet-space formalism. Evaluating Z at $F = 0$ shows that the integral curves of Z project to curves on the solution surface $x \to (x, u = u(x))$ which are integral curves of

$$\widetilde{Z} = R^i \frac{\partial}{\partial x^i} - S \frac{\partial}{\partial u}.$$

The PDE $F = 0$ can be rewritten as

$$\widetilde{\mathbf{Z}} \cdot \mathbf{n} = 0,$$

where the vector

$$\mathbf{n} = (\partial_1 u, \ldots, \partial_n u, -1)$$

is normal to the solution surface $u = u(x)$ in $\mathbb{R}^{n+1}$. Therefore $\widetilde{\mathbf{Z}}$ is tangent to this surface. The characteristic curves which foliate the solution surface are solutions to the system of ODEs:

$$\dot{x}^i = R^i(x, u) \quad \text{and} \quad \dot{u} = -S(x, u), \qquad i = 1, \ldots, n$$

(where $\dot{} = \partial/\partial z$) with the initial conditions given by the initial data for (C31):

$$x^i(0) = x^i(s_1, \ldots, s_{n-1}) \quad \text{and} \quad u(0) = u_0(s_1, \ldots, s_{n-1}).$$

The method breaks down if the initial data is not transverse of $\widetilde{Z}$. A surface tangent to $\widetilde{Z}$ is called characteristic. Thus initial data specified along a characteristic surface does not determine the solution uniquely.

- **Example.** Consider the initial value problem for the dispersionless KdV equation

$$u_t + u u_x = 0, \qquad u(x, 0) = f(x).$$

The characteristic equations are

$$\frac{dx}{dz} = u, \quad \frac{dt}{dz} = 1, \quad \text{and} \quad \frac{du}{dz} = 0.$$

The solution surface must contain the curve $\Sigma \subset \mathbb{R}^3$ which we parameterize as

$$s \longrightarrow (x(s), t(s), u(s)) = (s, 0, f(s)).$$

Using this as the initial condition for the characteristic ODEs yields

$$x(s, z) = f(s)z + s, \quad t(s, z) = z, \quad \text{and} \quad u(s, z) = f(s).$$

Eliminating (s, z) between these formulae yields the general solution in the implicit form

$$u = f(x - ut). \tag{C33}$$

This will be valid in the domain where the coordinates (s, z) are well defined and can be used instead of (x, t). In general one needs to analyse the Jacobian of the transformation to specify the domain of solution. In our case we can proceed as follows: The characteristic curves project to the straight lines $x(s) = f(s)t(s) + s$ in the domain of (x, t) in $\mathbb{R}^2$. These lines have different slopes for different values of s (say s_1 and s_2) and thus they can intersect. The intersection will take place at a point $(x, t) \in \mathbb{R}^2$ where

$$t = \frac{s_2 - s_1}{f(s_1) - f(s_2)}.$$

At this point the solution becomes multivalued, taking values $f(s_1)$ and $f(s_2)$. To understand it better, differentiate the implicit solution (C33) to find

$$u_x = \frac{f'(s)}{1 + tf'(s)}.$$

Hence if $f'(s) < 0$ the derivative u_x becomes infinite at the finite positive time $t = -[f'(s)]^{-1}$. At this time the solution experiences a gradient catastrophe.

C.6 Cartan–Kähler theorem

The Frobenius theorem (Theorem C.2.3) gives a criterion for the existence of integral manifolds for EDSs generated algebraically by one-forms. The Cartan–Kähler theorem (proved by Cartan for Pfaffian systems, and extended to the general case by Kähler) deals with arbitrary EDSs. Our brief presentation of the subject in this section follows [25].

The proof of the Frobenius theorem (Theorem C.2.3) was based on Picard's existence theorem for ODEs. Thus the Frobenius theorem works in the smooth category. The proof of the Cartan–Kähler theorem involves the Cauchy–Kowalewska existence theorem for PDEs. The Cauchy–Kowalewska theorem which we shall state below is valid in the real-analytic category.

Let $u : \mathbb{R}^{n+1} \to \mathbb{R}^N$. Thus the collection of functions $u^\alpha, \alpha = 1, \ldots, N$, depends on $(n + 1)$ independent variables $(x^i, t), i = 1, \ldots, n$. The system of PDEs in

Cauchy form is

$$\frac{\partial u^\alpha}{\partial t} = F^\alpha(x^i, t, u^\alpha, \frac{\partial u^\alpha}{\partial x^i}), \tag{C34}$$
$$u^\alpha|_{t=t_0} = g^\alpha(x^i).$$

Theorem C.6.1 (Cauchy–Kowalewska) *If the equation* (C34) *and the initial data are real-analytic then there exists a unique solution in the form of a power series*

$$u^\alpha(t, x) = g^\alpha(x^i) + g_1^\alpha(x^i)(t - t_0) + \frac{1}{2} g_2^\alpha(x^i)(t - t_0)^2 + \cdots$$

which converges on some domain containing $t = t_0$.

This theorem can be refined if the first derivatives of u with respect to t are specified only for the first r components of u, that is, if $\alpha = 1, \ldots, r < N$ in (C34). In this case the system is underdetermined as there are fewer equations than unknowns. The general analytic solution to (C34) depends on $(N - r)$ arbitrary functions. This is quite obvious, a choice of $(N - r)$ functions is needed to put the equation in the 'determined form' (C34) with $\alpha = 1, \ldots, N$.

Definition C.6.2 *A k-dimensional subspace $E \subset T_x M$ is an integral element of $\mathcal{I}$ if*

$$\theta(e_1, \ldots, e_k) = 0$$

for all $\theta \in \mathcal{I}^k$ and $e_i \in E, i = 1, \ldots, k$.

The set of all k-dimensional integral elements is denoted $V_k(\mathcal{I})$. It is clear that tangent space to any k-dimensional integral manifold is an integral element. We aim to answer the following:

- **Question.** When is an integral element tangent to an integral manifold?

Certainly not always, as obstructions can arise from the Frobenius theorem.

- **Example.** The EDS

$$\mathcal{I} = \langle dx \wedge dz, dy \wedge (dz - y dx) \rangle_{\text{diff}}$$

  has a two-dimensional integral element

$$\{\partial_x + y\partial_z, \partial_y\}$$

  at each point, but no two-dimensional integral manifolds as the vectors spanning E do not satisfy the Frobenius condition (C9).

If $E \subset V_k(\mathcal{I})$ and $G \subset E$ is a p-dimensional subspace of E then $G \subset V_p(\mathcal{I})$. Thus restrictions of integral elements are integral elements. But the converse is not true, and not every extension of integral element may be an integral element.

Definition C.6.3 *Let $E \subset V_k(\mathcal{I})$ be spanned by $\{e_1, \ldots, e_k\}$. The polar space of E is*

$$H(E) = \{v \in T_xM, \theta(v, e_1, \ldots, e_k) = 0, \forall \theta \in \mathcal{I}^{k+1}\} \subset T_xM.$$

The polar space is a vector space containing E, but it does not have to be an integral element. However if $v \in H(E)$ and v is not an element of E then the direct sum $E \oplus \text{span}\{v\}$ is a $(k+1)$-dimensional integral element. Thus $H(E)$ is the space of possible one-dimensional extensions of a given integral element. Constructing H from a given E comes down to solving a set of linear homogeneous equations for components of v. In practice to compute the polar space of a k-dimensional integral element E, contract all $(k+1)$-forms in the ideal with all vectors in E. The resulting one-forms should be annihilated by all vectors in $H(E)$. An integral element E is called regular if the dimension of the polar space is constant in a neighbourhood of E in $V_k(\mathcal{I})$. Moreover E is called ordinary if the intersection of $V_k(\mathcal{I})$ with an open neighbourhood of E is a smooth submanifold of the Grassmanian $\text{Gr}_k(TM)$ of all k planes in TM.

For a given $E \in V_k(\mathcal{I})$ define

$$r(E) = \dim\,[H(E)] - k - 1$$

to be the dimension of the set of $(k+1)$ integral elements that contain E with $r(E) = -1$ if there are no such elements.

- **Example [25].** Let

  $$\mathcal{I} = \langle dx \wedge dz, dy \wedge (dz - ydx)\rangle_{\text{diff}}$$

  be an EDS on $\mathbb{R}^3$. One-dimensional integral element E is spanned by $e_1 = a\partial_x + b\partial_y + c\partial_z$. The vector $v = f\partial_x + g\partial_y + h\partial_z$ is in $H(E)$ if two linear equations

  $$cf - ah = 0 \quad \text{and} \quad -ybf - (c - ya)g + bh = 0$$

  for (f, g, h) hold. If $c - ya \neq 0$ we get $H(E) = E$ and thus $r(E) = -1$. If $c - ya = 0$ then $\dim\,[H(E)] = 2$ and $r(E) = 0$. In particular E is not a regular integral element.

An integral manifold $S \subset M$ is called ordinary/regular iff all of its tangent spaces are ordinary/regular elements. For regular integral manifolds we define $r(S) = r(T_xS)$.

Theorem C.6.4 (Cartan–Kähler) *Let $(M, \mathcal{I})$ be a real analytic EDS and let $\Sigma \subset M$ be an n-dimensional analytic submanifold whose tangent spaces are regular integral elements such that $\dim[H(T_x\Sigma)] = n + 1$. Then there exists an open neighbourhood of $x \in \Sigma$ and a unique analytic $(n+1)$-dimensional integral manifold $S \subset U$ containing $\Sigma \cap U$.*

The Cartan–Kähler theorem states when an n-dimensional integral manifold can be thickened to an $(n+1)$-dimensional integral manifold. This theorem needs to be modified by introducing a so-called restraining manifold if the dimension of the polar space of $T_x\Sigma$ is greater than $(n+1)$. This is needed for uniqueness. The restraining manifold R is an analytic submanifold of M of co-dimension $r(\Sigma)$

such that $\Sigma \subset R$ and $T_x R \cap H(T_x\Sigma)$ has dimension $(n+1)$ for all $x \in \Sigma$. Then there exists a unique connected $(n+1)$ analytic integral manifold S which satisfies $\Sigma \subset S \subset R$. The reader is referred to [23] where this is discussed.

The proof of the Cartan–Kähler theorem is obtained by adopting local coordinates and reducing the problem to a solution of system of PDEs of the form (C34). This uses the Cauchy–Kowalewska theorem and so one needs to require real-analyticity. Again, consult [23] or [90] for details.

The integral manifolds can in principle be constructed successively using the Cartan–Kähler theorem. At each step the integral manifold is determined by a choice of restraining manifolds and the arbitrary functions in the maximal integral manifold parameterize these choices. We shall now discuss the Cartan test which gives a handle on how to calculate this freedom in the 'general solution'. Applying the Cartan–Kähler theorem successively, starting from one-dimensional integral manifolds gives a sufficient condition for the existence of an integral manifold tangent to a given integral element: If $E \subset V_n(\mathcal{I})$ contains a flag of subspaces

$$\{0\} = E_0 \subset E_1 \subset \cdots \subset E_n = E \subset T_x M,$$

where the integral elements $E_k \subset V_k(\mathcal{I})$ are regular, then there exists a real-analytic n-dimensional integral manifold $S \subset M$ passing through x and satisfying $T_x P = E$.

This corollary from Theorem C.6.4 is not of great practical significance, as the regularity assumption needs to be checked at each step. Also, it gives a sufficient condition which is not necessary as not all integral manifolds have tangent spaces which are final objects in a flag of regular integral elements.

To get around this, consider the integral flag $\mathcal{F} = (E_0, \ldots, E_n)$, not necessarily regular, and set

$$c(E_k) := \dim(T_x M) - \dim H(E_k), \qquad k = 1, 2, \ldots, n$$

and let $c(E_{-1}) = 0$. The Cartan characters of the flag $\mathcal{F}$ are the non-negative numbers defined by

$$s_k(\mathcal{F}) := c(E_k) - c(E_{k-1}).$$

Theorem C.6.5 (Cartan test) *Let $(M, \mathcal{I})$ be an EDS and let $\mathcal{F} = (E_0, \ldots, E_n)$ be an integral flag of $\mathcal{I}$. Then $V_n(\mathcal{I})$ has co-dimension at least*

$$c(\mathcal{F}) := c(E_0) + c(E_1) + \cdots + c(E_{n-1})$$

in the Grassmannian[8] *$Gr_n(TM)$ at E_n. Moreover $V_n(\mathcal{I})$ is a smooth submanifold of $Gr_n(TM)$ of co-dimension $c(\mathcal{F})$ iff the flag $\mathcal{F}$ is regular.*

[8] Recall that the Grassmannian $\mathrm{Gr}_k(E)$ is the set of k-dimensional subspaces of a vector space E. It is a smooth manifold of dimension $k[\dim(E-k)]$. The set of all k-dimensional subspaces in $T_x M$ as x varies over M is denoted $\mathrm{Gr}_k(TM)$. It is a manifold of dimension $\dim(M) + k[\dim(M-k)]$. Given a k-plane E in $T_x M$ on which $dx^1 \wedge \cdots \wedge dx^k \neq 0$ we can choose coordinates $(x^1, \ldots, x^k, u^1, \ldots u^s)$ on M such that E is spanned by vectors

$$\frac{\partial}{\partial x^i} + \sum_{\alpha=1}^{s} p_i^\alpha(E) \frac{\partial}{\partial u^\alpha}, \qquad i = 1, \ldots, k,$$

where $p_i^\alpha = p_i^\alpha(E)$ are coordinates on the fibres of $\mathrm{Gr}_k(TM) \to M$.

Performing Cartan test on a given flag is just a matter of linear algebra. If the flag passes the test and therefore is regular, the Cartan–Kähler theorem implies the existence of at least one real-analytic n-dimensional integral manifold $S \subset M$ such that $T_x S = E_n$ (there will be exactly one such manifold if the $r(E_{n-1}) = 0$. Otherwise the restraining manifold has to be chosen). Of course for a given integral element E_n there may be more than one flag which terminates at E_n. In practice it makes sense to choose the first element in a flag such that the first Cartan character s_k is as large as possible, then choose the second element such that the next character is as large as possible, etc. The sum $s_1 + s_2 + \cdots + s_n$ is fixed regardless of these choices. In what follows we shall drop the reference to the flag and write s_k instead of $s_k(\mathcal{F})$. The highest k such that $s_k \neq 0$ is called the Cartan character. Moreover let $c(E_n) = s$. Using the definitions of Cartan characters and $c_n = \dim(M - n)$ we can write

$$s_0 + s_1 + \cdots + s_k = c_k$$

and rewrite the inequality in Cartan test as

$$\dim [V_n(\mathcal{I})] - \dim (M) \leq s_1 + 2s_2 + \cdots + ns_n, \qquad \text{(C35)}$$

where the LHS is the fibre dimension of $V_n(\mathcal{I})$.

Given a flag $\mathcal{F}$ which passes the test it is possible to chose a coordinate system

$$(x^1, \ldots, x^n, u^1, \ldots, u^s)$$

centreed at $x \in U \subset \mathbb{R}^{n+s}$ such that E_k is spanned by

$$\{\frac{\partial}{\partial x^1}, \ldots, \frac{\partial}{\partial x^k}\}, \qquad 0 \leq k < n$$

and elements $H(E_k)$ are annihilated by the one-forms

$$\{du^1, \ldots, du^{c_k}\}.$$

Let $\mathcal{S}$ be the collection of real-analytic integral manifolds near S. This means that $\hat{S} \in \mathcal{S}$ if it can be represented by

$$u^\alpha = F^\alpha(x^1, \ldots, x^n),$$

where the analytic functions F^α are defined in the neighbourhood of $\mathbf{x} = 0$. Then the collection $\mathcal{S}$ depends on s_0 constants, s_1 functions of one variable, ..., s_n functions of n variables. Thus the integers $(s_0, s_1, \ldots, s_n)$ measure the arbitrariness of the general integral manifold.

- **Example.** A Lagrangian submanifold of $\mathbb{R}^{2n}$ is an integral manifold of the ideal generated by a symplectic structure

$$\theta = dx^1 \wedge du^1 + dx^2 \wedge du^2 + \cdots + dx^n \wedge du^n.$$

Choose a flag

$$\{0\} \subset \{\frac{\partial}{\partial x^1}\} \subset \{\frac{\partial}{\partial x^1}, \frac{\partial}{\partial x^2}\} \subset \cdots \subset \{\frac{\partial}{\partial x^1}, \frac{\partial}{\partial x^2}, \ldots, \frac{\partial}{\partial x^n}\}.$$

For this flag $H(E_0)$ is the whole tangent space, so $c_0 = 0$. Then $H(E_1)$ consists of all vectors which annihilate du^1 and more generally

$$H(E_k) = \{du^1, \dots, du^k\}^\perp.$$

Thus

$$c_0 = 0, c_1 = 1, c_2 = 2, \dots, c_n = n$$

which implies

$$s_0 = 0, \quad s_1 = s_2 = \cdots = s_n = 1.$$

To calculate the fibre dimension of $V_n(\mathcal{I})$ note that the vectors spanning E_n annihilate the one-forms du^k, $k = 1, \dots, n$. The nearby integral planes are given by

$$du^k = \sum_{j=1}^{n} p^{jk} dx^j, \quad k = 1, \dots, n,$$

and the total number of symmetric coefficients $p^{jk} = p^{kj}$ is the fibre dimension of $V_n(\mathcal{I})$ which appears on the LHS of the inequality (C35). This number is

$$\binom{n+1}{2}$$

which is also equal to the RHS of (C35). Thus we have equality and the flag is regular. The general integral manifold depends on one function of n variables and functions of lower number of variables. Explicitly

$$u^k = \frac{\partial f}{\partial x^k}, \qquad k = 1, \dots, n,$$

where $f = f(x^1, x^2, \dots, x^n)$.

- **Example.** Consider the EDS (C5) in $\mathbb{R}^5$

$$<\theta^1 = du - pdx - qdy, \, \theta^2 = dp \wedge dq - dx \wedge dy>_{\text{diff}}$$

for the Monge–Ampere equation. The four-dimensional space $V_1(\mathcal{I}) = \{\theta^1\}^\perp$ of one-dimensional integral elements is spanned by

$$\{\partial_x + p\partial_u, \partial_y + q\partial_u, \partial_p, \partial_q\}.$$

Pick $E_1 = \{\partial_p\}$ to be the first element in the flag. The two-forms in the ideal are

$$\theta^2, \quad d\theta^1 = dx \wedge dp + dy \wedge dq, \quad \text{and} \quad \theta^1 \wedge \gamma,$$

where γ is any one-form. Therefore the polar space $H(E_1)$ will consist of all vectors annihilating $\partial_p \lrcorner \theta_2$, $\partial_p \lrcorner d\theta^1$, and θ^1. Thus $H(E_1) = \{dq, dx, \theta^1\}^\perp$. This space is two-dimensional and there is a unique extension of E_1 to an integral element

$$E_2 = \{\partial_p, \partial_y + q\partial_u\}$$

(we would have got the same integral element E_2 if we had picked $E_1 = \{\partial_y + q\partial_u\}$). Contracting the two vectors in E_2 with all three-forms in the ideal shows that this cannot be further extended, that is, $H(E_2) = E_2$. Therefore we pick a flag

$$\{0\} \subset \{\partial_p\} \subset \{\partial_p, \partial_y + q\partial_u\} = E \subset T_x\mathbb{R}^5.$$

This flag has $c_0 = 5 - 4 = 1, c_1 = 5 - 2 = 3, c_2 = 5 - 2 = 3$ and so $s_0 = 1, s_1 = 2$. To perform the Cartan test we need to compute the co-dimension of $V_2(\mathcal{I})$ in the Grassmannian of two-planes. The two-planes close to E_2 are spanned by

$$v_1 = \partial_p + \alpha(\partial_x + p\partial_u) + \beta\partial_q + \gamma\partial_u \quad \text{and} \quad v_2 = \partial_y + q\partial_u + \delta(\partial_x + p\partial_u) + \epsilon\partial_q + \phi\partial_u$$

for some $(\alpha, \beta, \ldots, \phi)$. The conditions

$$\theta^1(v_1) = 0, \quad \theta^1(v_2) = 0, \quad d\theta^1(v_1, v_2) = 0, \quad \text{and} \quad \theta^2(v_1, v_2) = 0$$

give four linear equations

$$\gamma = 0, \quad \phi = 0, \quad \epsilon - \alpha = 0, \quad \text{and} \quad \beta + \delta = 0.$$

Thus the fibre co-dimension of $V_2(\mathcal{I})$ is four which is equal to $c_0 + c_1 + c_2$. The Cartan test is satisfied and the general solution to the Monge–Ampere equation depends on two functions of one variable.

In theory one could always reduce a problem to the analysis of a Pfaffian system (i.e. one where $\mathcal{I}$ is generated by one-forms) as any EDS can be prolonged to such system. If a Pfaffian system is generated by

$$<\theta^1, \ldots, \theta^N>$$

then the vectors $\{e_1, e_2, \cdots, e_k\}$ spanning E_k in a flag

$$\{0\} \subset E_1 \subset E_2 \subset \cdots \subset E_n \subset T_x M$$

are found by solving the system

$$e_i \lrcorner\, \theta^\alpha = 0 \quad \text{and} \quad e_i \lrcorner\, (e_j \lrcorner\, d\theta^\alpha) = 0, \qquad i, j = 1, \ldots, k, \quad \alpha = 1, \ldots, N.$$

This however comes at the price of introducing more variables and working in spaces of high dimension: For a system of r PDEs of order k for N functions of n unknowns

$$F^\rho\left(x^i, u^\alpha, \frac{\partial u^\alpha}{\partial x^i}, \ldots, \frac{\partial^k u^\alpha}{\partial x^{i_1}\partial x^{i_2}\ldots\partial x^{i_k}}\right) = 0, \qquad \rho = 1, \ldots, r,$$
$$\alpha = 1, \ldots N, \quad i = 1, \ldots, n$$

the Pfaffian system is generated by one-forms

$$du^\alpha - p_i^\alpha dx^i, \quad dp_i^\alpha - p_{ij}^\alpha dx^j, \quad \ldots, \quad dp^\alpha_{i_1 i_2\cdots i_{k-1}} - p^\alpha_{i_1 i_2\cdots i_k} dx^{i_k}$$

on the manifold M given by the zero locus

$$F^\rho\left(x^i, u^\alpha, p_i^\alpha, \cdots, p^\alpha_{i_1 i_2\cdots i_k}\right) = 0$$

in the kth jet space $J^k(\mathbb{R}^n, \mathbb{R}^N)$. There are however more economical tricks to reduce a problem to a Pfaffian system. The following example, modified from [17], shows one such trick.

- **Example.** Consider the Ricci-flat Kähler equation (C7) in four dimensions. We are interested in the real-analytic solutions, so we can complexify the dependent and independent variables and regard $(w, z, \bar{w}, \bar{z})$ as independent holomorphic coordinates on an open ball in $\mathbb{C}^4$. The equation (C7) is modelled by the ideal generated by one one–form and one four–form on $\mathbb{C}^9$

$$<d\Omega - pdw - \bar{p}d\bar{w} - qdz - \bar{q}d\bar{z},\ dp \wedge dq \wedge dw \wedge dz - d\bar{w} \wedge d\bar{z} \wedge dw \wedge dz>_{\text{diff}}$$

together with the independence condition $dw \wedge dz \wedge d\bar{w} \wedge d\bar{z} \neq 0$. To reformulate the problem as a Pfaffian system rewrite the vanishing of the four-form as

$$d(pdq - \bar{w}d\bar{z}) \wedge dw \wedge dz = 0.$$

The independence condition implies $dw \wedge dz \neq 0$. Thus locally there exist functions a, b, Σ such that

$$pdq - \bar{w}d\bar{z} = d\Sigma - adz - bdw$$

on integral manifolds. Conversely equation (C7) can be modelled as a Pfaffian EDS

$$\mathcal{I} = <\theta^1 = d\Omega - pdw - \bar{p}d\bar{w} - qdz - \bar{q}d\bar{z}, \theta^2 = d\Sigma - adz - bdw - pdq + \bar{w}d\bar{z}>_{\text{diff}}$$

in $\mathbb{C}^{12}$ with coordinates $(w, z, \bar{w}, \bar{z}, p, q, \bar{p}, \bar{q}, a, b, \Omega, \Sigma)$.
The space of one-dimensional integral elements $\{\theta^1, \theta^2\}^{\perp}$ is 10-dimensional, thus $c_0 = 2$ and $s_0 = 2$. Let $E_1 = \{e_1\}$. The polar space of E_1 is the eight-dimensional vector space

$$H(E_1) = \{\theta^1, \theta^2\ e_1 \lrcorner\ d\theta^1, e_1 \lrcorner\ d\theta^2\}^{\perp}.$$

Thus $c_1 = 4$ and $s_1 = c_1 - c_0 = 2$. Let $E_2 = \{e_1, e_2\}$. Then

$$H(E_2) = \{\theta^1, \theta^2, e_1 \lrcorner\ d\theta^1, e_1 \lrcorner\ d\theta^2, e_2 \lrcorner\ d\theta^1, e_2 \lrcorner\ d\theta^2\}^{\perp},$$

so $c_2 = 6$ and $s_2 = 2$. We continue looking for polar spaces and extending the integral elements. Let $E_3 = \{e_1, e_2, e_3\}$. This gives[9] $c_3 = 8, s_3 = 2$. Pick some $e_4 \in H(E_3)$ and set $E_4 = \{e_1, e_2, e_3, e_4\}$. Now

$$H(E_4) = \{\theta^1, \theta^2, e_i \lrcorner\ d\theta^1, e_i \lrcorner\ d\theta^2\}^{\perp}, \qquad i = 1, \dots, 4$$

and $\dim[H(E_4)] \leq 4$. However $E_4 \subset H(E_4)$ and we must have $H(E_4) = E_4$ and the integral element E_4 is not extendable. We have $c_4 = 12 - 4 = 8$ and $s_4 = 0$. Thus the maximal integral manifolds may be at most four-dimensional if we can pick a regular flag. We can verify the computations of Cartan characters by

[9] The flag must be chosen carefully for this to be true. The choice (C36) will do.

choosing the flag with

$$e_1 = \partial_w + \partial_{\bar{w}} + (p + \bar{p})\partial_\Omega + b\partial_\Sigma, \quad e_2 = \partial_p - \partial_{\bar{p}}, \quad e_3 = \partial_{\bar{q}} + \partial_a, \quad \text{and} \quad e_4 = \partial_{\bar{q}}. \tag{C36}$$

Then

$$H(E_1) = \{\theta^1, \theta^2, dp + d\bar{p}, db + d\bar{z}\}^\perp$$
$$H(E_2) = \{\theta^1, \theta^2, dp + d\bar{p}, db + d\bar{z}, dw - d\bar{w}, dq\}^\perp$$
$$H(E_3) = \{\theta^1, \theta^2, dp + d\bar{p}, db + d\bar{z}, dw - d\bar{w}, dq, d\bar{z}, dz\}^\perp$$
$$H(E_4) = E_4.$$

The co-dimension of $V_4(\mathcal{I})$ around $E = E_4$ can be now computed as in the last example. The Cartan test holds and thus the general real-analytic Ricci-flat Kähler metric in four dimensions depends on two arbitrary functions of three variables.

References

[1] Ablowitz, M. J., Ramani, A., and Segur, H. (1980) A connection between nonlinear evolution equations and ordinary differential equations of P-type. I, II, *J. Math. Phys.* **21**, 715–721 and 1006–1015.

[2] Ablowitz, M. J. and Clarkson, P. A. (1992) *Solitons, Nonlinear Evolution Equations and Inverse Scattering*, London Mathematical Society Lecture Note Series No 149, CUP, Cambridge.

[3] Ablowitz, M. J. and Fokas, A. S. (2003) *Introduction and Applications of Complex Variables*, 2nd edition, Cambridge University Press, Cambridge.

[4] Anand, C. (1997) Ward's solitons, *Geom. Topol.* **1**, 9–20 (electronic).

[5] Arnold, V. I. (1989) *Mathematical Methods of Classical Mechanics*, 2nd edition. Graduate Texts in Mathematics, **60**, Springer.

[6] Ashtekar, A. Jacobson, T., and Smolin, L. (1988) A new characterization of half flat solutions to Einstein's equations, *Commun. Math. Phys.* **115**, 631–648.

[7] Atiyah, M. F., Hitchin, N. J., and Singer, I. M. (1977) Deformations of instantons, Proc. Natl. Acad. Sci. U.S.A. **74**, 2662–2663.

[8] Atiyah, M. F. and Ward, R. S. (1977) Instantons and algebraic geometry, *Commun. Math. Phys.* **55**, 117–124.

[9] Atiyah, M. F., Hitchin, N. J., Drinfeld, V. G., and Manin, Y. I. (1978) Construction of Instantons, *Phys. Lett.* **A65**, 185–187.

[10] Atiyah, M. F., Hitchin, N. J., and Singer, I. M. (1978) Self-duality in four-dimensional Riemannian geometry, *Proc. Lond. Math. Soc. A* **362**, 425–461.

[11] Atiyah, M. F. and Hitchin, N. J. (1988) *The Geometry and Dynamics of Magnetic Monopoles*, Princeton University Press, Princeton, NJ.

[12] Baston, R. J. and Eastwood, M. G., (1989) *The Penrose Transform. Its Interaction with Representation Theory*, OUP, New York.

[13] Belinski, V. A., Gibbons, G. W., Page, D. N., and Pope, C. N. (1978) Asymptotically Euclidean Bianchi IX metrics in quantum gravity, *Phys. Lett.* **B76**, 433–435.

[14] Bielawski, R. (2001) *Twistor Quotients of Hyper-Kähler Manifolds*, World Scientific Publishing, River Edge, NJ.

[15] Bogdanov, L. V., Konopelchenko, B. G., and Martines Alonso, L. (2003) The quasiclassical $\bar{\partial}$-method: Generating equations for dispersionless integrable hierarchies, *Theor. Math. Phys.* **134**, 39–46.

[16] Bogdanov, L. V., Dryuma, V. S., and Manakov, S. V. (2007) Dunajski generalization of the second heavenly equation: Dressing method and the hierarchy, *J. Phys. A* **40**, 14383–14393.

[17] Boyer, C. P. and Plebański, J. F. (1977) Heavens and their integral manifolds, *J. Math. Phys.* **18**, 1022–1031.

[18] Boyer, C. (1988) A note on hyperhermitian four-manifolds, *Proc. Am. Math. Soc.* **102**, 157–164.

[19] Braaten, E., Curtright, T. L., and Zachos, C. K. (1985) Torsion and geometrostasis in nonlinear sigma models, *Nucl. Phys.* **B260**, 630–688.

[20] Branson, T. P., Cap, A., Eastwood, M. G., and Gover, A. R. (2006) Prolongations of geometric overdetermined systems, *Int. J. Math.* **17**, 641–664.

[21] Bredon, G. E. (1993) *Topology and Geometry*, Springer-Verlag, New York.

[22] Brieskorn, E. (1970) Singular elements of semisimple algebraic groups, *Actes Congres Int. Math.* **2**, 279–284.

[23] Bryant, R. L., Chern, S. S., Gardner, R. B., Goldschmidt, H. L., and Griffiths, P. A. (1991) *Exterior Differential Systems*, Mathematical Sciences Research Institute Publications, Vol. 18, Springer-Verlag, New York.

[24] Bryant, R. L. (2000) Pseudo-Riemannian metrics with parallel spinor fields and vanishing Ricci tensor. *Global Analysis and Harmonic Analysis*, Seminaires et Congrés, **4**, pp. 53–94, French Mathematical Society, Paris.

[25] Bryant, R. L. (2002) Lectures on Exterior Differential Systems.

[26] Bryant, R. L., Dunajski, M., and Eastwood, M. (2008) Metrisability of two-dimensional projective structures, `arXiv:0801.0300`, *J. Diff. Geom.*

[27] Burke, L. W. (1985) *Applied Differential Geometry*, CUP, Cambridge.

[28] Calderbank, D. M. J. and Pedersen, H. (2000) Selfdual spaces with complex structures, Einstein-Weyl geometry and geodesics, *Ann. Inst. Fourier (Grenoble)* **50**, 921–963.

[29] Calderbank, D. M. J. (2006) Selfdual 4-manifolds, projective structures, and the Dunajski-West construction, `math.DG/0606754`.

[30] Cartan, E. (1943) Sur une classe d'espaces de Weyl, *Ann. Sci. Ecole Norm. Supp.* **60**, 1–16.

[31] Chamseddine, A. H. and Nicolai, H. (1980) Coupling the SO(2) supergravity through dimensional reduction, *Phys. Lett. B* **96**, 89.

[32] Chruściel, P. T. and Nadirashvili, N. S. (1995) All electrovacuum Majumdar Papapetrou spacetimes with non-singular black holes, *Class. Quant. Grav.* **12**, L17.

[33] Chruściel, P. T., Reall, H. S., and Tod, K. P. (2006) On Israel-Wilson-Perjes black holes, *Class. Quant. Grav.* **23**, 2519.

[34] Corrigan, E. and Fairlie, D. B. (1977) Scalar field theory and exact solutions to a classical SU(2) gauge theory, *Phys. Lett.* **B67**, 69–71.

[35] Dai, B. and Terng, C. L. (2008) Bäcklund transformations, Ward solitons, and unitons, *J. Diff. Geom.* **75**, 57–108.

[36] Darboux, G. (1896) *Lecons sur la théorie generale des surfaces*, Vol. III, Chelsea Publishing.

[37] Derrick, G. H. (1964) Comments on nonlinear wave equations as models for elementary particles, *J. Math. Phys.* **5**, 1252.

[38] Dirac, P. A. M. (1931) Quantised singularities in the electromagnetic field, *Proc. R. Soc. Lond.* **A133**, 60.

[39] Donaldson, S. K. and Kronheimer, P. B. (1990) *The Geometry of Four-Manifolds*. Oxford Mathematical Monographs, Oxford Science Publications, The Clarendon Press, Oxford University Press, New York.

[40] Donaldson, S. K. (2002) *Floer Homology Groups in Yang-Mills Theory*, Cambridge Tracts in Mathematics, **147**, Cambridge University Press, Cambridge.

[41] Dorey, N., Hollowood, T., Khoze, V., and Mqttis, M. (2002) The calculus of many instantons, *Phys. Rep.* **371**, 231–459.

[42] Drazin, P. G. and Johnson, R. S. (1989) *Solitons: An Introduction*, Cambridge Texts in Applied Mathematics, Cambridge University Press, Cambridge.

[43] Dubrovin, B. A. Fomenko, A. T., and Novikov, S. P. (1985) *Modern Geometry–Methods and Applications. Part II. The Geometry and Topology of Manifolds*, Springer, New York.

[44] Dunajski, M. (1999) The twisted photon associated to hyperhermitian four manifolds, *J. Geom. Phys.* 30, 266–281.

[45] Dunajski, M. and Mason, L. J. (2000) Hyper-Kähler hierarchies and their twistor theory, *Commun. Math. Phys.* **213**, 641–672.

[46] Dunajski, M., Mason, L. J., and Tod, K. P. (2001) Einstein–Weyl geometry, the dKP equation and twistor theory, *J. Geom. Phys.* **37**, 63–92.

[47] Dunajski, M. (2002) Anti-self-dual four-manifolds with a parallel real spinor, *Proc. R. Soc. Lond. Ser.* **A458**, 1205–1222.

[48] Dunajski, M. and Tod, K. P. (2002) Einstein–Weyl spaces and dispersionless Kadomtsev-Petviashvili equation from Painlevé I and II, *Phys. Lett.* **303A**, 253–264.

[49] Dunajski, M. and Mason, L. J. (2003) Twistor theory of hyper-Kähler metrics with hidden symmetries, *J. Math. Phys.* **44**, 3430–3454.

[50] Dunajski, M. (2003) Harmonic functions, central quadrics and twistor theory, *Class. Quant. Grav.* **20**, 3427–3440.

[51] Dunajski, M. (2004) A class of Einstein–Weyl spaces associated to an integrable system of hydrodynamic type. *J. Geom. Phys.* **51**, 126–137.

[52] Dunajski, M. and Sparling G. (2005) A dispersionless integrable system associated to Diff(S^1) gauge theory. *Phys. Lett.* **A343**, 129–132.

[53] Dunajski, M. and Hartnoll, S. A. (2007) Einstein-Maxwell gravitational instantons and five dimensional solitonic strings, *Class. Quant. Grav.* **24**, 1841–1862.

[54] Dunajski, M. and Manton, N. S. (2005) Reduced dynamics of Ward solitons, *Nonlinearity* **18**, 1677–1689.

[55] Dunajski, M. and Plansangkate, P. (2007) Topology and energy of time dependent unitons, *Proc. R. Soc.* **A463**, 945–959.

[56] Dunajski, M. and Kaźmierczak, M. (2007) Moduli spaces with external fields, *J. Geom. Phys.* **57**, 1883–1894.

[57] Dunajski, M. and West, S. (2007) Anti-self-dual conformal structures from projective structures. *Commun. Math. Phys.* **272**, 85–118.

[58] Dunajski, M. and Plansangkate, P. (2009) Strominger–Yau–Zaslow geometry, Affine Spheres and Painlevé III, *Commun. Math. Phys.* 290, 997–1024.

[59] Dunajski, M. (2008) An interpolating dispersionless integrable system, *J. Phys.* **A41**, 315202.

[60] Eguchi, T. and Hanson, A. J. (1978) Asymptotically flat self–dual solutions to Euclidean gravity, *Phys. Lett.* **B74**, 249–251.

[61] Eguchi, T. and Hanson, A. J. (1979) Self-dual solutions to Euclidean gravity, *Ann. Phys.* **120**, 82–106.

[62] Eguchi, T., Gilkey, P., and Hanson, A. J. (1980) Gravitation, gauge theories and differential geometry, *Phys. Rep.* **66C**, 213–393.

[63] Ferapontov, E. V. and Khusnutdinova, K. R. (2004) The characterization of two-component (2 + 1)-dimensional integrable systems of hydrodynamic type, *J. Phys.* **A37**, 2949–2963.

[64] Ferapontov, E. V. and Khusnutdinova, K. R. (2004) On the integrability of (2 + 1)-dimensional quasilinear systems, *Commun. Math. Phys.* **248**, 187–206.

[65] Finley, J. D. and Plebański, J. F. (1979) The classification of all H spaces admitting a Killing vector, *J. Math. Phys.* **20**, 1938–1945.

[66] Fradkin, E. (1991) *Field Theories of Condensed Matter Systems*, Addison-Wesley. Redwood City, CA.

[67] Gardner, C., Green, J., Kruskal, M., and Miura, R. (1967) Method for solving the Korteweg-deVries equation, *Phys. Rev. Lett.* **19**, 1095–1097.

[68] Gauntlett, J. P., Gutowski, J. B., Hull, C. M., Pakis, S., and Reall, H. S. (2003) All supersymmetric solutions of minimal supergravity in five dimensions, *Class. Quant. Grav.* **20**, 4587–4634.

[69] Gauntlett, J. P. (2004) Branes, calibrations and supergravity, in *Strings and Geometry*, Clay Mathematical Proceedings, **3**, pp. 79–126, American Mathematical Society, Providence, RI.

[70] Gelfand, I. M. and Levitan, B. M. (1951) On the determination of a differential equation from its spectral function, *Izv. Akad. Nauk SSSR, Ser. Mat.* **15**, 309–360.

[71] Gibbons, G. W. and Hawking, S. W. (1978) Gravitational multi-instantons, *Phys. Lett.* **78B**, 430–432.

[72] Gibbons, G. W. and Hawking, S. W. (1979) Classification of gravitational instanton symmetries, *Commun. Math. Phys.* **66**, 291–310.

[73] Gibbons, G. W. (2003) Gravitational instantons, confocal quadrics and separability of the Schrödinger and Hamilton–Jacobi equations, *Class. Quant. Grav.* **20**, 4401–4408.

[74] Gross, D. J. and Perry, M. J. (1983) Magnetic monopoles in Kaluza-Klein theories, *Nuclear Phys.* **B226**, 29–48.

[75] Guven, R. (1992) Black p-brane solutions of $D = 11$ supergravity theory, *Phys. Lett.* **B276**, 49–55.

[76] Hartle, J. B. and Hawking, S. W. (1972) Solutions of the Einstein-Maxwell equations with many black holes, *Commun. Math. Phys.* **26**, 87.

[77] Hawking, S. W. (1977) Gravitational instantons, *Phys. Lett.* **A60**, 81–83.

[78] Hitchin, N. J. (1982) Monopoles and geodesics, *Commun. Math. Phys.* **83**, 579–602.

[79] Hitchin, N. (1982) Complex manifolds and Einstein's equations, in *Twistor Geometry and Non-Linear Systems*, LNM 970, eds. H. D. Doebner and T. D. Palev.

[80] Hitchin, N. J. (1983) On the construction of monopoles, *Commun. Math. Phys.* **89**, 145–190.

[81] Hitchin, N. J., Karlhede, A., Lindstrom, U., and Rocek, M. (1987) Hyper-Kähler metrics and supersymmetry, *Commun. Math. Phys.* **108**, 535–589.

[82] Hitchin, N. J., Segal, G. B., and Ward, R. S. (1999) *Integrable Systems. Twistors, Loop Groups, and Riemann Surfaces*, OUP, New York.

[83] Huggett, S. A. and Tod, K. P. (1994) *An Introduction to Twistor Theory*, LMS Student Texts, **4**, 2nd edition. CUP, Cambridge.

[84] Hydon P. E. (2000) *Symmetry Methods for Differential Equations: A Beginner's Guide*, CUP, Cambridge.

[85] Ince, E. L. (1956) *Ordinary Differential Equations*, Dover, New York.

[86] Ioannidou, T. (1996) Soliton solutions and nontrivial scattering in an integrable chiral model in (2 + 1) dimensions, *J. Math. Phys.* **37**, 3422–3441.

[87] Ioannidou, T. and Zakrzewski, W. J. (1998) Solutions of the modified chiral model in (2 + 1) dimensions, *J. Math. Phys.* **39**, 2693–2701.

[88] Ioannidou, T. and Manton, N. S. (2005) The energy of scattering solitons in the Ward model, *Proc. R. Soc.* **A461**, 1965–1973.

[89] Israel, W. and Wilson, G. A. (1972) A Class of stationary electromagnetic vacuum fields, *J. Math. Phys.* **13**, 865.

[90] Ivey, T. A. and Landsberg, J. M. (2003) *Cartan for Beginners: Differential Geometry via Moving Frames and Exterior Differential Systems*, AMS, Providence, RI.

[91] Jackiw, R., Nohl, C., and Rebbi, C. (1978) Classical and semiclassical solutions of the Yang–Mills theory, in Particles and Fields (Proceedings of the Banff Summer Institute, Banff, Alberta, 1977), pp. 199–258, Plenum, New York/London.

[92] John, F. (1938) The ultrahyperbolic differential equation with four independent variables, *Duke Math. J.* **4**, 300–322.

[93] Jones, P. and Tod, K. P. (1985) Minitwistor spaces and Einstein-Weyl spaces, *Class. Quant. Grav.* **2**, 565–577.

[94] Kobayashi, S. and Nomizu, K. (1969) *Foundations of Differential Geometry*, Vols. I and II, John Wiley & Sons, New York/London.

[95] Kodaira, K. (1963) On stability of compact submanifolds of complex manifolds, *Am. J. Math.* **85**, 79–94.

[96] Konopelchenko, B. G., Martinez Alonso, L., and Ragnisco, O. (2001), The $\bar{\partial}$-approach to the dispersionless KP hierarchy, *J. Phys.* **A34**, 10209–10217.

[97] Konopelchenko, B. and Martinez Alonso, L. (2002) Dispersionless scalar integrable hierarchies, Whitham hierarchy, and the quasiclassical $\bar{\partial}$-dressing method, *J. Math. Phys.* **43**, 3807–3823.

[98] Krichever, I. M. (1994) The τ-function of the universal Whitham hierarchy, matrix models and topological field theories, *Commun. Pure Appl. Math.* **47**(4), 437–475.

[99] Kronheimer, P. (1989) The construction of ALE spaces as hyper-Kähler quotient, *J. Diff. Geom.* **29**, 665.

[100] Kronheimer, P. (1989) A Torelli type theorem for gravitational instantons, *J. Diff. Geom.* **29**, 685–697.

[101] Kruglikov, B. (2008) Invariant characterisation of Liouville metrics and polynomial integrals. J. Geom. Phys, 58, 979–995.

[102] Landau, L. D. and Lifshitz, E. M. (1995) *Course of Theoretical Physics*, Vols. I and II, Butterworth-Heinemann.

[103] Lax, P. (1968) Integrals of nonlinear equations of evolution and solitary waves, *Commun. Pure Appl. Math.* **21**, 467–490.

[104] LeBrun, C. R. (1991) Explicit self-dual metrics on $CP^2\# \cdots \# CP^2$, *J. Diff. Geom.* **34**, 233–253.

[105] Lewandowski, J. (1991) Twistor equation in a curved spacetime, *Class. Quant. Grav.* **8**, L11–L17.

[106] Lindstrom, U. and Rocek, M. (1988) New hyper-Kähler metrics and new supermultiplets, *Commun. Math. Phys.* **115**, 21–29.

[107] Liouville, R. (1887) Sur une classe d'équations différentielles, parmi lesquelles, en particulier, toutes celles des lignes géodésiques se trouvent comprises, *Comptes rendus hebdomadaires des seances de l'Academie des sciences* **105**, 1062–1064.

[108] Magri, F. (1978) A simple model of the integrable Hamiltonian equation, *J. Math. Phys.* **19**, 1156–1162.

[109] Magri, R. and Morosi, C. (1983) On the reduction theory of the Nijenhuis operators and its applications to Gelfand-Diki equations, *Proceedings of the IUTAM-ISIMM Symposium on Modern Developments in Analytical Mechanics*, Vol. II.

[110] Manakov, S. V. and Zakharov, V. E. (1981) Three-dimensional model of relativistic-invariant field theory, integrable by the inverse scattering transform, *Lett. Math. Phys.* **5**, 247–253.

[111] Manakov S. V. and Santini P. M. (2006) The Cauchy problem on the plane for the dispersionless Kadomtsev–Petviashvili equation, *JETP Lett.* **83**, 462–466.

[112] Manakov, S. V. and Santini, P. M. (2007) A hierarchy of integrable partial differential equations in dimension $2+1$, associated with one-parameter families of vector fields, *Theor. Math. Phys.* **152**, 1004–1011.

[113] Manton, N. S. (1982) A remark on the scattering of BPS monopoles, *Phys. Lett.* **B100**, 54–56.

[114] Manton, N. S. and Sutcliffe, P. M. (2004) *Topological Solitons*, CUP, Cambridge.

[115] Marchenko, V. A. (1955) Reconstruction of the potential energy from the phases of scattered waves, *Dokl. Akad. Nauk SSSR* **104**, 695–698.

[116] Mason, L. J. and Newman, E. T. (1989) A connection between the Einstein and Yang-Mills equations, *Commun. Math. Phys.* **121**, 659–668.

[117] Mason, L. J. and Sparling, G. A. J. (1992) Twistor correspondences for the soliton hierarchies, *J. Geom. Phys.* **8**, 243–271.

[118] Mason, L. J. and Woodhouse, N. M. J. (1996) *Integrability, Self-Duality and Twistor Theory*, LMS Monograph New Series, **15**, OUP, Oxford.

[119] Mikhailov, A. V. (1981) The reduction problem and the inverse scattering method, *Physica 3D* **1** and **2**, 73–117.

[120] Mimura, Y. and Takeno, H. (1962) *Wave Geometry*, Sci. Rep. Res. Inst. Theor. Phys. Hiroshima Univ. No. 2.

[121] Morrow, J. and Kodaira, K. (2006) *Complex Manifolds*, AMS bookstore, Providence, RI.

[122] Novikov, S., Manakov, S. V., Pitaevskii, L. P., and Zakharov V. E. (1984) *Theory of Solitons: The Inverse Scattering Method*, Consultants Bureau, New York.

[123] Nurowski, P. and Sparling, G. A. J. (2003) Three-dimensional Cauchy-Riemann structures and second-order ordinary differential equations, *Class. Quant. Grav.* **20**, 4995–5016.

[124] Olver, P. J. (1993) *Applications of Lie Groups to Differential Equations*. Springer-Verlag, New York.

[125] Ooguri, H. and Vafa, C. (1991) Geometry of N = 2 strings, *Nucl. Phys.* **B361**, 469–518.

[126] Page, D. N. (1979) Green's functions for gravitational multi-instantons. *Phys. Lett.* **B85**, 369–372.

[127] Pavlov, M. V. (2003) Integrable hydrodynamic chains, *J. Math. Phys.* **44**, 4134–4156.

[128] Pedersen, H. and Tod, K. P. (1993) Three-dimensional Einstein–Weyl geometry, *Adv. Math.* **97**, 74–109.

[129] Penrose, R. (1967) Twistor algebra, *J. Math. Phys.* **8**, 345–366.

[130] Penrose, R. (1969) Solutions of the zero-rest-mass equations, *J. Math. Phys.* **10**, 38–39.

[131] Penrose, R. (1976) Nonlinear gravitons and curved twistor theory, *Gen. Rel. Grav.* **7**, 31–52.

[132] Penrose, R. and Rindler, W. (1986) *Spinors and Space-Time*, Vols. **1** and **2**, CUP, Cambridge.

[133] Perjes, Z. (1971) Solutions of the coupled Einstein-Maxwell equations representing the fields of spinning sources, *Phys. Rev. Lett.* **27**, 1668.

[134] Piette, B., Stokoe, I., and Zakrzewski, W. I. (1988) On stability of solutions of the U(N) chiral model in two dimensions, *Z. Phys. C* **37**, 449–455.

[135] Plebański, J. F. (1975) Some solutions of complex Einstein Equations, *J. Math. Phys.* **16**, 2395–2402.

[136] Plebański, J. F. and Robinson, I. (1976) Left-degenerate vacuum metrics, *Phys. Rev. Lett.* **37**, 493–495.

[137] Pontecorvo, M. (1992) On twistor spaces of anti-self-dual hermitian surfaces, *Trans. Am. Math. Soc.* **331**, 653–661.

[138] Prasad, M. K. and Sommerfield, C. M. (1975) Exact classical solution for the t'Hooft monopole and the Julia–Zee dyon, *Phys. Rev. Lett.* **35**, 760–762.

[139] Przanowski, M. (1983) Locally Hermite–Einstein, self-dual gravitational instantons, *Acta Phys. Polon.* **B14**, 625–627.

[140] Przanowski, M. (1991) Killing vector fields in self-dual, Euclidean Einstein spaces with $\Lambda \neq 0$, *J. Math. Phys.* **32**, 1004–1010.

[141] Rajaraman, R. (1982) *Solitons and Instantons: An Introduction to Solitons and Instantons in Quantum Field Theory*, North-Holland Publishing Co., Amsterdam, the Netherlands.

[142] Rubin, H. and Ungar, P. (1957) Motion under a strong constraining force, *Commun. Pure Appl. Math.* **10**, 65–87.

[143] Sacks, J. and Uhlenbeck, K. (1981) The existence of minimal immersions of 2-spheres, *Ann. Math.* **113**, 1–24.

[144] Schiff, L, I (1969) *Quantum Mechanics*, 3rd edition, McGraw-Hill.

[145] Schuster, H. G. (1988) *Deterministic Chaos: An Introduction*, 2nd edition, VCH Publishers, New York.

[146] Sibata, T. and Morinaga, K. (1935) A complete and simpler treatment of wave geometry, *J. Sci. Hiroshima Univ.* **5**, 173–189.

[147] Skyrme, T. H. R. (1962) A unified field theory of mesons and baryons, *Nucl. Phys.* **31**, 556–569.

[148] Sorkin, R. D. (1983) Kaluza-Klein monopole, *Phys. Rev. Lett.* **51**, 87–90.

[149] Sparling, G. A. and Tod, K. P. (1981) An example of an *H*-space, *J. Math. Phys.* **22**, 331–332.

[150] Spencer, D. C. (1969) Overdetermined systems of linear partial differential equations, *Bull. Am. Math. Soc.* **75**, 179–239.

[151] Spivak, M. (1999) *A Comprehensive Introduction to Differential Geometry*, Vol. 5, 3rd edition.

[152] Stephani, H., Kramer, D., MacCallum, M., Hoenselaers, C., and Herlt, E. (2003) *Exact Solutions of Einstein's Field Equations*, CUP, Cambridge.

[153] Stokoe, I. and Zakrzewski, W. (1987) Dynamics of solutions of the $CP^{(N-1)}$ models in $(2+1)$ dimensions, *Z. Phys. C* **34**, 491–496.

[154] Tafel, J. (1989) A comparison of solution generating techniques for the self-dual Yang-Mills equations, *J. Math. Phys.* **30**, 706–710.

[155] Takasaki, K. (1989) An infinite number of hidden variables in hyper-Kähler metrics, *J. Math. Phys.* **30**, 1515–1521.

[156] t'Hooft, G. Unpublished.

[157] Tod, K. P. (1983) All metrics admitting supercovariantly constant spinors, *Phys. Lett. B* **121**, 241.

[158] Tod, K. P. (1995) Scalar-flat Kähler and hyper-Kähler metrics from Painlevé-III class, *Quant. Grav.* **12**, 1535–1547.

[159] Tod, K. P. (1997) The SU(∞)-Toda field equation and special four-dimensional metrics, in *Geometry and Physics* (Aarhus, 1995), Lecture Notes in Pure and Applied Mathematics, **184**, pp. 307–312, Dekker, New York.

[160] Tod, K. P. (1995) Cohomogeneity-one metrics with self-dual Weyl tensor, in *Twistor Theory*, Lecture Notes in Pure and Applied Mathematics, **169**, pp. 171–184, Dekker, New York.

[161] Tod, K. P. (2001) Indefinite conformally-ASD metrics on $S^2 \times S^2$, in *Further Advances in Twistor Theory*, Vol. III, Chapman & Hall/CRC, pp. 61–63. Reprinted from *Twistor Newslett.* **36**, 1993.

[162] Tong, D. (2005) TASI Lectures on Solitons arXLV: hep-th/0509216.

[163] Trautman, A. (1977) Solutions of the Maxwell and Yang-Mills equations associated with hopf fibrings, *Int. J. Theor. Phys.* **16**, 561–565.

[164] Tzitzéica, G. (1908) Sur une nouvelle classe de surfaces, *Rendiconti del Circolo Matematico di Palermo* **25**, 180–187.

[165] Uhlenbeck, K. (1982) Removable singularities in Yang–Mills fields, *Commun. Math. Phys.* **83**, 11–29.

[166] Uhlenbeck, K. (1989) Harmonic maps into Lie groups, *J. Diff. Geom.* **30**, 1–50.

[167] Veblen, O. and Thomas, J. M. (1926) Projective invariants of affine geometry of paths, *Ann. Math.* **27**, 279–296.

[168] Weinberg, S. (2005) *The Quantum Theory of Fields*, Vol. II. *Modern Applications*, Cambridge University Press, Cambridge.

[169] Ward, R. S. (1977) On self-dual gauge fields, *Phys. Lett.* **61A**, 81–82.

[170] Ward, R. S. (1980) Self-dual space-times with cosmological constant, *Commun. Math. Phys.* **78**, 1–17.

[171] Ward, R. S. (1985) Integrable and solvable systems and relations among them, *Phil. Trans. R.* Soc. **A 315**, 451–457.

[172] Ward, R. S. (1985) Slowly-moving lumps in the CP^1 model in $(2+1)$ dimensions, *Phys. Lett.* **158B**, 424–428.

[173] Ward, R. S. (1988) Soliton solutions in an integrable chiral model in $2+1$ dimensions, *J. Math. Phys.* **29**, 386–389.

[174] Ward, R. S. (1989) Twistors in $2+1$ dimensions, *J. Math. Phys.* **30**, 2246–2251.

[175] Ward, R. S. and Wells, R. (1990) *Twistor Geometry and Field Theory*, CUP, Cambridge.

[176] Ward, R. S. (1988) Integrability of the chiral equations with torsion term, *Nonlinearity* **1**, 671–679.

[177] Ward, R. S. (1990) Classical solutions of the chiral model, unitons, and holomorphic vector bundles, *Commun. Math. Phys.* **128**, 319–332.

[178] Ward, R. S. (1995) Nontrivial scattering of localized solitons in a (2 + 1)-dimensional integrable system, *Phys. Lett.* **A208**, 203–208.

[179] Ward, R. S. (1998) Twistors, geometry, and integrable systems, in *The Geometric Universe. Science, Geometry and the Work of Roger Penrose*, eds. S. A. Huggett et al., Oxford University Press, New York.

[180] Weinberg, E. J. (1979) Parameter counting for multimonopole solutions, *Phys. Rev.* **D20**, 936–944.

[181] Wess, J. and Zumino, B. (1971) Consequences of anomalous Ward identities, *Phys. Lett.* **37B**, 95–97.

[182] Whitt, B. (1985) Israel-Wilson metrics, *Ann. Phys.* **161**, 241–253.

[183] Witten, E. (1981) A new proof of the positive energy theorem, *Commun. Math. Phys.* **80**, 381–402.

[184] Witten, E. (1984) Nonabelian bosonization in two dimensions, *Commun. Math. Phys.* **92**, 455–472.

[185] Witten, E. (2004) Perturbative gauge theory as a string theory in twistor space, *Commun. Math. Phys.* **252**, 189–258.

[186] Woodhouse, N. M. J. (1985) Real methods in twistor theory, *Class. Quant. Grav.* **2**, 257–291.

[187] Woodhouse, N. M. J. (1987) *Introduction to Analytical Dynamics*, OUP, Oxford.

[188] Yang, C. N. and Mills, R. L. (1954) Conservation of isotopic spin and isotopic gauge invariance, *Phys. Rev.* **96**, 191–195.

[189] Yau, S-T (1977) Calabi's conjecture and some new results in algebraic geometry, *Proc. Natl. Acad. Sci.* **74**, 1798–1799.

[190] Yuille, A. L. (1987) Israel-Wilson metrics in the Euclidean regime, *Class. Quant. Grav.* **4**, 1409–1426.

[191] Zaharov, V. E. and Shabat, A. B. (1979) Integration of the nonlinear equations of mathematical physics by the method of the inverse scattering problem. II. *Funct. Anal. Appl.* **13**, 13–22.

[192] Zaharov, V. E. (1994) Dispersionless limit of integrable systems in 2 + 1 dimensions, in *Singular Limits of Dispersive Waves*, eds. N. M. Ercolani et al., Plenum Press, New York, pp. 165–174.

[193] Zakrzewski W. J. (1989) *Low-Dimensional Sigma Models*, Adam Hilger, Bristol, UK.

Index